Palgrave Studies in the History of Science and Technology

James Rodger Fleming (Colby College) and Roger D. Launius (National Air and Space Museum), Series Editors

This series presents original, high-quality, and accessible works at the cutting edge of scholarship within the history of science and technology. Books in the series aim to disseminate new knowledge and new perspectives about the history of science and technology, enhance and extend education, foster public understanding, and enrich cultural life. Collectively, these books will break down conventional lines of demarcation by incorporating historical perspectives into issues of current and ongoing concern, offering international and global perspectives on a variety of issues, and bridging the gap between historians and practicing scientists. In this way they advance scholarly conversation within and across traditional disciplines but also to help define new areas of intellectual endeavor.

Published by Palgrave Macmillan:

Continental Defense in the Eisenhower Era: Nuclear Antiaircraft Arms and the Cold War
By Christopher J. Bright

Confronting the Climate: British Airs and the Making of Environmental Medicine
By Vladimir Jankovic´

Globalizing Polar Science: Reconsidering the International Polar and Geophysical Years
Edited by Roger D. Launius, James Rodger Fleming, and David H. DeVorkin

Eugenics and the Nature-Nurture Debate in the Twentieth Century
By Aaron Gillette

John F. Kennedy and the Race to the Moon
By John M. Logsdon

A Vision of Modern Science: John Tyndall and the Role of the Scientist in Victorian Culture
By Ursula DeYoung

Searching for Sasquatch: Crackpots, Eggheads, and Cryptozoology
By Brian Regal

Inventing the American Astronaut
By Matthew H. Hersch

The Nuclear Age in Popular Media: A Transnational History
Edited by Dick van Lente

Exploring the Solar System: The History and Science of Planetary Exploration
Edited by Roger D. Launius

The Sociable Sciences: Darwin and His Contemporaries in Chile
By Patience A. Schell

The First Atomic Age: Scientists, Radiations, and the American Public, 1895–1945
By Matthew Lavine

NASA in the World: Fifty Years of International Collaboration in Space
By John Krige, Angelina Long Callahan, and Ashok Maharaj

Empire and Science in the Making: Dutch Colonial Scholarship in Comparative Global Perspective
Edited by Peter Boomgaard

Anglo-American Connections in Japanese Chemistry: The Lab as Contact Zone
By Yoshiyuki Kikuchi

Eismitte in the Scientific Imagination: Knowledge and Politics at the Center of Greenland
By Janet Martin-Nielsen

Climate, Science, and Colonization: Histories from Australia and New Zealand
Edited by James Beattie, Matthew Henry and Emily O'Gorman

The Surveillance Imperative: Geosciences during the Cold War and Beyond
By Simone Turchetti and Peder Roberts

Post-Industrial Landscape Scars
By Anna Storm

Voices of the Soviet Space Program: Cosmonauts, Soldiers, and Engineers Who Took the USSR into Space
By Slava Gerovitch

After Apollo?: Richard Nixon and the American Space Program
By John M. Logsdon

After Apollo?

Richard Nixon and the American Space Program

John M. Logsdon

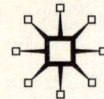

AFTER APOLLO?
Copyright © John M. Logsdon, 2015.

All rights reserved.

First published in 2015 by
PALGRAVE MACMILLAN®
in the United States—a division of St. Martin's Press LLC,
175 Fifth Avenue, New York, NY 10010.

Where this book is distributed in the UK, Europe and the rest of the world, this is by Palgrave Macmillan, a division of Macmillan Publishers Limited, registered in England, company number 785998, of Houndmills, Basingstoke, Hampshire RG21 6XS.

Palgrave Macmillan is the global academic imprint of the above companies and has companies and representatives throughout the world.

Palgrave® and Macmillan® are registered trademarks in the United States, the United Kingdom, Europe and other countries.

ISBN: 978–1–137–43852–2

Library of Congress Cataloging-in-Publication Data

Logsdon, John M., 1937–
   After Apollo? : Richard Nixon and the American space program / John M. Logsdon.
     pages cm—(Palgrave studies in the history of science and technology)
   Includes bibliographical references and index.
   ISBN 978–1–137–43852–2 (hardback : alk. paper)
    1. Astronautics and state—United States—History—20th century.
2. United States. National Aeronautics and Space Administration—Planning—History—20th century. 3. Project Apollo (U.S.)—History. 4. Nixon, Richard M. (Richard Milhous), 1913–1994—Influence. I. Title.
TL789.8.U5L6385 2014
629.4500973′09047—dc23   2014035724

A catalogue record of the book is available from the British Library.

Design by Newgen Knowledge Works (P) Ltd., Chennai, India.

First edition: March 2015

10 9 8 7 6 5 4 3 2 1

Printed in the United States of America.

This book is dedicated to my grandchildren—Jacob, Sara, and Aaron—with the hope that they will see the space dream become reality. I was present when men first left for the Moon in 1969. I hope that they will see men and women return to the Moon and leave for Mars—or maybe even make the trip themselves.

Contents

Preface and Acknowledgments   ix

Overture   1

Act 1   No More Apollos

Chapter 1
Richard Nixon and *Apollo 11*   7

Chapter 2
Setting the Post-Apollo Stage   31

Chapter 3
After the Moon, Mars?   55

Chapter 4
Space and National Priorities   83

Chapter 5
The Nixon Space Doctrine   103

Chapter 6
The End of the Apollo Era   117

Intermission   125

Act 2   What Next?

Chapter 7
A New Cast of Characters   131

Chapter 8
The Space Shuttle Takes Center Stage   143

Chapter 9
National Security Requirements Drive Shuttle Design   161

Chapter 10
A Time of Transitions   173

Chapter 11
A Confused Path Forward   197

Chapter 12
Debating a Shuttle Decision   213

Chapter 13
Which Shuttle to Approve?   239

Chapter 14
A "Space Clipper"   255

Finale   271

Epilogue
Richard Nixon and the American Space Program   277

Notes   303

Bibliography   339

Index   343

Preface and Acknowledgments

This study has had a very long genesis. When my first book, *The Decision to Go to the Moon: Project Apollo and the National Interest*, was published by the MIT Press in summer 1970, I gave a copy to NASA Deputy Administrator George Low. By that time there had been two successful landings on the Moon—*Apollo 11* and *Apollo 12*—and one near-tragedy—*Apollo 13*. Low told me that NASA at that point in time was in the midst of a confused process of dealing with Richard Nixon's White House with respect to what the space agency should do after Apollo. He suggested that I take a look at that post-Apollo decision-making process similar to the one that had led to my Apollo study, and provided a modest NASA grant to facilitate such an effort. That suggestion set me on the lengthy and winding path that 44 years later has resulted in this book.

Working with NASA chief historian Gene Emme and especially Nat Cohen of NASA's policy office, during late 1970 and 1971 I carried out a series of interviews with many of the key actors in the post-Apollo debate; these interviews took place as NASA was struggling to get White House approval for developing the space shuttle as the central focus of its efforts for the 1970s. Those interviews are one basis for the current study; they provide an "at the moment" look at what was on the minds of those trying to decide what kind of post-Apollo space program was in the nation's, and President Nixon's, interest. In 1973, I wrote up but never published an initial account of post-Apollo decision making, and put that draft and transcripts of the supporting interviews in the NASA Historical Reference Collection at NASA Headquarters in Washington, DC; other researchers have drawn on that material over the years. I continued on a sporadic basis over the following years to interview individuals involved in post-Apollo decisions; the last of those interviews was with top Nixon assistant John Ehrlichman in 1983. I published several articles on the space shuttle, most notably a controversial analysis titled "The Space Shuttle Program: A Policy Failure?" that appeared in the journal *Science* a few months after the January 1986 *Challenger* accident. But the press of teaching and administrative responsibilities was a barrier to completing the book-length study needed to tell the full post-Apollo story.

It was only in 2008 as I left after 38 years the active faculty at the Space Policy Institute, part of The George Washington University's Elliott School of International Affairs, that I could turn my full attention to my backlog of policy history work. First up was a relook at President John Kennedy's 1961 decision to send Americans to the Moon and a fresh examination of what he did to turn that decision into reality. The result was published by Palgrave Macmillan in December 2010 as *John F. Kennedy and the Race to the Moon*. One of those reading an early copy of the Kennedy manuscript and providing a book jacket endorsement was Bill Anders. Bill had flown around the Moon in December 1968 on the *Apollo 8* mission and had taken the iconic "Earthrise" photograph, then came to Washington to be executive secretary of the National Aeronautics and Space Council, the organization set up in 1958 to provide White House level space policy coordination. Anders was thus a participant in post-Apollo policy discussions from fall 1969 through the decision to approve the space shuttle, and he encouraged me to continue my research and writing to present a full account of space decision making during the Nixon administration. Bill backed his encouragement both with continued involvement as the study progressed, commenting on chapter drafts, and with crucial financial support from the Anders Family Foundation. That support helped me visit various archives during my research and avoid other compensated activity so I could focus on my writing. I thus owe a strong "thank you" to Bill Anders for all his effort in helping bring this book into existence.

If I had completed my study of post-Apollo decision making on its original schedule, it would have been a far less rich account. The availability of books by senior White House staff and of the Nixon administration papers at the Richard Nixon Presidential Library in Yorba Linda, California, and the release of Nixon's tape recordings, which can be accessed at a variety of websites (I used www.nixontapes.org) were all essential to a full narrative. At the Nixon Library, the staff of the research room was extremely researcher-friendly. I owe particular thanks there to audio-visual archivist Jon Fletcher, who was very responsive in my search for fresh images to include in the book. Freelance researcher Alicia Fernandez provided useful help in tying up some last-minute loose ends.

I also consulted the papers of Caspar Weinberger, Clay Thomas Whitehead, and Tom Paine at the Manuscript Division of the Library of Congress; George Low's papers at Rensselaer Polytechnic Institute; James Fletcher's papers at the University of Utah (at an early stage in my research); material in the Johnson Space Center Historical Collection at the University of Houston Clear Lake; and interviews available in the Archives Division of the National Air and Space Museum. The staffs at all these venues were very helpful; I am grateful to them all but owe particular thanks to Jean Grant at Clear Lake for provide a large amount of useful material. The NASA Historical Reference Collection is a treasure trove for researchers into NASA's history and was absolutely crucial to my work, and I owe thanks to the NASA history office staff, particularly its director, Bill Barry, chief archivist Jane

Odom, and archivists Colin Fries and Liz Suchow for their help. I have put the documents and interviews that form the basis for this study on deposit at the NASA Historical Reference Collection as "Logsdon Source Notes."

As I completed the study I was able to interview a number of those involved in the 1969–1972 events, including Bill Anders, Don Rice of the Office of Management and Budget (OMB), and former astronaut and Nixon adviser on space Frank Borman. Russell Drew of the White House Office of Science and Technology, Dan Taft of OMB, and original shuttle program manager Bob Thompson provided useful comments on chapter drafts. In addition, Frank Borman, Richard Speier, Chuck Friedlander, James Dewar, and Jim Behling were good enough to share material from their personal files, and Paul Shawcross gave me access to the few files on the shuttle decision that had been retained at OMB.

I owe a particular debt of gratitude to "space shuttle guru" Dennis Jenkins. Dennis shared material from his voluminous files and read and perceptively commented on drafts of every chapter. My book is *not* a history of the early evolution of the space shuttle; rather it is an account of the decisions made by the Nixon White House and the NASA leadership in Washington that made the shuttle central to what the United States has done in space for over four decades. I hope that when I do discuss the early years of the shuttle program, I make no major errors. When it becomes available in 2015, Dennis Jenkins's three volume compendium on the totality of the space shuttle program will be the definitive work.

My former student and colleague Andre Bormanis also read every chapter with an eagle eye, catching my many typos while providing thoughtful substantive comments. Other colleagues who commented on chapter drafts include Roger Launius, Teasel Muir-Harmony, Russ Drew, Dan Taft, Dwayne Day, and L. Parker Temple III. I must thank Scott Pace, my successor as director of GW's Space Policy Institute, for his hospitality in providing an aging professor emeritus continuing work space at the university. There have been a number of people at GW who helped in the early stages of my research on post-Apollo decisions, but frankly I cannot remember any specific names. If any of those individuals happen to read this book, I thank you for your help and apologize for my poor recall. More recently, student assistants Caitlan Dowling helped with archival research and retyping some of the early interviews, Luis Suter took on the unenviable task of trying to transcribe the often-garbled conversations on the Nixon tapes, and Gaurav Dhiman helped get the manuscript in shape for submission. Rachel Nishan of Twin Oaks Indexing did an extremely thorough job of compiling the book's index, and she and Dwayne Day provided invaluable "second eyes" in reviewing the study's page proofs.

I am appreciative of Roger Launius's interest in having this book be part of the Palgrave Series in the History of Science and Technology that he and Jim Fleming co-edit, and to editors Chris Chappell and his successor Kristin Purdy, editorial assistant Mike Auperach, and production editor Erin Ivy at Palgrave Macmillan for seeing the book through to publication.

The time taken in completing this study covers most of my professional career—38 years on the active GW faculty and six as an emeritus professor. I tell people that I have not retired, and offer *John F. Kennedy and the Race to the Moon* and this book as evidence. There are likely to be more books to follow, both in terms of policy history and perhaps also a collection of my own insights and opinions over the years. In those same 44 years, my two sons have grown to be outstanding men and three delightful grandchildren have been born. I am dedicating this book to Jacob, Sara, and Aaron Logsdon with the hope that they will see a future in space with more purpose and payoffs than the one created by the Nixon administration decisions chronicled in this work. Throughout these 44 years, and even before, my wife Roslyn has provided the loving foundation of my life. Maybe now that this long-running opus is finished we can find more time to enjoy life together.

Needless to say, I am responsible for all errors of fact (including what was actually being said on the Nixon tapes!) and interpretation. I am sure that many people will not agree with my assessment of the Nixon space heritage, especially with respect to the space shuttle, and my characterization of the recent and current state of the U.S. space program. After a career devoted to that program, I regret that my conclusions are so downbeat. I can only remain hopeful that better days are ahead.

JOHN M. LOGSDON
January 2015

Overture

On July 20, 1969, U.S. astronaut Neil Armstrong took "one small step for a man, one giant leap for mankind," as he became the first human to set foot on the Moon. The success of the *Apollo 11* mission satisfied the goal that had been set by President John F. Kennedy just over eight years earlier—"before this decade is out, landing a man on the moon and returning him safely to earth."[1] Inevitably, it also raised the question "What do you do next, after landing on the Moon?"

It fell to President Richard M. Nixon, sworn into office exactly six months before Armstrong's historic moonwalk, to answer this question. The following account traces in detail how Nixon and his associates in the 1969–1972 period went about developing their response. The decisions made then have defined the U.S. program of human space flight well into the twenty-first century. Those choices have thus had a much more lasting impact than did John Kennedy's 1961 decision to go to the Moon. The factors leading to Kennedy's decision are well understood, but that is not the case with respect to space policymaking under President Nixon. The goal of this study is to provide that understanding, and thus to fill in the details of a crucial period in the history of the U.S. space program, and particularly its human space flight element. The Nixon administration also made influential decisions with respect to space science and applications efforts, but those decisions will not be discussed here.

The process of deciding what the United States should do in space after Apollo is presented here as a "play in two acts." In the first act, unfolding in chapters 1–6, decisions were made on what *not* to do—not to continue during the 1970s a fast-paced, high-priority, Apollo-like effort aimed at rapid development of new space capabilities and leading to human missions to Mars in the early to mid-1980s. Nixon soon after taking office chartered a top-level review to recommend post-Apollo space goals and programs. That review took place even as *Apollo 11* gained worldwide acclaim; Richard Nixon made sure that he would bask in the glow of that achievement. But when presented with a recommendation for an ambitious post-Apollo space effort, Nixon decided that the nation neither wanted nor could afford such an undertaking. In March 1970 the president spelled out a policy that assigned to the space program reduced priority among the many demands on the federal budget. The refrain "after the Moon, Mars" did not resonate with the Nixon White House, even though the president himself identified with American astronauts and was intrigued with a future in space exploration that included eventual Martian journeys.

The second act of the drama, discussed in chapters 7–14, involved answering the question, "if not an ambitious post-Apollo program centered on human space flight, then what?" Options evaluated during the 1970–1972 period ranged from focusing the nation's space capabilities on Earth-bound problems, and perhaps even transforming the space agency to a general-purpose technology organization, to a modestly paced effort using surplus Apollo hardware, to developing a fully or partly reusable space shuttle.

During 1970, the future development that had had highest priority in 1969, developing a long duration orbital outpost—a space station launched by the Saturn V Moon rocket and serviced by the space shuttle—fell from favor, and thus other rationales for developing a shuttle had to be articulated. A wide variety of shuttle designs were assessed, with the president's technical and budget advisers arguing for a far less ambitious system than that advocated by the National Aeronautics and Space Administration (NASA). Factors such as aerospace unemployment and its impact on the 1972 presidential election entered into consideration, as did the message the United States would send to the world if it were to decide not to continue to seek space leadership. All involved believed that Richard Nixon wanted to continue some type of human space flight program, even as he personally tried to cancel the final flights to the Moon to avoid the possibility of the kind of near-fatal accident that had threatened the *Apollo 13* crew.

Out of this complex mix of influences came the decision, announced by President Nixon on January 5, 1972, "to revolutionize transportation into near space by routinizing it."[2] By approving NASA's plans for a large space shuttle, Nixon put the shuttle at the center of U.S. space efforts without proposing clear strategic goals that it would serve. Because the shuttle would be flown by a two-person astronaut crew and on most missions would carry additional astronauts, it met Nixon's desire to keep the human space flight program alive. The belief was that, by reducing the cost of space launch, the shuttle would open up space to a wide variety of activities. By providing capabilities for satellite deployment, in-orbit servicing, in-orbit assembly, and return of payloads to Earth, NASA hoped that the shuttle would usher in a new era of space operations. There were suggestions of innovative, potentially provocative, national security missions made possible by the new capabilities that the shuttle would offer.

The decision to develop a space shuttle was the culmination of the drama of post-Apollo space policymaking. The decision carried with it NASA's intent, once the shuttle entered operations, to seek presidential support for developing a space station launched in separate elements by the shuttle and assembled in orbit. Those two activities—developing and flying the space shuttle, then developing, assembling, and utilizing the space station—have dominated U.S. human space flight efforts for four decades after the last American astronaut left the Moon in December 1972. As *Apollo 17* lifted off the lunar surface on December 14, 1972, President Nixon issued a statement saying "this may be the last time in this century that men will walk on the Moon."[3] By the decisions he made between 1969 and 1972, Richard Nixon ensured that his forecast would come true.

Act 1
No More Apollos

Chapter 1

Richard Nixon and *Apollo 11*

President-elect Richard Nixon, like most Americans, was thrilled by the December 1968 *Apollo 8* mission, the first space flight to leave Earth orbit with humans aboard. *Apollo 8* sent Frank Borman, Jim Lovell, and Bill Anders into orbit around the Moon on December 24. In his *Memoirs,* Nixon recalled that on that Christmas Eve, he "was a happy man." At his retreat on Key Biscayne, Florida, "a wreath hung on the front door and a beautifully trimmed Christmas tree stood in the living room...Far out in space Apollo VIII orbited the moon while astronaut Frank Borman read the story of the Creation from the Book of Genesis.* Those days were rich with happiness and full of anticipation and hope."[1]

The afterglow of the bold *Apollo 8* mission was still bright as Richard Milhous Nixon was sworn in as the thirty-seventh president of the United States on January 20, 1969. References to that mission and to space exploration in general appeared throughout the new president's inaugural address:

- "In throwing wide the horizons of space, we have discovered new horizons on earth."
- "We find ourselves rich in goods, but ragged in spirit; reaching with magnificent precision for the moon, but falling into raucous discord on earth."
- "As we explore the reaches of space, let us go to new worlds together—not as new worlds to be conquered, but as a new adventure to be shared."
- "Only a few weeks ago we shared the glory of man's first sight of the world as God sees it, as a single sphere reflecting light in the darkness. As the Apollo astronauts flew over the moon's grey surface on Christmas Eve, they spoke to us of the beauty of earth."
- "In that moment of surpassing technological triumph, men turned their thoughts toward home and humanity—seeing in that far perspective that

---

*Actually, Borman was joined in the reading by his astronaut colleagues Lovell and Anders.

man's destiny on earth is not divisible; telling us that however far we reach into the cosmos, our destiny lies not in the stars but on earth itself."[2]

As he assumed the presidency, Richard Nixon was well aware that the success of the *Apollo 8* mission meant that the United States during his first year in the White House almost surely would achieve the lunar landing goal set by Nixon's long-time nemesis John F. Kennedy eight years earlier. He also knew that in his first year in office he would face significant space policy decisions, choices that would set the path in space for the United States for the coming decade and beyond. But there was no sense of urgency within the Nixon administration with respect to defining what the United States would do in space after landing on the Moon; the space program was not high on Nixon's policy agenda. More important in the short run was making sure that the lunar landing program was a success and that Richard Nixon was closely identified with that success.

Preparing for a Lunar Landing

To Nixon, "the most exciting event of the first year of my presidency came in July 1969 when an American became the first man to walk on the moon." Not only was the historic *Apollo 11* mission to the Moon personally exciting to the president, it also provided him an ideal vehicle to promote many of the themes he hoped would characterize his time in the White House, particularly America's global leadership. In addition, by linking himself closely with the message left on the Moon—"We came in peace for all mankind"— Richard Nixon could portray himself as a peacemaker, eager to reduce the tensions that had led to conflict among nations in the years since World War II. To Nixon, the American spirit, as exemplified by the Apollo missions to the moon, was "the most important psychological weapon that could be used in building the generation of peace." Nixon had decided that the lunar landing "was (a) a necessary shot in the arm to the American body politic, (b) a lift to the spirit of a war-weary people, (c) a boost for technology that was being unfairly derided by environmentalists—and (d), (e), and (f)—that he was going to be an enthusiastic part of it."[3]

Project Apollo had in fact been intended from its 1961 approval by President Kennedy to be a large-scale effort in "soft power," sending a peaceful but unmistakable signal to the world that the United States, not its Cold War rival the Soviet Union, possessed preeminent technological and organizational power, and that the American way of life provided an example other nations should admire and aspire to follow. In his May 25, 1961, address to a joint session of Congress in which he proposed setting as a national goal sending Americans to the Moon, Kennedy had said "if we are to win the battle for men's minds, the dramatic achievements in space...should have made clear to us all...the impact of this adventure on the minds of men everywhere who are attempting to make a determination of which road they should take."[4] Although he was extremely reluctant to acknowledge that the *Apollo 11* mission would be the culmination of the pledge Kennedy had made eight years earlier, Richard Nixon agreed with Kennedy's rationale for the lunar landing effort. Even after the dismal events of the 1960s—assassinations, urban riots, and seemingly endless U.S. involvement in a war in Southeast Asia—landing Americans on the Moon, thought Nixon, was an achievement that could help both communicate to the rest of the world an extremely positive image of U.S. leadership and power and restore national morale.

Nixon and the Apollo Astronauts

According to his senior advisor John Ehrlichman, the Apollo astronauts were to Nixon "very wonderful people. There was just not enough the country can do for these guys, and they are doing an enormous amount for the country...He would always be enormously stimulated by contact with these folks. And there was an element of hero worship on his part." Nixon "liked heroes. He thought it was good for this country to have heroes."[5] *Apollo 8* commander Frank Borman suggested that the president believed that the Apollo astronauts were "something special—not as individuals so much as for what we represented." According to veteran *Time/Life* correspondent Hugh Sidey, whenever Nixon met with one or more of the Apollo astronauts, "the color comes to his face and the bounce to his step." Sidey suggested that Nixon saw the astronauts as "the sons he never had...They are the distillers of what Nixon considers to be the best in this country."[6] Nixon saw the Apollo astronauts as exemplars of the best characteristics of Americans and was eager to use them both overseas and in the United States as role models for what humans could achieve with positive intent and sufficient determination. Nixon's attitude toward the Apollo astronauts led to a judgment on the part of those planning post-Apollo space efforts that he would never accept a proposal to end U.S. human space flights; any future NASA program would have to keep Americans flying in space.

While Nixon may have had positive feelings toward all of the Apollo astronauts, he developed a continuing relationship with only one of the group—Frank Borman. The *Apollo 8* commander was invited to Nixon's inaugural; to Borman, the invitation suggested that "Nixon was not only genuinely interested in space, but seemed to have embraced me personally as the space program's symbolic representative."[7] By the time of the inauguration Borman was already scheduled to go on a three-week European "goodwill" tour. One of the first decisions of the incoming Nixon administration was to give its approval to the trip; Nixon's secretary of state, William Rogers, later told Borman "we clearly made a wise decision."

The *Apollo 8* crew was invited to the White House on January 30, as the president announced that "it is very appropriate for Colonel Borman to go to Western Europe and to bring...not only the greetings of the people of the United States, but to point out what is the fact: that we in America do not consider that this is a monopoly, these great new discoveries that we are making; that we recognize the great contributions that others have made and will make in the future; and that we do want to work together with all peoples on this earth in the high adventure of exploring the new areas of space." Upon his return to Washington, Borman reported that "space technology in Europe lags behind American achievement by a considerable amount" and suggested that the United States "immediately request an international agency to select a certain number of qualified scientists from different nations of the earth to join our program to participate as scientists/astronauts in future earth-orbital space stations." This suggestion interested Richard Nixon; in the months to come he would press his associates to find ways to fly non-U.S. individuals on future U.S. space flights.[8]

Borman was surprised by "the extent to which Richard Nixon accepted me." Indeed, until he left NASA and government service in mid-1970, Borman served as Nixon's "in-house astronaut," frequently consulted on space policy and personnel issues as well as serving as liaison between the White House and NASA during the *Apollo 11* and *Apollo 13* missions. Borman in early 1969 and again in fall 1970 might even have become head of NASA if he had been so inclined. With respect to his relationship with President Nixon, Borman recalls that "I liked him, I really did...I know he was terribly shy, even ill at ease with people he didn't know, and when it came to making small talk he was a disaster." However, "we never had to engage in small talk; at every meeting I had with him, we always discussed important matters on a one-on-one basis. He took advice—and sometimes it was advice that he either didn't want to hear or that was contrary to what his advisers had told him." Borman was "sure that he trusted me personally and he trusted my judgment in areas in which he knew I had some knowledge."[9]

Planning for Presidential Involvement

In the five months after Richard Nixon was sworn in as president on January 20, 1969, there were two Apollo missions, both of which had to be successful in order for the July *Apollo 11* flight to be the first try at a lunar landing. Both did succeed, clearing the path to the Moon. *Apollo 9* (March 3–13) was an Earth-orbit test of the lunar module. *Apollo 10* (May 18–26) was the dress rehearsal that performed all elements of the lunar landing mission except the landing itself.

Several of Nixon's immediate staff, including chief of staff H. R. "Bob" Haldeman and appointments secretary Dwight Chapin, had worked in the advertising agency J. Walter Thompson, and they applied that expertise to making sure that the *Apollo 11* mission and its aftermath would communicate the messages important to the president and in the process burnish Nixon's image as a world leader. On May 28, two days after *Apollo 10* splashed down, Chapin and Peter Flanigan, Nixon's assistant with specific responsibility for space issues, met with NASA Administrator Tom Paine "to go over the Apollo 11 activities which could conceivably involve the President, either directly or indirectly."[10] Nixon, briefed on these discussions, quickly suggested that NASA assign Borman to the White House to help manage activities "with relation to this shot and subsequent congratulation of the astronauts." Borman recognized that the *Apollo 11* mission was "obviously going to be one of the most epochal events in history if it succeeded, and by the same token an unparalleled catastrophe if the crew didn't survive." Those within NASA close to Project Apollo, like Borman, realized just how risky missions to the Moon were, and thus were very conscious of the possibility of failure in the first landing attempt.[11]

Haldeman, Chapin, and Flanigan had their own ideas on how best to portray the president in the most positive possible light, and they did not trust Paine and other top NASA officials to give the president's interests top priority in the run up to *Apollo 11*. Paine had been selected as the NASA administrator only after several candidates preferred by the White House had turned down the position. Paine was a holdover from the Johnson administration; as a liberal Democrat, he was an unlikely choice as Nixon's top space official. (His selection is discussed in chapter 2.) After discussing Paine's suggestions with Nixon, Haldeman told Chapin that "the President is intrigued with having a very big dinner" after the *Apollo 11* crew was released from quarantine; the dinner would include all U.S. astronauts and the widows "of the three that were burned." [This was a reference to the deaths of *Apollo 1* astronauts Gus Grissom, Ed White, and Roger Chaffee when a fire broke out in their spacecraft during a launch pad test on January 27, 1967.] Nixon first considered having the dinner at the White House, then thought "it ought to be bigger." After considering both New York and Chicago as venues, Nixon "ended up being primarily intrigued with the possibility of Los Angeles, doing it at the Century Plaza." Nixon proposed charging $100 a person for the dinner and "using the income for space scholarships for underprivileged kids." (This proposal was later dropped.) He "definitely wants to go ahead with plans to visit the Cape for the shoot" and "liked the idea of watching the launch from aboard a ship." Nixon wanted to make sure that any prelaunch reception "would clearly be the President's affair—not NASA's." Nixon had been told that it would be possible to talk on split-screen television with the astronauts while they were on the Moon; he was "extremely anxious to pursue the television participation idea." The president, reported Haldeman, "still feels he probably should go to the carrier for the pick up," but "we can talk him out of that." A week letter, the idea of President Nixon having dinner with *Apollo 11* crew—Neil Armstrong, Edwin Aldrin, Jr., better known as "Buzz," and Michael Collins—the night before their launch had been added to the list of possibilities.

Nixon's interest in going to the recovery carrier had been communicated to NASA, which was skeptical of the desirability of such an undertaking. NASA's top public relations official, Julian Scheer, told the White House that Nixon could not greet the astronauts personally, but only "talk with the Apollo 11 crew through a porthole (two feet by two feet in size)" in the isolation quarters in which they would stay for two weeks after their return from the Moon to avoid the remote possibility that they were carrying alien organisms. Even so, after meeting with Nixon on June 10, Ehrlichman ended his meeting notes with the question "splash down—DO WE GO?" Richard Nixon's answer was "yes."[12]

Another White House idea for putting *Apollo 11* in a broader cultural and historical context was asking poet Archibald MacLeish, who had written the stirring words with respect to the *Apollo 8* mission that Nixon had quoted in his inaugural address, to compose something similar in connection with *Apollo 11*. MacLeish had initially responded positively to an informal inquiry asking whether he would accept such a request, so on July 1, Nixon, noting that there was "no precedent for such a request by a President in office," wrote MacLeish, asking him "to write a poem commemorative of this event, examining the meaning and portent of the achievement," which Nixon noted should be viewed "not only as a great adventure, but in the perspective of the search for truth and a quest for peace." However, even before receiving the president's letter, MacLeish changed his mind; apparently he "thought twice about doing anything with Nixon connected with it." On June 26 he called Henry Kissinger, Nixon's national security adviser and a former faculty colleague at Harvard, indicating that his "artistic creativity" could not be marshaled on request. MacLeish did write such a poem, but rather than providing it to President Nixon, it was published on the front page of *The New York Times* on the morning after the Moon landing. According to Nixon speechwriter Safire, "this slap in the face did not go unnoticed, and was an episode to recall and mutter about when we were criticized for not considering the spiritual meaning of the moon landing."[13]

By mid-June, Frank Borman had arrived at the White House and had begun to work with Flanigan and Chapin on *Apollo 11* activities. He relayed the information that the *Apollo 11* crew was "very pleased the President will accept their invitation to dinner." He recommended that Nixon "should *not* stay" for the next morning's launch, since "there is the possibility of last minute delays." Borman felt that the dinner with the crew would "set the stage" and "the President's activity will build—with the television from the moon and the events thereafter." The decision that Nixon would be present as the crew splashed down in the Pacific had been made by this time, and "plans are being made aboard the carrier for the President and his party—up to a total of 30." After the crew's release from quarantine in August, the White House was planning "a swing to New York City, Chicago and back to Los Angeles for the dinner in the evening." Borman had objected to this plan, suggesting that the crew travel only to Los Angeles, but he was overruled. Nixon wanted a nationwide celebration of the mission's success.[14]

What were supposed to be final plans for the president's involvement were in place by July 1. Nixon would fly to Cape Kennedy on July 15 for an early dinner with the *Apollo 11* crew, who had to get up at 4:00 a.m. on launch day, and then return to Washington after dinner. He would watch the launch from the White House. On July 20, the day the astronauts would land on the Moon, there would be a White House church service with a large attendance of members of Congress, NASA officials, and other dignitaries. Shortly after the crew members began their walk on the Moon, at that point scheduled for the early morning of July 21, they would unveil a plaque on the lunar module saying "Here Men from the Planet Earth First Set Foot Upon the Moon, July 1969, A.D. We Came in Peace for All Mankind." The plaque would bear the signatures of Armstrong, Aldrin, and Collins, the three men who had actually journeyed to the Moon—and that of Richard Nixon. Adding Nixon's signature was a late decision on NASA's part, without White House urging, reflecting the space agency's interest in making the president positively disposed toward NASA's post-Apollo plans.

The final wording on the plaque was a White House responsibility, after NASA had prepared a first draft of the text. Initially it was to read "first landed," but there were Central Intelligence Agency reports that the Soviet Union might land a robotic spacecraft on the lunar surface before the astronauts arrived, so "landed" was changed to "set foot." Safire, who was reviewing the text for the plaque, changed "we come in peace" to "we came in peace." He thought the former phrase sounded like "a stereotyped salute from white settlers to Hollywood Indians." NASA's adding "A.D." to the date, noted Safire, was "a shrewd way of sneaking God in"; it would "tell space travelers eons hence that earthlings in 1969 had a religious bent." Safire recalls that "the one item we did not bother to discuss was the signature of the President" on the plaque, since "the President, whoever he is, always signs a new Federal bridge or post office," so "we took it for granted he would sign his name to the moon project." Safire added, "we were insensitive to the sensitivity of old Kennedy hands," who interpreted Nixon's signature as "trying to horn in on a Kennedy project." The president was given two alternatives for the last line on the plaque: "A New Dawn for the Human Spirit" and "A New Dawn of Peace for All Mankind." Nixon decided to stay with "We Came in Peace for All Mankind." He gave his personal approval to the wording of the plaque, writing "OK" on a June 16 memorandum communicating the text.[15]

Nixon also decided in June to make his long flight to the *Apollo 11* splashdown on July 24 the first stop on a round-the-world diplomatic tour that would have as its theme "The Spirit of Apollo." In this way Nixon could use his long trip to be present at the mission's end as a springboard for broader diplomatic purposes. In particular, Nixon was eager to visit Romanian head of state Nicolae Ceausescu, who had indicated that he could serve as a communication channel to Chinese Premier Chou En-Lai for a Nixon initiative to begin the process of normalizing the U.S.-Chinese relationship. This planning assumed the mission's success, which was certainly not guaranteed, and thus represented significant risk-taking on Nixon's part.[16]

Richard Nixon got much of his information about what was going on in the world from assiduously reading his "daily news summary," a digest of stories from around the world, usually prepared by his young staff assistant Patrick Buchanan. The July 7, 1969, news summary reported that NASA medical officials were "extremely upset by the President's plans to have dinner with the APOLLO 11 astronauts the night before they blast off." The source of the reported concern turned out to be NASA's Dr. Charles Berry, who billed himself as the astronaut's personal physician, although according to Mike Collins, "we seldom saw him." Berry apparently was worried that the president might be carrying germs that could affect the crew's health during the mission. The *Apollo 11* astronauts thought that this concern was absurd, given that they were in daily contact with a number of others not under quarantine restrictions, and would have dinner a few days before the flight with NASA Administrator Paine who, noted Collins sarcastically, "was apparently germ-free." Borman called Berry's warning "totally ridiculous" and "dammed stupid," but advised Nixon to cancel the planned dinner because "if anyone sneezes on the Moon, they'd put the blame on the president." As the story gained wide circulation, Nixon's staff accepted Borman's advice and decided it had no choice but to cancel the president's prelaunch dinner with the crew. Armstrong, Collins, and Aldrin on July 9 sent a telegram to the president, expressing their "deepest regrets over the unfortunate circumstances that precluded your coming...You are welcome in our quarters at any time." Instead of dining with the *Apollo 11* crew on July 15, Richard Nixon called them as they were having dinner and sent them a telegram saying: "On the eve of your epic mission, I want you to know that my hopes and my prayers—and those of all Americans—go with you...It is now your moment."[17]

Apollo 11, Richard Nixon, and John F. Kennedy

There was little inclination on Richard Nixon's part to acknowledge President John Kennedy's role in initiating the lunar landing program as the launch of *Apollo 11* approached. Indeed, throughout the many celebrations of the *Apollo 11* achievement, Nixon never once publicly spoke Kennedy's name.

This visceral aversion to sharing credit for Apollo became evident as Nixon's Special Assistant for Urban Affairs Daniel P. Moynihan, who was among the more liberal of Nixon's White House staff and who had earlier served as an assistant to President Kennedy, received a request from another Kennedy alumnus, Bill Moyers. Moyers, in 1969 the publisher of the Long Island, New York, newspaper *Newsday*, on June 4 forwarded a column he had written suggesting that the *Apollo 11* spacecraft be commissioned "The John F. Kennedy" in recognition of the late president's role in initiating Project Apollo. Moyers told Moynihan "you knew John Kennedy even better than I did; can't you influence your friends there to take up this suggestion?" Moynihan forwarded the suggestion to Haldeman, saying that "the *Newsday* proposal has a certain gallant quality to it. I imagine this would be interesting to the President, and I strongly suspect it would be to his advantage." Haldeman had the proposal circulated among other senior staff members. Counselor Arthur Burns, at that point Nixon's top advisor on domestic policy, "heartily" endorsed the idea, saying that "such an act of graciousness is justified by history and would be, I think, good politics besides."

Presidential science advisor Lee DuBridge thought that the proposal would be "a fitting tribute indeed to the man who, against great opposition, initiated this bold project." In contrast, White House communications director Herb Klein "strongly" recommended against the proposal, saying "the Kennedy angle will get major play anyway. We would get more mileage with a gracious Presidential mention of Kennedy's vision." Congressional relations assistant Bryce Harlow noted that it was President Eisenhower who initiated the U.S. space program and remarked that "we have gone far enough in 'Kennedyizing' the mission." Senior advisor John Ehrlichman pragmatically noted that "such an action would win us neither friends in Congress nor votes in 1972," suggesting "fall prey to this and the next step will be renaming the moon because NBC thinks it would be a good idea." After receiving these diverse views, Haldeman directed that "any plan to commission the Apollo 11 shot John F. Kennedy be abandoned"; in initialing the memorandum recording this decision, he added in bold handwriting with double underlining, "positively!!"[18]

There is no evidence in the written record that President Nixon knew of this episode, although it is hard to imagine that Haldeman in his frequent and extended meetings with Nixon did not raise the matter. At any rate, Haldeman's decision meant that there was no obstacle to the *Apollo 11* crew themselves choosing the names for their spacecraft, as had become the tradition. The crew announced at their last prelaunch press conference on July 5 that their command and service module would be christened *Columbia* and their lunar lander, *Eagle*.

Negative Press Reactions

While the White House debate over the Moyers proposal was out of the public view, such was not the case as both *The Washington Post* and *The New York Times* published editorials critical of Richard Nixon's granting himself a central role in celebrating the lunar landing. Nixon was deeply suspicious of the media, and especially the elite Eastern newspapers; less than a month into his presidency, he had told one of his speechwriters "they are waiting to destroy us." In this case, he had reasonable cause for his anger. The *Post* objected "with special sarcasm" to the fact that Richard Nixon's signature was on the plaque that would be left on the Moon, saying "how dare the space program be treated as some run-of-the-mill public works project!" A rather snarky *Times* editorial was captioned "Nixoning the Moon." It noted that "Mr. Nixon's attempt to share the stage with the three brave men on Apollo 11 when they attain the moon appears to us to be rather unseemly." It criticized the plan to have the president "share a split television screen with the two lunar pioneers" and noted that an "unnecessary" presidential conversation with the astronauts as they walked on the moon would cut into the "extremely precious time" available to Armstrong and Aldrin to carry out their scientific program. The *Times* concluded that such a "publicity stunt" was "unworthy of the President of the United States." Richard

Nixon learned of this editorial at his Camp David presidential retreat; typically angry and vindictive, he "wanted action" in response to the *Time*'s criticism, directing Haldeman to "ban" the *Times* from the White House and to organize attacks on the newspaper's views. Nixon assistant Buchanan was asked, in coordination with Borman, to stimulate letters to the editor of the *Times* critical of the paper's position.[19]

The rejected idea of naming the Apollo spacecraft "John F. Kennedy" may have caused confusion among some subsequent accounts of the *Apollo 11* mission. On occasion, it has been suggested in books and documentary films that NASA requested the White House to assign the newly commissioned aircraft carrier *John F. Kennedy* as the recovery ship to be in the central Pacific as the *Apollo 11* crew splashed down after their historic journey, and that the Nixon White House rejected that request. For example, Craig Nelson in his book *Rocket Men* states that "NASA had asked for aircraft carrier *USS John F. Kennedy* to take part [in the recovery] as a tribute to the president's original vision; the Nixon White House gave them *USS Hornet* instead."[20] Nelson gives no evidence for this claim, and the research associated with this book did not reveal either a request for the *Kennedy* from NASA or a denial (which surely would have come) from the White House of such a NASA request. In addition, the carrier *Kennedy* and her battle group were on a just-begun deployment as part of the Sixth Fleet in the Mediterranean Sea in mid-1969; it would have taken a major effort to re-deploy the *Kennedy* to the Pacific Ocean for the sole purpose of being the recovery ship for *Apollo 11*. So the notion that the *Kennedy* might have served as the *Apollo 11* recovery ship if not for Nixon White House ill-will is almost certainly one of the long-standing inaccuracies in the history of *Apollo 11*. (The worst, of course, being that the mission never happened and that there has been since 1969 a well-orchestrated conspiracy to conceal this reality.)

Final Preparations

With most preparations for President Nixon's involvement with *Apollo 11* in place, Frank Borman in early July made a quick visit to the Soviet Union. He had met Soviet ambassador Anatoly Dobrynin in January, and Dobrynin had followed that meeting with an invitation for Borman and his family to visit Moscow. Borman informed Nixon and his national security adviser Kissinger of the invitation, and they urged him to accept. Borman remembers that Nixon "was already intrigued" with the idea of U.S.-U.S.S.R. cooperation in a joint space mission, and he viewed the Borman visit as an "opening wedge" in the process of defining such a mission. Borman was the first U.S. astronaut to visit the Soviet Union, and his trip received positive press coverage there. In a formal meeting with the president of the Soviet Academy of Sciences, Mstislav Keldysh, who was the senior publicly acknowledged official in the Soviet space program, Borman raised the possibility of the United States and the Soviet Union increasing their space cooperation, and got a positive response. On his return to the White House, Borman reported to the president that he had not "gathered much technical information on the Soviets' space program," but had gotten the impression that "the Soviets would be receptive to a joint space mission." The July 1969 Borman visit can thus be seen as a first step leading to the 1975 joint U.S.-Soviet Apollo-*Soyuz* mission with its "handshake in space."[21]

The good relations created by Borman on his trip had an immediate payoff. On July 13, three days before the Apollo launch, the Soviet Union launched the *Luna-15* robotic probe, with the intent of first orbiting, then landing on, the Moon, scooping up some lunar soil, and bringing it back to Earth. There was some concern that the trajectory of the Soviet mission might intersect with *Apollo 11* while both were in lunar orbit, resulting in a collision. At NASA's request, Borman used the White House–Kremlin "hot line" to send a message to Keldysh requesting the orbital parameters of the Soviet probe. On July 17, Keldysh replied with the requested information, saying that "the orbit of probe *Luna-15* does not intersect the trajectory of Apollo-11 spacecraft." Never before had the Soviet Union provided such detailed information on one of its ongoing space missions. While *Luna-15* did reach lunar orbit, it crashed onto the Moon on July 21 as the *Apollo 11* crew was preparing to lift off of the lunar surface.[22]

By July 14, Borman was back from his trip to the Soviet Union; he would stay involved with President Nixon until the *Apollo 11* astronauts were safely back on Earth on July 24. One action Borman took at the president's request was to prepare brief profiles of the *Apollo 11* crew for Nixon and similar profiles of the crew's wives for Mrs. Nixon. With respect to Neil Armstrong, Borman told Nixon that the mission commander was a "quiet, perceptive, thoroughly decent man, whose interests still turn to flying," and that he "follows the stock market actively." Armstrong was "a little reserved, but when you get to know him, he has a very warm personality." Buzz Aldrin was described as "very athletic, aggressive, hard charging," an "almost humorless, serious personality," and "very concerned about social problems." Michael Collins was in "superb physical condition." Collins was "in some sense skeptical, more inclined toward the arts and literature rather than engineering" and a "devoted family man." With respect to the astronauts' wives, Borman described Jan Armstrong as "quite composed and very factual." Joan Aldrin was "more demonstrative than either of the other wives, and perhaps more apt to show her concern." Pat Collins "tends toward the intellectual; [is] very interested in current events"; and "enjoys evenings that include candlelight and wine for dinner."[23]

NASA had sent to the White House proposed remarks for President Nixon to use as he spoke with the astronauts on the Moon. From Borman's perspective, "the gist of those remarks was that the current administration was responsible for Apollo 11's success...The statement was pure politics, an exercise in self-congratulations." Borman advised Nixon not to use NASA's input. He told the president "look, Mr. President, you really don't have anything to do with *Apollo 11*. You're just the fortunate or unfortunate recipient of this mission...If it fails, you'll get tarred with it, and if it succeeds you'll get *some* of the credit. But for you to say what NASA is suggesting—that in effect you were the father of the space program—is just plain wrong." Rather, suggested Borman, the president should say "something very simple and nonpartisan, a few words of congratulations, and then get off the air." Borman also advised against the plan of playing the national anthem as Armstrong and Aldrin stood next to the American flag during the telecast conversation involving the president. This "would force the crew to stand at attention for some two and one-half minutes. This time, plus the time allocated to unveiling the plaque and mounting the flag, would add up to a significant portion of the time on the lunar surface which is non-productive from a scientific or exploration viewpoint."[24]

President Nixon met with Haldeman, Flanigan, Chapin, and Borman on July 14 to discuss plans for his involvement. According to Haldeman, Nixon "was really intrigued with his participation in the whole thing." The plan at this point was for the president to go to either the Manned Spacecraft Center in Houston or the Kennedy Space Center in Florida for his phone call to the astronauts on the Moon; Nixon's long-time personal secretary Rose Mary Woods suggested that the call should instead come from the Oval Office, and the president agreed. Going into the meeting, Nixon was "cranked up" about playing the Star-Spangled Banner when the American flag was placed on the Moon, but he accepted Borman's reservations about that idea, also recognizing "possible adverse reaction to overnationalism."[25]

One more important detail had to be attended to in the final days before the launch: what to do in case of a mission failure involving astronaut deaths, particularly if Armstrong and Aldrin could not lift off the Moon to rendezvous with Michael Collins in lunar orbit. NASA had prepared a disaster contingency plan and sent it to the White House. In addition, Flanigan's assistant Jonathan Rose reviewed with Borman and Safire a "rain plan" in the event of an *Apollo 11* disaster, suggesting the need for a presidential statement and phone calls to the crew's widows, and then a "National Day of Mourning" after the president returned from his around-the-world trip. Borman had earlier urged the president's speechwriters to think about "what to say to the widows," and Safire had prepared a statement in the event that Armstrong and Aldrin were stranded on the Moon. The suggested remarks began by saying: "Fate has ordained that the men who went to the moon to explore in peace will stay on the moon to rest in peace. These brave men, Neil Armstrong and Edwin Aldrin, know that there is no hope for their recovery. But they also know that there is hope for mankind in their sacrifice." The message added: "Others will follow, and surely will find their way home." After the president's statement, at the point when NASA cut off communications with the astronauts, "a clergyman should adopt the same procedure as a burial at sea, commending their souls to the 'deepest of the deep.'"[26] Fortunately, this statement was not needed.

One Small Step

Armstrong, Aldrin, and Collins were launched toward the Moon at 9:32 a.m. (all times are Eastern Daylight Time) on July 16, 1969.* President Nixon watched the launch in the White House together with Borman. Soon after the third stage of the Saturn V booster fired to send the crew on a trajectory that would bring them to the Moon three days later, the White House issued a presidential proclamation designating July 21 as a "National Day of Participation." The statement declared "Apollo 11 is on its way to the moon...Never before has man embarked on so epic an adventure." It noted that "in past ages, exploration was a lonely enterprise. But today the miracles of space travel are matched by the miracles of space communication...Television brings the moment of discovery into our homes, and makes all of us participants." Indeed, the *Apollo 11* mission was the first event to be televised globally; the communications satellite required to complete a global network had been put into orbit over the Indian Ocean only a few days earlier. Nixon ordered all federal government offices to be closed on July 21; he urged "the Governors of the States, the mayors of cities, the heads of school systems, and other public officials to take similar action" and "private employers to make appropriate arrangements so that as many of our citizens as possible will be able to share in the significant events of that day." While Armstrong and Aldrin were scheduled to land on the Moon on the afternoon of July 20, their mission timeline called for a sleep period before emerging from *Eagle* for their historic moonwalk sometime after 2:00 a.m. on the morning of July 21. One purpose of declaring July 21 as what amounted to a national holiday was to allow as many as possible to stay up well past midnight to watch the first steps on the Moon without having to worry about getting up to go to work the same morning.

On the morning of July 20, President Nixon presided over an interdenominational church service in the East Room of the White House. The service was attended by some 300 people, including cabinet secretaries, members of Congress, and the diplomatic corps. Borman read the same verses from the Bible that he and his crew had read as they circled the Moon on Christmas Eve, and a Quaker minister provided the sermon.[27]

After a virtually trouble-free voyage, the *Apollo 11* spacecraft went into orbit around the Moon on July 19, and at 1:44 p.m. on July 20 the lunar module *Eagle* separated from the command and service module *Columbia* to begin its descent to the lunar surface. After a hair-raising final few moments which saw Neil Armstrong take over manual control of *Eagle* to pilot the spacecraft to a safe landing spot, *Apollo 11* landed on the Moon at 4:18 p.m. A few seconds later, Armstrong reported "Houston, Tranquility Base here. The *Eagle* has landed." Accompanied by Borman, Nixon watched the landing on television in his hideaway office in the Executive Office Building next to the White House.

---

* The author had the good fortune to be present at the *Apollo 11* launch.

Even before *Apollo 11* lifted off, the crew and mission planners back in Houston had agreed that if all was going well, Armstrong and Aldrin would skip their scheduled rest period and start their extra-vehicular activity on the lunar surface as soon as they were ready. Within an hour after landing, Armstrong received permission to begin the crew's moonwalk at approximately 9:00 p.m. Informed of this change in plans, President Nixon arrived in the White House office area just before 9:00 p.m., only to be advised that preparations were running more slowly than expected. Almost two hours later, Armstrong stood on the outside of the lunar module, ready to climb down to the surface of the Moon. A worldwide audience watched his ghost-like image descend the module's ladder; then, Armstrong announced that he was ready to step off the lunar module. He took his historic "one small step for [a] man, one giant leap for mankind" at 10:56 p.m. on July 20, 1969. (In the excitement of the moment, Armstrong did not fully articulate the "a" in his statement, although some later acoustic analyses suggested that he had indeed included the article in what he said. In retrospect, Armstrong himself was typically enigmatic, saying to his biographer "I would hope that history would grant me leeway for dropping the syllable and understand that it was certainly intended, even if it wasn't said—and it actually might have been.[28]) Aldrin soon followed Armstrong to the lunar surface, stepping off the lunar module at 11:15 p.m.

President Richard Nixon talks to Neil Armstrong and Buzz Aldrin on the surface of the Moon, July 20, 1969. (NASA photograph GPN-2000-1672)

President Nixon watched the historic first steps on the Moon on a small television in his private office in the White House, next to the more formal Oval Office. Borman and Haldeman were with him. According to Haldeman, Nixon was "very excited by the whole thing. Was fascinated by the moon walk." The president then went into the Oval Office, where from 11:45 to 11:50 p.m., in the dispassionate words of the his official "Daily Diary," he "held an interplanetary conversation with the Apollo 11 astronauts Neil Armstrong and Edwin Aldrin on the Moon." The conversation was shown on split-screen television and seen live around most of the world, but not in the Soviet Union.[29]

Nixon had available to him for this conversation two different versions of prepared remarks, one written by lead speechwriter Ray Price and the other by William Safire, but he used neither version. Borman says that he and Safire composed the actual comments, while Haldeman suggests that Nixon "wrote his own remarks." Safire recalls that he was watching the preparations for the moonwalk from his home and was struck by the idea that the president should work the theme of "tranquility" into his remarks, given that *Eagle* had landed on the Moon's Sea of Tranquility. Safire called the White House and asked that his thought be relayed to the president as he prepared for his *Apollo 11* phone call. Whatever the source of the rhetoric, what the president said reflected the themes—pride, power, and peace—that Nixon had from the start of his preparations wanted to associate with the lunar landing. Nixon told Armstrong and Aldrin as they stood beside the American flag on the lunar surface:

> Hello Neil and Buzz, I am talking to you by telephone from the Oval Room at the White House, and this certainly has to be the most historic telephone call ever made from the White House.
>
> I just can't tell you how proud we all are of what you have done. For every American this has to be the proudest day of our lives, and for people all over the world I am sure that they, too, join with Americans in recognizing what an immense feat this is.
>
> Because of what you have done the heavens have become a part of man's world, and as you talk to us from the Sea of Tranquility, it inspires us to redouble our efforts to bring peace and tranquility to earth.
>
> For one priceless moment in the whole history of man all the people on this earth are truly one—one in their pride in what you have done and one in our prayers that you will return safely to earth.

Armstrong replied to the president: "It is a great honor and privilege for us to be here representing not only the United States, but men of peaceable nations, men with an interest and a curiosity, and men with a vision for the future."[30]

The president's phone call came as a complete surprise to Aldrin, who found it "awkward" and decided not to respond. Armstrong had been alerted before launch that there might be a "special communication" while the two astronauts were on the Moon, but he was not told that it would be

President Nixon on the line. Armstrong did not share this "heads up" with Aldrin. Armstrong later suggested that "If I'd known it was going to be the president, I might of tried to conjure up some appropriate statement." Armstrong's not sharing his advance information with Aldrin was typical of the relationship between the members of the *Apollo 11* crew, described by Collins as "amiable strangers."[31]

On the morning of July 21, the front page of the *The New York Times* in a 96-point banner headline announced "Men Walk on Moon." (In the early edition of the paper, sent to press before Aldrin had joined Armstrong on the lunar surface, the headline had been singular—"Man Walks on Moon.") The newspaper also included on its front page the poem Archibald MacLeish had composed to commemorate the occasion, titled "Voyage to the Moon."[32]

Eagle with Armstrong and Aldrin and 49 pounds of lunar samples aboard lifted off of the Moon's surface at 1:54 p.m. on July 21, first to rendezvous in lunar orbit with *Columbia*, where Collins had been patiently waiting, and then to head back for an early morning splashdown in the South Pacific on July 24. The crew had little to do on the return trip, and reverted to characteristics that Borman had noted in his July 14 memo to Nixon. Armstrong asked mission control for a report on the stock market, and Collins rummaged around the various storage areas of the spacecraft, hoping, with tongue in cheek, that someone had surreptitiously smuggled aboard a small supply of cognac.[33]

Welcome Back to Earth

President Nixon and a large entourage left Washington on the evening of July 22 to begin the trip to the *Apollo 11* splashdown and then to undertake the president's round-the-world diplomatic mission. After spending the night in San Francisco, on July 23 they flew to Johnston Island, a small atoll 750 miles west of Hawaii. During that flight, Nixon, NASA Administrator Paine, and national security adviser Kissinger spent some time discussing the president's desire to increase international participation in the U.S. space program; Paine remembers that "we made a great deal of progress in laying out the plan for international cooperation."[34] Borman was also aboard Air Force One, and met separately with the president and Kissinger, also to discuss international space cooperation.

The president's party arrived on Johnston Island at 5:00 p.m. local time. Those of the group that would view the *Apollo 11* splashdown then boarded helicopters for the hour and a half trip to the aircraft carrier *Arlington*, where they would spend the evening. As he had earlier met with Ehrlichman to plan his trip to meet the returning *Apollo 11* astronauts, President Nixon had attempted to stage manage his trip to the splash down. He recognized that Secretary of State William Rogers and Kissinger would have to be part of the diplomatic trip, but he did not want them to accompany him to the recovery; instead, Nixon declared, they would stay on Johnston Island awaiting his return. Nixon did not want to share the event with

President Nixon, *Apollo 8* astronaut Frank Borman (right) and Admiral John McCain (left) watch as the recovery carrier *Hornet* approaches the *Apollo 11* capsule after it splashed down in the Pacific Ocean on July 24, 1969. (NASA photograph 6900598)

a large entourage; his presence as the crew returned from the Moon was "to be *his* triumph, not *theirs*." Nixon told Ehrlichman that "no staff—no Dr. [Doctor]—only two SS [Secret Service]—no press pool—nobody" was to ride on his helicopter to the recovery carrier. Ehrlichman described these directives as an example of the "forlorn and impossible wishing game he liked so well." He added "as he knew it would, Nixon's entourage at the splashdown included the full complement of bodyguards, a vast press contingent, the President's doctor, Haldeman, Haldeman's aide, and, of course, both Rogers and Kissinger."[35]

Haldeman in his diary described in vivid detail both the trip from the *Arlington* to the smaller recovery carrier *Hornet* to view the splashdown and the event itself:

> Up at 4:00 for 4:40 departure. It was beautiful on the flight deck, absolutely dark, millions of stars, plus the antenna lights on the ship. Borman said it looked more like the sky on the back side of the moon than any he had ever seen on earth. Helicopter left in the dark and flew over the ocean to the *Hornet*. Landed and went through quick briefings on the decontamination setup and the recovery plan. Then waited on the bridge for the capsule to appear.
>
>   It did, in spectacular fashion. We saw the fireball (like a meteor with a tail) rise from the horizon and arch through the sky, turning into a red ball,

then disappearing. Waited on the bridge for an hour or so until we could see the helicopters over the capsule and raft in the sea. We steamed toward them. Watched the pickup, first through binoculars, then with naked eye. P [the president] was exuberant, really cranked up, like a little kid. Watched everything, soaked it all up.

Then the pickup helicopter landed on deck. P ordered band to play "Columbia the Gem of the Ocean."...Then down to the hangar deck for P chat with the astronauts in quarantine chamber. Great show. He was very excited, personal, perfect approach. Then prayer and "Star Spangled Banner." Then "Ruffles and Flourishes" and "Hail to the Chief," and we left.[36]

The *Apollo 11* command module *Columbia* splashed down on target at 5:51 a.m. local time (12:51 p.m. EDT) on July 24, 13 miles from the *Hornet*. After donning their "biological containment garments," Armstrong, Aldrin, and Collins were helped from their spacecraft into a raft, then lifted into a waiting helicopter. By now, the *Hornet* was only a quarter of a mile away, and the helicopter carrying the *Apollo 11* crew landed on its deck at 6:57 local time.

The astronauts had an hour before interacting with the president, first undergoing a quick medical examination, then taking a shower and changing into comfortable clothing. Armstrong later reflected "there were the Nixon ceremonial activities to attend to. We needed to do that and get it behind us so we could celebrate." Collins added that after showering and shaving, "we were looking for something to do, and it's not long in coming." The crew was summoned to the end of the quarantine facility and "parting the curtains we see that the hangar deck has arranged for some sort of ceremony—the first of many, I would guess." After the band played *Ruffles and Flourishes*, "in marches none other than President Nixon, looking very fit and relaxed as he stands by a microphone just outside our window." As he spoke with the crew, Nixon demonstrated his lack of facility with small talk, attempting to joke that his conversation with the crew while they were on the Moon was a collect call, pointing out Frank Borman standing nearby, and asking whether the crew knew the results of the baseball All Star game and whether they were fans of the American or National League. One of Nixon's biographers suggested that his conversation "set some sort of record for inappropriateness." He told the astronauts that he had spoken to their wives—"three of the greatest ladies and most courageous ladies in the whole world today"—and had invited them to a dinner on August 13. He asked the crew "Will you come?" Demonstrating his "penchant for hyperbole and weakness for gross exaggeration," Nixon "came out with the all-time Nixonism," telling the crew that "this is the greatest week in the history of the world since the Creation, because as a result of what happened in this week, the world is bigger, infinitely" and "as a result of what you have done, the world has never been closer together before."[37]

Shortly after 9:00 a.m., President Nixon and his party boarded their helicopters for the return trip to Johnston Island; by early that afternoon, they were on their way to Guam, the first stop in a tour that would bring

President Nixon jokes with the *Apollo 11* crew in their mobile quarantine facility. (Photo courtesy of Milt Putnam, the Navy photographer who recorded the recovery of the *Apollo 8, 10, and 11* crews after their return from the Moon.[38])

Nixon to the Philippines, Indonesia, Thailand, Vietnam, India, Pakistan, Romania, and the United Kingdom before returning to Washington on August 3. At each stop on the journey, Nixon evoked "the spirit of Apollo 11." For example, when he landed in Manila, the president said "as we think of that great venture into space, as we think of the first man setting foot on the moon, we realize the meaning that that has, clearly apart from the technical achievement, we realize that if man can reach the moon, that we can bring peace to the earth. And that should be the great lesson of that great space journey for all of us." In Romania, Nixon added "mankind has landed on the moon. We have established a foothold in outer space." He added "but there are goals that we have not reached here on earth. We are still building a just peace in the world. This is a work that requires the same cooperation and patience and perseverance from men of good will that it took to launch that vehicle to the moon. I believe that if human beings can reach the moon, human beings can reach an understanding with each other on the earth." As had been planned from the start of the Nixon administration, and indeed from 1961 as President Kennedy had laid out his rationale for sending Americans to the Moon, the *Apollo 11* triumph was used by President Nixon as a powerful tool in Earth-bound diplomacy.[39]

Missing in Richard Nixon's communications during the *Apollo 11* mission and his subsequent world tour was any mention of John F. Kennedy. However, some in NASA did recognize President Kennedy's role. As the *Apollo 11* spacecraft splashed down in the Pacific Ocean, one of the video screens in the front of the mission control room at NASA's Manned Spacecraft Center in Houston had displayed Kennedy's 1961 challenge, while another screen noted simply: "Task Accomplished."[40]

On the evening of August 10, 21 days after leaving the surface of the Moon, the *Apollo 11* crew members were released from their Houston quarantine. Early on the morning of August 13 they left Houston on what promised to be an exhausting day. The crew and their wives and children were flown by Air Force Two to New York City for a ticker-tape parade. According to Armstrong's biographer, "not even the revelry at the end of World War II or the parade for Lindbergh in 1927 matched in size" the crowd watching the crew's parade through Manhattan; one estimate of the turnout was 4 million people. Then on to Chicago, where the crowds were "even wilder." Finally the astronauts arrived in Los Angeles for the huge dinner celebrating their mission.

Richard Nixon acted as master of ceremonies for the evening. The assemblage included representatives of 83 countries, governors from 44 states, 14 members of the president's cabinet ("More members of the Cabinet than are usually present at a Cabinet meeting," joked Nixon), the chief justice of the Supreme Court, 50 members of Congress, a bevy of Hollywood stars, NASA officials and astronauts, aerospace industry executives, and the man who Nixon had defeated in the contest to be president, former Vice President Hubert Humphrey. At the culmination of the evening, Vice President Spiro Agnew presented the Medal of Freedom, the nation's highest civilian award, to each of the *Apollo 11* crew members. Then the astronauts spoke. Michael Collins said "here stands one proud American, proud to be a member of the Apollo team, proud to be a citizen of the United States of America which nearly a decade ago said that it would land two men on the moon and then did so, showing along the way, to the world, both the triumphs and the tragedies—and proud to be an inhabitant of this most magnificent planet." Buzz Aldrin added, "There are footprints on the moon. Those footprints belong to each and every one of you, to all of mankind, and they are there because of the blood, the sweat, and the tears of millions of people. These footprints are a symbol of the true human spirit." Neil Armstrong hoped that "this is the beginning of a new era, the beginning of an era when man understands the universe around him, and the beginning of the era when man understands himself." President Nixon closed the evening, saying "It has been my privilege in the White House, and also in other world capitals, to propose toasts to many distinguished people, to emperors, to kings, to presidents, to prime ministers...Tonight, this is the highest privilege I could have, to propose a toast to America's astronauts." Reflecting on the event the next day, Haldeman suggested that the "dinner was a truly smashing success...Highly emotional and patriotic evening that completely succeeded in meeting all the P's objectives. Well worth all the work."[41]

"Giant Step": the *Apollo 11* World Tour

Although both NASA and the White House certainly expected that at some point after their mission the *Apollo 11* crew would embark on an international tour, there were no concrete plans for such a junket in place at the time of the *Apollo 11* gala dinner. One characteristic of the Nixon White House evident early on was the intent to exercise close control over executive agency activities of direct interest to the president; there was little trust in the career bureaucracy. By early August, the White House was becoming increasingly impatient to hear from NASA regarding plans for the *Apollo 11* tour. On August 6, three days after the president returned from his round-the-world trip, Nixon's assistant Peter Flanigan wrote to NASA's Julian Scheer, saying "No doubt you will be arranging for international trips for the Apollo 11 astronauts." Flanigan requested that "before any specific schedule has been agreed upon, we would appreciate an opportunity to have the chief scheduler sit down here at the White House with the appropriate members of the White House, the National Security Council and the State Department [so] that we can coordinate the proposed schedule." Five days later, Flanigan again wrote Scheer, this time saying "the President has again asked that he personally have an opportunity to review the Apollo 11 astronauts' foreign travels. He has some strong opinions on this matter and wants to make sure he can express those opinions before any commitments are made." Flanigan added "he is also anxious that there be some movement along this line, so I would appreciate hearing NASA's thoughts with regard to the schedule in the near future." On August 14, Nixon told Haldeman that the White House should control the tour schedule, with "no countries included w/o WH [without White House] approval." As a result, Flanigan on August 15 wrote NASA administrator Paine, saying "the President is most anxious that the Apollo 11 astronauts commence their world-wide trip as soon as possible."[42]

On August 15, the same day that Flanigan wrote Paine, Scheer finally replied, sending Flanigan a plan for the crew in the United States, to include an appearance before a joint session of Congress, as well as suggested "operational guidelines for the overseas tour" and a proposed itinerary. Scheer noted that the plan was put together "with the guidance of U. Alexis Johnson of the Department of State." Johnson was a veteran diplomat, then undersecretary of state for political affairs, who had long involvement in space policy matters and was at the time part of the White House review of post-Apollo space plans. With respect to the proposed itinerary, Scheer noted that "it was more than advisable: 51 days, 28 countries and 30 cities. We would like to reduce this by 10 days." With respect to the trip's guidelines, Scheer suggested that "the Apollo 11 astronauts represent the President on a Presidential 'Spirit of Apollo' world trip." He noted that "a Presidential aircraft, such as Air Force 2/3, is important for image purposes overseas." Scheer proposed that

NASA supply both the "Chief of Mission" and the "Mission Director," with a supporting staff of 11 additional NASA people; there would be four people from the U.S. Information Agency and only one from the Department of State in the traveling party.[43]

Little in what NASA was proposing was acceptable to the White House, which wanted a "highly political and carefully choreographed" tour designed to "reward friends, snub foes" and to produce "a flood of positive foreign headlines." Nixon, reflecting his August 14 decision to take over from NASA the responsibility for planning the astronaut trip, told Kissinger "if you leave things in their [government bureaucrats] hands like this, they come out with an utter disaster." Flanigan told Scheer on August 23 that "the President was dismayed at the proposed foreign schedule for the astronauts," believing that "it went to too many countries, many of which were unimportant, while leaving out others of considerably greater importance." Flanigan announced to NASA in no uncertain terms that "the President has given the White House staff the responsibility for reconstructing this schedule" and that "as soon it is completed it will be sent to you." To make sure his point was clear, Flanigan added "Please be sure that all interested parties know that this is now a White House responsibility."[44]

On August 26, completing the White House takeover of the trip planning, Flanigan informed Administrator Paine that the astronauts would indeed "tour the world as his [the President's] representatives." Rather than NASA managing the tour, Nicholas Ruwe, a senior Department of State protocol officer, was designated "Chief of Mission" and would be "responsible to the President for its successful completion." Both NASA and the State Department would provide staff, but only "as requested by the White House."[45]

NASA was not at all pleased by the White House intervention in the tour arrangements; tension between Scheer, particularly, and the White House ran high. Ruwe on September 23, a week before the tour was to commence, reported to Kissinger "NASA and I are at complete loggerheads with regard to the execution of the Apollo 11 trip." Dissatisfaction with tour planning extended to Armstrong, Aldrin, and Collins themselves. On September 17, the day after they had addressed a joint session of Congress, the three were briefed at the State Department with respect to tour preparations. The astronauts had set as their objectives for the trip "to demonstrate goodwill to all people in the world and to stress that what we had done was for all mankind." According to Aldrin, they were not impressed when they perceived from their briefing that an important objective of the tour was "to visit the American embassies anxious to score social coups." The crew's response was "we would take care of Americans in America."[46]

The *Apollo 11* tour was code-named "Giant Step." It departed on September 29, with the first stop being Mexico City. The day before, Nixon, reflecting his personal concern that the tour serve his broader purposes, called Armstrong to give him some final thoughts. Using talking points prepared by Borman, Nixon urged Armstrong to convey to the leaders in each of the countries visited that the *Apollo 11* flight and the astronauts' tour represented "the interest of the United States in maintaining space exploration as a project of peaceful benefits for all nations of the world." He suggested that Armstrong might repeat what the president had said during his post-mission trip—that "the success of the Apollo XI mission belongs to all the people of the earth and not just the people of the United States."[47]

The crew visited 27 cities in 24 countries over 39 days. They returned to Washington on November 5. Neil and Jan Armstrong and Mike and Pat Collins enjoyed most of the exhausting trip; Collins remembers that "despite the fatigue and the repetitive nature of the ceremonies," the tour "was the rarest of opportunities, to cram in slightly over a month's time visits with the Queen of England, Marshal Tito, the Pope, the Emperor of Japan, the Shan of Iran, Generalissimo Franco, Badouin King of the Belgians, King Olaf of Norway, Queen Wilhemina of the Netherlands, the King and Queen of Thailand, and dozens of Presidents, Prime Ministers, ambassadors, and lesser lights." In contrast, Buzz Aldrin found the trip extremely stressful, and became increasingly depressed as the tour continued; he and his wife were at times not on speaking terms.[48]

When the crew arrived back in Washington, they went by helicopter directly to the South Lawn of the White House. There they were welcomed by President Nixon, speaking "for all of the American people in expressing the heartfelt thanks of this Nation to the Armstrongs, the Aldrins, and the Collinses for what I think is the most successful goodwill trip in the history of the United States of America... Certainly the first men ever to land on the moon have demonstrated that they are the best possible ambassadors America could have on this earth." That evening, President and Mrs. Nixon hosted a White House dinner; the only other people present were the crew members and their wives. Aldrin remembers a "friendly, warm evening." The president told the crew that he had used his stop in Romania in his around-the-world tour to send a secret message to China's leaders that he was open to normalizing U.S.-Chinese relations and said that opportunity had "paid for everything we spent on the space program." He asked each crew member what they wanted to do next. While Armstrong and Aldrin were non-committal, Collins expressed interest in continuing work in public diplomacy. In a conversation with NASA Administrator Paine even before leaving on the "Giant Step" tour, Collins had learned that Secretary of State Rogers had expressed interest in Collins becoming the assistant secretary of state for public affairs. Collins told Nixon of his interest in that position. The president immediately called Rogers, telling him that Collins would be an excellent fit for the job. After dinner, Pat Nixon led a tour through the White House and the Executive Office Building next door. When the crew had interacted with the First Lady at the August 13 banquet, they had found her distant and stiff. Now, she was "charming," a "delightful, warm hostess who really tried to make us feel at home"; the tour was carried off "with unexpected enthusiasm and a beautiful informality." The three astronauts and their wives then spent the night at the White House. A few weeks later, "Giant Step" would be resumed for a two-day trip to Canada, but the White House evening provided a satisfying conclusion to the mission of *Apollo 11* and its immediate aftermath. According to Collins, Mrs. Nixon's hospitality "made our stay at the White House the real highlight of our around the world trip."[49]

Now What?

The excitement of *Apollo 11* had barely begun to diminish when on September 15 President Nixon received the report of the "Space Task Group" he had created in February 1969 to recommend the course of the post-Apollo space program. That report laid out an ambitious plan, culminating in human trips to Mars sometime in the next 15 years. The president was soon to decide that the nation neither wanted nor could afford that kind of ambition in space. But this "deceleration" of the U.S. space program was still in the future as Richard Nixon and his associates made sure that the president was closely identified with the success of Apollo 11, even though he had only the good fortune to be the occupant of the White House when the lunar landing occurred. One way of emphasizing the linkage between the president and the mission's success was a purposeful ignoring in Nixon's statements related to *Apollo 11* of the role of the two presidents actually responsible for Apollo—Lyndon B. Johnson, who had provided steady support for the project during his five years in the White House, and especially John F. Kennedy, who had the original vision of using a mission to the Moon as an instrument of U.S. grand strategy and then had backed up that vision with a massive commitment of human and financial resources. Richard Nixon was able to harvest the fruits of Kennedy's and Johnson's nurturing of Apollo without any additional commitment of tangible resources on his part. His major, and not insignificant, contribution was linking the prestige of the office of the president of the United States to the Apollo achievement. He did so skillfully, personally orchestrating his engagement with the lunar landing and its aftermath. Nixon took some significant risks along the way. If there had been a mission failure at some point or if the *Apollo 11* crew members had not been so successful in their unaccustomed role as global diplomats, the "spirit of Apollo" that President Nixon so effectively used to signal U.S. determination to maintain global leadership might not have been so potent a symbol. But NASA delivered extraordinary results in carrying out the first landing on another celestial body, and Richard Nixon was able to leverage that success to a major strategic triumph for the United States.

Chapter 2

Setting the Post-Apollo Stage

While Richard Nixon's involvement with the *Apollo 11* mission provided the background to the first steps in the process of deciding what the United States would do in space after reaching the Moon, it did not create the positive momentum needed to overcome both skepticism on the part of those advising the new president about the value of continuing a fast-paced and expensive program of space activities after Apollo and the reality that NASA was ill-prepared to face its future. All involved recognized that there was a need for decisions on what would follow Apollo, but they approached that imperative with widely differing perspectives. It took almost a year to make and announce an initial judgment—that the United States would *not* continue an Apollo-like program of space development and exploration. The confused process of reaching this outcome is described in this and the following four chapters, which together constitute the first act of the post-Apollo drama.

Candidate Nixon and Space

Richard Nixon would face his decisions on the future in space with some background in space policy, particularly in comparison to John Kennedy as he became president eight years earlier. Then, a leading journalist had observed "of all the major problems facing Kennedy when he came into office, he probably knew and understood least about space."[1] Nixon as Dwight Eisenhower's vice president had an early impact on the organization of the U.S. space effort. In a February 4, 1958, meeting in which President Eisenhower discussed how the United States should organize its response to the October and November 1957 launches of *Sputniks 1* and *2* by the Soviet Union, Nixon had suggested that "our posture before the world would be better if non-military research in outer space were carried forward by an agency entirely separate from the military." Nixon judged that having a separate agency for "peaceful" research projects would also make possible a broader range of internationally cooperative space activities. Eisenhower accepted this advice, which came not only from Nixon but from other sources; the result was the president's April 1958 proposal to create the National Aeronautics and Space Administration (NASA) as a civilian agency. Nixon's 1968 transition task force on space noted that "separation of the space program into a part directed towards military applications in the DOD and a largely unclassified part without strong military coloring in NASA has, we believe, been an eminently wise policy."[2] Richard Nixon was an early advocate of that policy.

One account of President Eisenhower's measured response to Sputnik notes that Nixon "was far more attuned than Eisenhower to the political ramifications of space." In White House discussions, Nixon suggested "we can make no greater mistake than seeing this as just a Soviet stunt. We've got to pull up our socks and get with it and make sure we maintain our leadership." This account suggests that, had he been elected president in 1960, Nixon "would have pursued a [space] policy more active and flashy than Eisenhower's." Nixon agreed with this assessment; in his *Memoirs* he suggested that in cabinet and National Security Council meetings in the final years of the Eisenhower administration, he "strongly advocated a sharp increase in our...space program." Once he was in the White House, however, Nixon did not follow this path, instead continuing the reductions in NASA's budget that had begun under Lyndon Johnson. To Nixon, in a theme that he would frequently repeat in his White House years, "when a great nation drops out of the race to explore the unknown, that nation ceases to be great"; like many Nixon pronouncements, this was more an empty rhetorical statement than a guide to his policy and budget decisions.[3]

There was little or no Nixon involvement in space issues between his defeat in the 1960 presidential election and his selection as the Republican nominee for president in August 1968. However, a few days after his February 1, 1968, announcement that he would be a candidate for that nomination, Nixon told a space-interested audience in Washington that "the United States must remain competitive in this field, and we must support a space program which is second to none. That's looking at it in long-term objectives." But in the shorter term, Nixon added "I believe that space is one of the areas that will have to be in the [next] President's recommendations for budget-cutting...With the immense financial crisis which currently confronts the United States, we will have to make some cuts." These views foreshadowed the approach to space issues that Nixon would actually pursue as president, but they were articulated before the glare of campaign attention had begun. As candidate for president, Richard Nixon was much more bullish, telling audiences in Texas and Florida that the "space program was indispensable and of major importance to our country," that in space "we must do all that we can," that the space program was "a national imperative," and that the United States "must be first in space." How candidate Nixon's general statements on space might translate into specific decisions was not made clear. As one observer commented after Nixon's election in November 1968, his statements during the campaign "provide few clues as to what he will really do"; the president-elect's views of the future of the space program were "as obscure...as his intentions across the spectrum of national problems."[4]

NASA Not Ready for Success

While Richard Nixon came to the White House knowing that he would soon have to make choices regarding the future of the United States in space, the NASA leadership was not well prepared to present the new president with attractive options for that future. At what should have been a moment of great triumph, with the spectacular success of the bold *Apollo 8* mission and with the first landing on the Moon just months in the future, the top officials of NASA in January 1969 did not have a clear sense of what might best follow Apollo. According to one of those officials, "the general atmosphere [among NASA's leaders] in terms of decisiveness, purpose, dynamics—a feeling that you were in an agency moving forward—that was not there." Those at the helm of NASA did not accurately perceive the broad societal changes that would influence political decisions on what space future was sustainable; "the dramatic political, cultural, and socioeconomic changes of the tumultuous decade of the 1960s" had left NASA, focused on the Cold War goal of beating the Soviet Union to the Moon, "in a time warp not completely of its own making." Apollo's message of America's technological power stemming from the concerted actions of government and industry "ran up against a powerful shift in American culture that was beginning to push in the opposite direction, and which ultimately undermined the very premise (and promise) of the manned space program."[5] Decisions on the post-Apollo space program would be made in a very different context than that existing as John F. Kennedy in 1961 decided to send Americans to the Moon.

NASA Resistance to Facing Its Future

James Webb had been NASA administrator from 1961 until he resigned in October 1968. Webb had seen as his overriding responsibility making sure that the Kennedy commitment to a lunar landing was carried out. With this as his focus, Webb had resisted agency-wide planning for what NASA should undertake in the post-Apollo period. According to Willis Shapley, one of Webb's close associates at NASA, Webb "refused to the extent possible to recognize the importance" of post-Apollo planning. Webb did believe, as a "fundamental tenet," that "we could not or should assume that the Apollo program would be a total success, and certainly not assume that it would be a total early success." Webb felt "that nothing should be allowed to dilute the focus of the program we had taken on already, and that we should not start dreaming about what would take place after that."[6] (Shapley as NASA associate deputy administrator had a major role during the period examined in this study in developing NASA's strategy and policies and articulating them to the White House and Congress. He was a prime example of a "faceless bureaucrat" who plays a key behind-the-scene policy role, in this case with respect to the nation's civilian space program.) Webb's perspective also reflected political reality. President Lyndon B. Johnson had made sure that the NASA budget remained adequate to assure Apollo's success, but faced with spiraling costs of the Vietnam War and of his Great Society programs as well as with widespread domestic unrest, he was unwilling to approve a NASA budget at a level that could support major new space initiatives. NASA itself was a badly divided organization, with its Office of Manned Space Flight and its human space flight centers in Houston, Texas and Huntsville, Alabama planning their own course for the future, while its Office of Space Science and Applications worked with the external scientific community to define a different preferred future, one which would redress the perceived imbalance between human and robotic space missions. As a result of Webb's resistance, agency-wide planning for the post-Apollo period began only in early 1968, and its early results were disappointing, reflecting the divisions within the organization.

An Unhappy Webb Leaves NASA

James Webb had insisted from the early years of Apollo that the undertaking was about much more than landing men on the Moon. Rather, its purpose was "to become preeminent" in all areas of space activity, and to do so "in such a manner that our emerging scientific, technological, and operational competence in space is clearly evident." To Webb (and John Kennedy), the space program was an instrument of national power, not an enterprise driven by the human desire to explore. In order to make sure that there was enough equipment to achieve the lunar landing goal, NASA ordered 15 Saturn V Moon rockets, 15 lunar landing spacecraft, and 20 command and service module spacecraft. The expectation was that most of this hardware would be necessary to assure Apollo's success; it seemed likely that a number of attempts would have to be made to achieve the various milestones in the lunar landing program.[7]

At the peak of the Apollo buildup in fiscal year (FY) 1965, NASA's budget was $5.25 billion; just four years later, the budget had shrunk by some 20 percent, to $3.99 billion, and NASA had only a few approved human space flight missions for the 1970s. Clearly NASA needed new objectives if it were to maintain the skilled workforce assembled for Apollo and other elements of its rapid 1960s buildup and to make use of the facilities and capabilities in which the nation had invested billions of dollars.

Given this lack of future large missions, Webb on August 1, 1968, refused to approve a request to begin procurement of "long-lead-time" items for the Saturn V Moon rocket, beginning the process of shutting down the booster's production line. This decision was deeply disappointing to Webb. It represented "only the most recent in a series of cutbacks that constitute what may be called a national decision." To Webb, that decision was "that the United States is not pursuing, for the time being at least, its goal of 'preeminence' in space."[8]

By mid-1968, James Webb was "noticeably very, very tired." Webb had for some time planned to retire from NASA before the 1968 presidential election. On September 16, 1968, he went to the White House to discuss the timing of his resignation with President Johnson. Given Webb's unhappiness with

Johnson's recent lack of support for NASA, it is likely that he made his disappointment known to the president. Johnson himself was eager to escape from the burdens of the presidency, and he was not very receptive to Webb's concerns. Somewhat to Webb's surprise, Johnson immediately accepted Webb's resignation, effective on Webb's 62nd birthday, October 7, and sent Webb to the White House press room to announce that action. Asked by a reporter to comment on the status of the space program, Webb responded "I am not satisfied with the program. I am not satisfied that we as a nation have not been able to go forward to achieve a first position in space." Commenting on Webb's departure, *The Washington Post* noted that he was leaving NASA without its having "a set mission beyond landing on the moon...The fading American taste for competition with the Russians in space and the rising competition of other claimants for Federal funds explains NASA's uncertain estate." The situation was "hardly his fault," but for Webb, "it is a bitter pill."[9]

Enter Tom Paine

Even before going to the White House press room after his meeting with Johnson, James Webb had made a quick call to NASA Deputy Administrator Thomas O. Paine, telling Paine that his resignation was about to be announced and that the president wanted Paine to serve as acting NASA administrator. This shift in command marked a new era for NASA; Tom Paine had a markedly different personality than James Webb. Where Webb was a consummate Washington insider, skilled in forging political coalitions in support of NASA's programs but careful not to get out in front of what in his judgment was politically acceptable, Paine was a Washington outsider, naive in political dealings, ebullient, and a technological visionary. He had been a submarine officer during World War II and had a fascination with all things naval. Paine had a doctorate in physical metallurgy from Stanford and had spent his whole professional career with General Electric. Since 1963 he had been the manager of the General Electric "think tank" called TEMPO; there he was exposed to a wide variety of innovative technological ideas in both the civilian and national security sectors. He had had no particular exposure to the space program prior to coming to NASA. Paine had decided that some Washington experience would be good for his career and had put his name on file with the Civil Service Commission as a person interested in a high-level government position; it was there that NASA found him in January 1968 as it searched for a replacement for Deputy Administrator Robert Seamans.[10]

In his early months as NASA deputy administrator Paine told senior NASA managers that he saw the position of the United States in space "as somewhat analogous to that of the Atlantic Coast of Europe in the 15th century. We have small ships and crude but usable navigational systems and life-support techniques." The question for the future, he thought, was "how should we structure our efforts to build navigation capability and conduct exploration?" Paine saw NASA as analogous to the Portuguese "Research Institute for Navigation" that had been established in 1418 by Prince Henry

Thomas O. Paine, NASA Acting Administrator and Administrator, 1968–1970. (NASA photograph)

the Navigator. That "maritime NASA" was "probably as significant as the later dramatic and successful Portuguese voyages of discovery," Paine suggested, because "it provided a central focus for the best European cartographers, astronomers, navigators, shipwrights, riggers, gunners, coopers, and other medieval scientists, technologists, and skilled workers." This emphasis on maritime technology, he noted, was "the base on which the Spanish and later the British, French, and Dutch empires were founded, spreading European seacoast culture, technology, and languages around the world." Paine wondered whether the United States could have "an analogous opportunity in space."[11] It would have been hard to conceive of Jim Webb pursuing this line of thought.

As Paine took over the direction of the space agency in October 1968, he urged people at NASA to be bolder in their thinking than they had been while Webb was administrator. New in Washington, believing strongly in the historical importance of the space program, and optimistic that he could convince others of that importance, Paine faced the incoming Nixon administration with anticipation, telling a reporter soon after the presidential election that he would present the new president "with an ambitious agenda for future man-in-space flights."[12]

What to Do after Apollo?

By early 1968, James Webb had grudgingly come to accept the need for NASA to begin to plan for its future. He first commissioned an internal study led by one of NASA's most senior people, director of NASA's Langley Research Center Floyd Thompson, and involving other experienced NASA leaders. This "Post-Apollo Advisory Group" reported to Webb in July 1968 that "objectives for manned space flight in earth orbit for the period immediately ahead must focus on deepening our understanding of man's capabilities and needs in a weightless space environment for extended periods of time." This advice led inexorably to identifying some form of orbital outpost—a space station—as the most appropriate post-Apollo program. A space station had been part of NASA's planning even before the lunar landing program was begun, and there had been a number of NASA studies of space station concepts during the 1960s. To serve as the crew transportation vehicle for a space station, the group thought that initially the three-person Apollo command and service modules could be used but, as crew size increased and capabilities for a land landing and spacecraft reuse were developed, a modified Gemini spacecraft launched by an expendable rocket was the appropriate choice to carry later crews to a space station.[13]

NASA's Associate Administrator for Manned Space Flight George Mueller was not part of the Thompson study team. A hard-charging, brilliant, tough-minded individual, Mueller since arriving at NASA in September 1963 had become almost autonomous in his management of NASA's human space flight efforts. He had a different idea with respect to what should be NASA's top post-Apollo priority. In an August 1968 speech to the British Interplanetary Society, he noted that "the exploitation of space is limited in concept and extent by the very high cost of putting payload in orbit, and the inaccessibility of objects once they have been launched." This reality, said Mueller, led him to conclude that "the next major thrust in space will be the development of an economical launch vehicle for shuttling between Earth and the installations, such as the orbiting space station, which will soon be operating in space." Mueller characterized such a vehicle as a "space shuttle." Over the next three years, Mueller's idea would become central to NASA's plans for the future.[14]

Webb in early 1968 also selected Homer Newell, who had been involved in NASA's space science activities since the agency's inception and who at the time headed NASA's Office of Space Science and Applications, to be the NASA associate administrator, the agency's number three position. Newell's primary responsibility was to design and manage what was characterized as an "experiment" in NASA-wide long range planning. Newell organized the planning effort in a very bureaucratic manner. There was little progress during 1968 in achieving an integrated approach to NASA's long-range plans. The results of the planning experiment, Newell admitted, "were not up to the standards of boldness and imagination expected...or worthy of our first decade in space." NASA had become "so conditioned to retreat over the past two years that an intellectual conservatism pervaded the planning...The total effect in terms of forward motion was pedestrian, even timid." One major issue with respect to the planning experiment was the limited participation of Mueller's Office of Manned Space Flight. As Newell commented, "the problem with manned space flight was that they were in the habit of going it alone, they wanted to go it alone, and they intended to go it alone."[15]

A Holding Action

As he took over the leadership of NASA in October 1968, one of Tom Paine's first tasks was to submit to the White House Bureau of the Budget (BOB) a NASA budget request for FY1970, which would begin on July 1, 1969. As acting administrator, Paine was not in a strong position, but that did not deter him from an aggressive posture with respect to NASA's future. The BOB had given NASA a budget target for FY1970 of $3.6 billion, continuing the downward trend in the NASA budget that had started four years earlier. Paine called the target "a going-out-of-business projection, certainly not a viable program." Paine argued that a budget at the BOB target level would immediately after Apollo bring "to a halt the great program that was built at such a great cost." Paine's arguments did not convince the BOB staff. In a paper commenting on NASA's request, the staff noted "the resource requirements of the Viet Nam war and of pressing domestic needs, coupled with an apparent acceptance of the Soviet presence in space, have tended to push the civil space program down the scale of national priorities." The paper recognized that "major decisions must be made in the 1970 and 1971 budgets." The BOB staff was skeptical of the value of human space flight, suggesting that "the case for a continuation of a manned space flight effort after Apollo is one of continuing to advance our capability to operate in space on a larger scale, for longer duration, for ultimate purposes that are unclear."[16]

Based on a judgment that an outgoing administration should not make decisions with long-term budget implications, BOB Director Charles Zwick told Paine that he would recommend a budget of only $3.9 billion to President Johnson. This was not acceptable to Paine; he insisted that he and Zwick meet with the president to allow Paine to argue his case for a higher budget. As Paine correctly saw it, Zwick's proposed budget would provide only "the minimum levels of funding required to preserve for the next Administration the option, in the next two years, to decide whether and in what areas to move ahead in aeronautics and space."

When Paine and Zwick met with Lyndon Johnson, the president supported BOB's position. Lyndon B. Johnson had been a major supporter of the NASA program as a senator, as vice-president, and in the first few years of his presidency. In his 1971 memoir, Johnson would speak of his hope that the United States could build on Apollo to develop "laboratories in space," "an Antarctica-type station on the moon," "a spacecraft that can be reused," and would eventually "move out to other planets." But in his last weeks in the White House, weary from the turmoil of the late 1960s, he was unwilling to do anything but pass the question of the future of the United States in space to Richard Nixon.[17]

Getting Ready for the New President

Paine, like most of the Washington space community, thought it unlikely that he would be kept on as NASA administrator by the incoming Nixon administration. He was a liberal Democrat, and his wife had campaigned for Nixon's opponent, Vice President Hubert Humphrey. But it was not in Paine's character to sit back in a caretaker role until his successor was named. On December 23 he briefed the space transition team that had been set up by the president-elect on NASA's future aspirations. He spent much of his time in the first weeks of 1969 trying to develop a more compelling argument than what was coming out of the Newell planning effort for developing a space station, the program that NASA had chosen to be the centerpiece of its post-Apollo efforts.

There was a problem in developing that argument—the various elements of NASA were not in agreement on what kind of space station the agency should be developing. The BOB had agreed that the FY1970 budget would contain modest funds for studies of a space station by the aerospace industry, and as 1969 began NASA was struggling to outline for potential contractors the characteristics of the station they should study. What had emerged from NASA's internal planning was a station with a six-to-nine astronaut crew capable of resupply and crew rotation. The goals of such a station were both to qualify astronauts and their equipment for long-duration flights in Earth orbit and beyond and to demonstrate the ability of astronauts to carry out useful engineering and science experiments in the microgravity environment of space.[18]

Paine found this station concept neither sufficiently ambitious nor exciting enough, and on January 27, 1970, called his top managers to Washington for a meeting on what kind of space station NASA should be proposing. By the time of this meeting, Richard Nixon was already president and NASA had received the expected request from Nixon's new budget director Robert Mayo to reexamine its FY1970 budget proposal, primarily to identify places where it could be reduced. Paine also knew that the White House was considering several candidates to be his replacement as Nixon's NASA administrator. Even so, Paine continued his push for bolder thinking. He told those invited to the meeting that there was a "need to outline bold objectives for the Space Station program. Modest goals...are not worthy successors to those of Apollo. They will neither challenge our people nor draw the support of the nation to retain a space effort of the present size and capability." These two objectives—developing a technologically challenging program for the NASA workforce and gaining enough public and political support to allow NASA to continue to operate in an Apollo-like mode—were underpinnings of Paine's approach to the future of NASA.[19]

At the January 27 meeting, Paine discovered that he was not alone in seeking a more ambitious post-Apollo goal. The director of the Marshall Space Flight Center, émigré German engineer and space visionary Wernher von Braun, observed that NASA should spell out "what we foresee as the ultimate—the long range—the dream—station." Then, he suggested, NASA could define a first-generation station "as a *core* facility in orbit from which the *ultimate* 'space campus' or 'space base' can grow." Director of the Manned Spacecraft Center Robert Gilruth suggested that NASA should be looking "at a step more comparable in challenge to that of Apollo after Mercury."[20] Paine found von Braun's and Gilruth's advice very much to his liking. Commenting on the space station meeting, he said "We're trying to get the best talent in NASA focused on setting the right course for the future." He added that "the Space Station looms very large in post-Apollo manned space flight, but we've not yet adequately planned for this."

Soon after the January 27 meeting, the trade publication *Aviation Week and Space Technology* reported that "all previous concepts have been retired from active competition in favor of a large station," with the goal of a "100-man earth-orbiting station with a multiplicity of capabilities" and with the first step the launch "of the first module of a large space station, with perhaps as many as 12 men, by 1975."[21] Paine would soon try to sell to the new Nixon administration an ambitious space station program as the initial large-scale post-Apollo space effort. It would prove to be a tough sell.

Space and the Presidential Transition

On December 3, 1968, President-elect Nixon created a transition task force on space, chaired by Nobel Prize–winning physicist Charles Townes of the University of California at Berkeley. This task force was one of 17 such panels established by the president-elect; their creation followed the model that had been originated by John Kennedy in 1960.[22] The members of the space transition task force in addition to Townes were Spenser Beresford, Lewis Branscomb, Francis Clauser, Harry Hess, Norman Horowitz, Samuel Lenher, Ruben Mettler, Charles O'Dell, Alan Puckett, Walter Roberts, Robert Seamans, and James van Allen. Seamans had been a senior NASA official from 1960 to 1968 and during the transition became Nixon's choice for Secretary of the Air Force; he seems to have had a particularly strong impact on the conclusions. Of the other members, Beresford was a Washington lawyer with experience on space issues as a Congressional staffer. Lenher, Mettler, and Puckett were leaders in the aerospace industry; Branscomb, Clauser, Hess, Horowitz, O'Dell, Roberts, and van Allen were well-known scientists. All had had some significant exposure to space issues prior to their transition team service. Thus they spent little time in fact-finding and never met as a group; rather, they worked by exchanging draft inputs to prepare what was intended to be a consensus report. In addition, Townes was also a member of the Space Science and Technology Panel of the President's Science Advisory Committee (PSAC); Branscomb was chair of that panel. The PSAC panel had met in December 1968 and prepared a report "with malice aforethought" that fed into the transition team activity. Townes met with president-elect Nixon on January 8 to brief him on the task force conclusions.[23]

The task force's report identified a number of issues and presented related recommendations. Among them were the following.

Is any significant change required in thrust or content of the present space program? A new look is required at the balance between the manned and unmanned segments of the NASA space program.

What should be the objectives and scope of the manned program? While this issue is complex, and the function of man in space not yet clear, a considerable majority of the task force believes there is a substantial role for man in the long term, and that a continued manned flight program, including lunar exploration, is justified at present.

What are the program items and their urgency for the immediate future? Various items needing special consideration are

- a. A manned space station. We are against any present commitment to the construction of a large space station.
- b. [omitted]
- c. Lunar exploration. Lunar exploration after the first Apollo landing will be exciting and valuable. But additional work needs to be initiated this year to provide for its full exploitation.
- d. Planetary exploration... The great majority of the task force is not in favor of a commitment now to a planetary lander or orbiter.

Cost Reduction and "Low Cost" Boosters. The unit costs of boosting payloads into space can be substantially reduced, but this requires an increased number of flights, or such an increase coupled with an expensive development program. We do not recommend initiation of such a development, but study of the technical possibilities and rewards.

International Affairs. Space operations put in a new light many international questions and also lead naturally toward some areas of international cooperation. We believe these offer opportunities for initiatives and some progress towards world cooperation and stability, and the U.S. should exploit these opportunities with both care and vigor.[24]

With respect to NASA, the task force estimated that a $4 billion annual budget, "about ¾ of one per cent of the GNP, does not seem excessive in view of the importance of the space developments to the nation." This figure included $2 billion annually for human space flight; the majority of the task force members accepted that the United States would have a continuing human space flight program into the indefinite future. The task force believed that because "a considerable number of boosters and space vehicles will remain after the first lunar landing, it is possible to have an active and successful manned program for several years while at the same time steadily decreasing the level of funding for manned space flight to perhaps $1.25 billion by fiscal 1972." While it accepted in principle the existence of a post-Apollo human space flight effort, the panel thought that "it would be undesirable to define at this time a new goal that is both very ambitious in scope and highly restrictive in schedule, for example a manned landing on Mars before 1985, even though such a goal might be achievable. Such a commitment, adopted now, might inhibit our ability to establish a proper balance between the manned space program and the scientific and application programs."[25] These findings and recommendations closely foreshadowed the approach the Nixon administration would take in its post-Apollo space decisions, including continuing human space flight, not setting another ambitious space goal, not approving space station development, and giving higher priority to international space cooperation than had been the case during the 1960s.

NASA did not receive a copy of the Townes report until sometime in March 1969, and it took another two months to generate its response. Tom Paine's response to the report, not surprisingly given his bullish approach to NASA's future, took umbrage at the report's tone, while welcoming its endorsement of the need for a vigorous U.S. space program. But, Paine asked, "What do we mean by the word 'vigorous'?" If "one associates vigor with youth, with growth, and with the promise of future accomplishments, one can only view the state of affairs in our space program with serious concern for the future." Paine also objected to the report's opposition to a near-term commitment to any major future space undertaking, suggesting that this posture was a continuation of the situation in which NASA had found itself in the final years of the Johnson administration. He complained that "we have been frustrated too long by a negativism that says hold back, be cautious, take no risks, do less than you are capable of doing."[26]

NASA's frustration was understandable. Given the uncertainty that accompanies the arrival of any new president, combined with the recognition that the 1961 commitment to a lunar landing by the end of the decade would soon lose its potency as the central focus around which NASA could organize its efforts, the fact that the Townes report took a "go slow" approach to the future in space meant that NASA, as it approached humanity's first steps on a celestial body other than Earth, had little sense of what might lie ahead. It was squarely up to the new administration of President Richard Nixon to chart America's future course in space. If the recommendations of the Townes report were to be the foundation of the Nixon space policy, that course would be a very different one than NASA had been following and hoped to continue to pursue.

Organizing the Nixon White House

Even after "having brooded, dreamed and schemed for the Presidency for the last sixteen of his fifty-five years," Richard Nixon on January 20, 1969, was not well prepared to take over the reins of government. Nixon had an "encyclopedic" understanding of foreign affairs, but there were "deep and obvious gaps...in his knowledge of the federal government and the Congress." As Nixon began his transition to the White House, there was "an appalling vacuum of advance planning on how to organize and operate one of the biggest and most intricate governments in the world." Nixon could "count on fewer close associates to help him run the government than any recent predecessor." His "handful of trusted [campaign] lieutenants and advisors would, of course, take up key positions in the White House and the administration," but "almost to a man, they were sadly inexperienced in the ways of Washington." To supplement his few close associates in filling key White House and administration positions, Richard Nixon had "to call on outsiders that would make his, at the beginning, an administration of strangers."[27] It took more than a year for the Nixon White House operation to settle into place; during its first year in office there was a great deal of policy, budget, and personnel confusion. This confusion had more of an influence on NASA, as its future plans were being debated, than on many other government agencies.

Choosing the Senior Staff

Fundamental to understanding how decisions were made with respect to space is thus the approach Richard Nixon took to assembling his senior White House staff. First in significance and power among Nixon's immediate associates was Harry Robbins "Bob" Haldeman, whom soon after the election Nixon designated as the White House chief of staff. Haldeman's background was in advertising; he had worked for the giant advertising company J. Walter Thompson for 20 years, taking time off during Nixon's 1960 presidential and 1962 gubernatorial campaigns. Haldeman and his staff controlled all papers flowing into and out of the Oval Office and controlled access to the president for all but a very few individuals who had "walk-in rights."[28]

Haldeman presented himself as being overridingly concerned with the process of making policy choices rather than their substance; he was dedicated to making sure that Nixon received all plausible policy options before reaching a decision. There was one important fact that Haldeman kept secret from Nixon—that he was compiling a detailed day-by-day account of the Nixon White House. He marked the daily entries "Top Secret" and stored them in a White House safe. Twenty years after leaving the White House under the cloud of the Watergate scandal, Haldeman, believing that his diary would "provide valuable insights for historians, journalists, and scholars," decided to make it public. A book containing some 40 percent of the 750,000 words in the diaries was published in 1994, after Haldeman's death.[29]

Although at the outset of the Nixon administration John Ehrlichman had a secondary role among the president's advisors, during 1969 he quickly became together with Haldeman a powerful member of Richard Nixon's inner circle. Ehrlichman was Bob Haldeman's college classmate, then got a law degree, and began a successful practice in Seattle. He, like Haldeman, was a veteran of Nixon's prior political campaigns. At the start of the Nixon administration, both Haldeman and Ehrlichman "were almost wholly ignorant of major national issues, the federal government, and politics in its broadest sense...That positions of such power and influence should be filled by men of such slight experience in public affairs" was described as "the single most extraordinary aspect of the early Nixon White House."[30]

Of Nixon's innermost circle, it was Ehrlichman who over the next few years would get most involved in space-related issues.

Haldeman, Ehrlichman, and other senior Nixon advisers acquired sizeable staffs to assist them in their responsibilities. Many of these staff members were under 30 years in age—much more so than in previous White House staffs. They were chosen primarily for their "pugnacity and proven loyalty," and were equally as inexperienced in actually managing the federal government as were Haldeman and Ehrlichman. During 1969 and 1970, a young staff assistant several layers down in the White House hierarchy, Clay Thomas "Tom" Whitehead, would have a great deal of influence in shaping decisions on post-Apollo space activities.

A third member of Nixon's inner circle was his national security adviser, Henry Kissinger. His choice was somewhat surprising; Kissinger as a Harvard professor had long been a protégé of New York governor and potential rival for the 1968 Republican presidential nominee Nelson Rockefeller. Nixon did not know Kissinger well before his election, but soon afterward the two met and found they thought along very similar lines with respect to international issues. Kissinger was quickly offered the national security advisor position and after consulting Rockefeller and others in the East Coast Republican establishment accepted Nixon's invitation to join his administration.

The relationship between Nixon and his three senior advisers was strictly professional. Leonard Garment, one of Nixon's law partners during the 1960s who came to Washington with Nixon in January 1969 and served in the White House through almost all of the Nixon administration, suggests that "the relationships among Haldeman, Ehrlichman, Kissinger, and Nixon were singularly devoted to the breeding and tending of power. They were not friends, not even a little. Indeed, if the members of Nixon's German general staff shared an emotion, it was an intense dislike of Nixon, which he returned." Garment notes that this "strange quartet" after 1969 was

H. R. "Bob" Haldeman (left) and John Ehrlichman (right), President Richard Nixon's top advisers on domestic policy and politics. (Photographs WHPO 6106–6 and WHPO 1040–22A, courtesy of the Richard Nixon Presidential Library & Museum)

increasingly able to centralize control over executive branch activities until the forces of Watergate scandal tore them apart.[31]

There was an important shift in the context within which the civilian space program was viewed by the Nixon administration compared to the approach since 1957; that earlier approach had seen space as primarily a foreign policy and national security issue. The primary rationale for the kind of space program that the United States had pursued during the 1960s was as a peaceful symbol of national power and as a foreign policy tool in the Cold War U.S.-Soviet competition. While Nixon recognized the continuing foreign policy salience of space achievements, by the time he entered the White House he had concluded that more domestically oriented rationales for what the United States would do in space after Apollo, such as applying space capabilities to problems on Earth and seeing the space program as a stimulus to technological innovation and as a way of maintaining a qualified aerospace industrial and employment base, would have priority in shaping his space policy. The race to the Moon was on the verge of being won, and Nixon saw no compelling reason to continue the space program at a racing pace. By treating space as primarily a domestic rather than a national security and foreign policy issue, the Nixon administration changed the calculus by which the benefits of a post-Apollo space effort would be measured. It was thus individuals on the Nixon White House staff with responsibility for domestic policy issues who had particular influence on Richard Nixon's space policy choices. This choice also meant that Nixon himself, who was far more interested in foreign policy than domestic issues, would view space policy as a matter of secondary concern.

The senior member of Nixon's staff with direct oversight responsibility with respect to NASA was thus Assistant to the President Peter Flanigan. Flanigan's other policy responsibilities were issues related to the U.S. financial community and international trade, to the 15 independent regulatory agencies that were then part of the executive branch, and to other technical government agencies like the National Science Foundation and the Atomic Energy Commission. At the outset of the Nixon administration, this position had been filled by former Congressman Robert Ellsworth. But Ellsworth had hoped for a more responsible position, and soon was ready to leave the White House to become ambassador to the North Atlantic Treaty Organization. He was replaced in April 1969 by Flanigan, described by Ehrlichman as a "young prince of Wall Street."[32] Flanigan was an investment banker and also a veteran of Nixon campaigns in 1960 and 1962. He had served the Nixon 1968 presidential campaign as its link to the financial community. As he assumed his White House position in April, Flanigan inherited from Ellsworth's staff the previously mentioned Tom Whitehead as one of his staff; Whitehead was Flanigan's primary assistant for NASA issues. Whitehead held a doctorate in management from MIT, where he had first majored in engineering. He during the 1960s had spent time at the Rand Corporation, a think-tank steeped in a systems analysis approach to assessing policy issues. Flanigan and Whitehead were to play key policy roles

Nixon assistants Peter Flanigan (left) and Clay Thomas Whitehead (right). (Photographs WHPO 1092–21 and MUG-W-322, courtesy of the Richard Nixon Presidential Library & Museum)

in shaping the approach that the Nixon administration would take to the post-Apollo space program.

At the center of this small group of individuals sat Richard Nixon, "a loner, seated in an Oval Office as hushed and solemn as a hermitage." Nixon designed his approach to governance to isolate himself "from the demands of the hated bureaucracy while ensuring that power was centralized in the White House." Much of Nixon's communication with his immediate staff was through notes he scribbled on the memorandums and on daily news summaries he read in the evenings as he sat alone. Nixon was an "improbable president" who "didn't particularly like people...lacked charm or humor or joy," and was "virtually incapable of small talk." Nixon was "insecure, self-pitying, vindictive, suspicious...and filled with long-nursed anger and resentments." This study will not probe deeply into the Nixon psyche. There are many other accounts of this "peculiar man" that analyze the way his personality influenced his conduct as president; on occasion, however, it will be clear how some of his peculiarities affected his space decisions.[33]

Because Nixon and his advisers were unfamiliar with how the process of governing actually worked and suspicious of career government bureaucrats, they seem to have underestimated the importance of the "institutional presidency" lodged in the Executive Office of the President. With respect to space issues, the Bureau of the Budget (BOB) and the Office of Science and Technology (OST) were particularly important. While the president could appoint the heads of these offices, the staffs of both were career government employees, more dedicated to supporting the institution of the presidency than to supporting any particular president. In order to make sure that these offices served the priorities of a particular president, in this case Richard Nixon, the individuals he appointed to lead these offices had to be strong managers, able to transmit the president's policy priorities to the permanent staff and able to see that they were reflected in specific recommendations and decisions. This did not happen at the start of the Nixon administration. Nixon selected as director of BOB a Chicago banker named Robert Mayo, whom he did not know. Mayo was suggested by Nixon's nominee for secretary of the treasury, David Kennedy, another Chicago banker, for whom Mayo had worked. Mayo turned out to an individual with whom Nixon found it unpleasant to deal; he was a weak BOB director and would leave the administration in 1970. Nixon selected as his science advisor and director of OST Lee DuBridge, the retiring president of the prestigious California Institute of Technology. Nixon had known DuBridge for over 20 years, but he also soon discovered that DuBridge was neither a strong leader nor someone to whom Nixon could turn for advice reflecting the president's interests. By the end of 1969 DuBridge found himself increasingly marginalized in the policy process, and he too would leave the White House in 1970. But it was DuBridge and his OST staff and Mayo and his BOB staff who would join with Peter Flanigan and Tom Whitehead to deal with space issues on a continuing basis during 1969.

First Steps on Space

There were both parallels and differences with respect to the status of the space program at the time John F. Kennedy entered the White House in January 1961 and the arrival of Richard M. Nixon eight years later. Both men as presidential candidates had spoken of the importance of U.S. space leadership. Both had commissioned a transition task force on space that had been skeptical regarding a presidential commitment to a major new space effort, especially one involving human space flight. During both transitions, NASA had ambitious plans for the future, but also was operating with high uncertainty with respect to whether the new man in the White House would embrace those ambitions. NASA at the start of both the Kennedy and the Nixon administrations was being led by an acting administrator, and the new president was having difficulty in finding a person to head the space agency on a permanent basis. In both 1961 and 1968, the new president faced important decisions in his first months in office with respect to the future of the U.S. space effort.

A major difference in the two situations was that while in January 1961 the United States was still four months away from the launch of its first astronaut, Alan Shepard, on a 15-minute suborbital flight, in January 1969 NASA had just sent three astronauts around the Moon and was preparing to make the initial attempt to land Americans on the lunar surface. Once the lunar landing was achieved, there was no clear next step for human space flight. Without such new missions, the U.S. program of human space flight would come to an end in the 1973–1975 period, after Apollo lunar landings missions through *Apollo 20* had been carried out and astronaut visits to an already approved orbital workshop based on Apollo hardware, later named Skylab, were completed. At the time of the Kennedy transition, NASA was a relatively small organization with a modest contractor support network; in 1969, as a result of the Apollo buildup, NASA had over 34,000 employees supported by over 200,000 contractors from the aerospace industry. Deciding what to do with this "space industrial complex" and the capabilities it represented was a rather more difficult problem for the Nixon administration than John F. Kennedy had faced as he decided to race to the Moon.

Organizing a Review of the U.S. Space Program

The incoming Nixon administration was advised that there was a need for a focused review of the future options for the U.S. civilian and national security space programs. Arthur Burns, an economist and long-time Nixon associate whom Nixon had appointed as his top domestic policy advisor, had reviewed the reports of the 17 Nixon transition task forces and had extracted from them recommendations for President Nixon's early attention. With respect to space, Burns had identified three items:

1. Opportunities for increasing the amount and broadening the character of international cooperation in space;
2. Opportunities for significant reduction in the costs of space launches;
3. The need for a comprehensive review of the nation's space programs.

The second and third of these items were quickly incorporated into February 4 memos from President Nixon to science adviser DuBridge. With respect to lowering launch costs, Nixon told DuBridge "I would appreciate having by February 10, 1969, your assessment of this matter, and also of the recommendation that the Department of Defense and NASA be directed to coordinate studies in this area." With respect to the overall program review, Nixon noted that "there is general agreement that our space efforts should continue, although there are notable differences of opinion in regard to specific projects and the amount of annual funding." Burns had proposed "the establishment of an interagency committee which would include you [DuBridge], the Administrator of NASA, and a senior official from the Department of Defense. The primary function of this committee would be to furnish recommendations to me [Nixon] on the scope and direction of our Post Apollo space program." Nixon also asked for an assessment of this proposal by February 10.[34] A similar presidential memorandum regarding the first of Burns's recommended items for attention, international space cooperation, was sent to Secretary of State William Rogers only on February 21.

NASA learned of the plans for the White House space review only by accident. The agency's public affairs office had noticed a news item in a Florida newspaper saying that the president had asked his science adviser to evaluate ways of achieving lower costs in the space program. NASA contacted DuBridge to learn what was going on. While the story had to do with the transition task force's suggestion that it might be possible to lower launch costs, when DuBridge talked to Paine, he was confused, and began to explain to Paine his not-yet-final plans for the overall space review. He told Paine that what he had in mind was a steering committee composed of DuBridge as chairman and including Paine from NASA, either Deputy Secretary of Defense David Packard or Secretary of the Air Force Robert Seamans from the Department of Defense, and Vice President Spiro Agnew in his role as chairman of the National Aeronautics and Space Council, the high-level interagency group set up in 1958 to develop a national perspective on space issues. DuBridge suggested that after this group had examined the space program he would integrate their views and would prepare a summary document that he would present to President Nixon. Paine "dissented strongly" from this proposal, saying that "it was not proper for the President's Special Assistant for Science and Technology to put himself in a position superior to the Vice President, the Secretary of Defense, and the Administrator of NASA, all of whom report directly to the President." DuBridge suggested that Paine's objections were a "question of protocol." Paine disagreed; to him, the issue was "a basic question of executive authority, organization, and responsibility." DuBridge closed their conversation by telling the NASA chief he would be in contact with a new proposal that he hoped would meet Paine's objections.[35]

DuBridge's apparent intent in organizing the post-Apollo review, with himself as its chair and his OST staff and the President's Science Advisory Committee (PSAC) playing key roles, was to make sure that the review "covered all the necessary bases and got all the necessary points of view exposed for the president." There was a concern within OST that if NASA controlled the review the science adviser "would be called upon to rubber stamp a NASA document." Paine's negative reaction was aimed at preserving NASA's direct access to the president; Paine feared that what DuBridge had in mind "might result in some diminution of NASA's authority...because you never want one bunch of guys to do the planning and another bunch to carry it out." NASA was also concerned about DuBridge having the key role in the review, given his reported skepticism regarding the value of human space flight.[36]

After two days, DuBridge came back to Paine with a new proposal. It met many of Paine's objections. One change was making Vice President Agnew the chair of the review. Paine asked DuBridge about "the delicate matter" of whether the White House really wanted to put Agnew in such an important role; even three weeks into the Nixon administration, it was clear that Agnew would not be part of Richard Nixon's inner circle. DuBridge assured Paine "that he had discussed this question with both the President and the Vice President and this was their decision." With this assurance and word that the White House did not want to wait until a permanent NASA administrator was selected to begin the review, Paine agreed to DuBridge's new proposal.[37]

Later that day, DuBridge sent a memorandum to the president suggesting a "Task Group" composed of the acting administrator of NASA, the secretary of defense, the chairman of the National Aeronautics and Space Council, and the director of the Office of Science and Technology (DuBridge himself) to oversee the review, with Vice President Spiro Agnew in his role as Space Council head as chair of the Task Group. DuBridge still proposed to reserve to himself the key role of "staff officer" and coordinator of the staff studies that the Task Group would review. He earlier had suggested that the separate review of space launch cost reductions be folded into the general review of the space program. DuBridge noted that "there is some urgency in proceeding with this review because of the very long lead time for space projects" and suggested a September 1, 1969, date for submitting the group's recommendations. DuBridge attached to his report a draft memorandum for presidential signature.

Richard Nixon on February 13 signed that memorandum. It said that "it is necessary for me to have in the near future definitive recommendation on the direction which the U.S. space program should take in the post-Apollo period." Thus was created what came to be known as the Space Task Group (STG). Over the next seven months, the STG would be the forum for debate over the American future in space.[38]

Why Spiro Agnew?

Richard Nixon's vice president, Spiro T. Agnew, was not an obvious choice to chair a review of the U.S. space program. Agnew had been elected governor of Maryland in 1966; before then he was a local Maryland politician. He had no prior exposure to space issues, or indeed to most national issues. Agnew had first supported Nelson Rockefeller as the Republican nominee for president in early 1968. But Rockefeller, much to Agnew's surprise, in March 1968 had announced he would not enter presidential primaries or otherwise campaign for the Republican nomination. (He later reversed this position and competed with Nixon to be the Republican nominee.) Richard Nixon met with Agnew for the first time two weeks later; Nixon was "impressed with his intelligence and poise." Nixon's campaign asked Agnew to be one of Nixon's nominators at the Republican convention; this put him among the leading candidates to be Nixon's choice for the vice presidential nomination. After two of Nixon's closest advisers turned down the vice presidential possibility, Nixon informed Agnew that he was his choice as vice-presidential candidate. Nixon noted in his *Memoirs* that Agnew at his first press conference admitted that his name was not exactly "a household word," and assured the press "that he would work to change that situation." In ways likely not intended, Agnew succeeded in that objective.[39]

There was a straightforward reason for involving Vice President Agnew in space affairs. The vice president by law was the chairman of the National Aeronautics and Space Council, the White House organization set up by the 1958 Space Act to provide presidential-level coordination of space policy. At its origin, the president chaired the Space Council, which included as members the administrator of NASA, the secretary of defense, the secretary of state, and the chairman of the Atomic Energy Commission. When John

F. Kennedy became president in 1961, he asked the Congress to change the law to make the vice president the council chair. Kennedy recognized that Vice President Lyndon B. Johnson had been deeply involved in space matters in the Senate, and he wanted to give Johnson some specific responsibilities during the Kennedy administration. Johnson in his role as Space Council chair had played an important part in developing the recommendations that led Kennedy to set a lunar landing within the decade as a national goal, but in the remaining 30 months of the Kennedy administration he had limited influence on space choices. Johnson did accumulate a sizeable staff for the Space Council. Once Johnson became president and chose Hubert Humphrey as vice president and thus council chair, the Space Council during the rest of the Johnson administration had become almost dormant, even while it retained its large staff.

The Nixon transition task force on space had discussed what to do with the Space Council. It observed that "the Space Council has not been very effective" and observed that President Nixon could ask Congress to abolish it. But, "as long as the Council exists...it should be made effective. For that purpose, there should be a strong staff and the President should be Chairman." As he considered how best to organize the post-Apollo space review, science adviser DuBridge also considered what to do with the council. One option, suggested Russell Drew, the space specialist on DuBridge's staff, was to "strengthen the Space Council," with a "vigorous and knowledgeable person as Executive Secretary." The Executive Secretary was the presidentially appointed, Senate-confirmed top staff person for the Space Council and ran the day-by-day operations of its staff. If the president were to replace the vice president as chair of the Space Council, then the council staff could logically become part of the presidential science adviser's office and the executive secretary could report to the president through DuBridge. (It is likely that OST staffer Russell Drew aspired to the position.) The other alternative was to abolish the Space Council, but this would be likely to run into vice presidential opposition, since it would mean that he would lose a large number of dedicated staff positions.[40]

There was no serious consideration at the start of the Nixon administration given to making the president the Space Council chair. However, over the course of 1969, there were attempts to revitalize the Space Council. One step in that direction was the May 1969 selection of 34-year-old *Apollo 8* astronaut Bill Anders as the Space Council's new executive secretary. NASA Administrator Paine was instrumental in Anders's selection, seeing an opportunity to place someone positively disposed toward human space flight in a senior White House position, counterbalancing the skepticism of OST and OMB. Anders could not take on the job immediately, since he was part of the *Apollo 11* backup crew; this meant that the council staff would not become engaged in the work of the STG. Anders had become convinced that he was unlikely to get a role on a later Apollo flight that would give him the opportunity to walk on the Moon, and so was ready to take on a new and very different challenge with the Space Council position. He was told

Vice President Spiro T. Agnew introduces his choice as executive secretary of the National Aeronautics and Space Council, *Apollo 8* astronaut Bill Anders. Anders is accompanied by his wife Valerie. (National Archives photo WHPO-1044-8)

by Agnew and Paine that once he came to Washington he would have the opportunity to reinvigorate the Space Council and its staff so that they could play a more influential role in space policy development.[41]

But that was in the future; by coming into the vice presidency with the Space Council as one of his assigned responsibilities, Spiro Agnew in February 1969 became the titular leader of the effort to define the U.S. future in space. Few could have predicted at the time that he would become perhaps the program's leading cheerleader within the Nixon administration.

Selecting a NASA Administrator

In 1961, a large number (anywhere from 15 to 24, according to various accounts) of individuals were considered for NASA administrator before President Kennedy and Vice President Johnson on January 30 finally settled on James Webb as their choice. Webb was one of the last Kennedy nominees for a high position. The Nixon administration also considered a (smaller) number of candidates to replace Acting Administrator Tom Paine. During the transition, the position was offered to retired Air Force general Bernard Schriever, who declined, saying "he had too many obligations" to take on a full-time administration job. The position was reportedly also offered to Simon Ramo, head of the aerospace industry firm TRW. President Nixon on January 28 personally offered the job to Patrick Haggerty, chairman of Texas Instruments. Haldeman recorded that Nixon was "very impressed by his obvious brain power, and with his concept of institutionalizing innovation." In fact, Nixon told Haggerty that his work in this respect "maybe was a more important contribution to the nation than actual federal service." Haggerty agreed. He wrote Nixon on February 4, saying that "your invitation to join your Administration as Director of NASA both did me a great honor and faced me with an extremely difficult decision." However, wrote Haggerty, he had decided, as the president had suggested, that finishing his effort at Texas Instruments to institutionalize innovation took priority. He thus turned down the president's invitation.[42]

Almost by default, Tom Paine thus became Richard Nixon's choice as NASA administrator. After Haggerty turned down the job, science adviser DuBridge recommended that Paine be kept on; the space trade press noted that "more and more sentiment was growing among space insiders to keep Dr. Paine." President Nixon himself would have preferred to offer the position to *Apollo 8* commander Frank Borman, with whom he had become impressed in the first weeks of his administration. Near midnight on February 24, the second day of his initial European tour as president, Nixon, after

President Richard Nixon, with Vice President Spiro Agnew looking on, introduces Thomas Paine as his choice as NASA administrator on March 5, 1969. (NASA photograph GPN-2000-001669)

returning from dinner with Prime Minister Harold Wilson at Chequers, the prime minister's country home, met with Haldeman in his room at the posh Claridge Hotel in London. Haldeman reported that he and Nixon, "in his pajamas and pretty well out... discussed the NASA appointment briefly. He said go ahead on Paine, the Deputy, unless I thought we could do Borman." Haldeman did not think that Borman would take the job, and thus the choice of Paine as the head of NASA was made.

President Nixon announced the nomination on March 5 as he presented a trophy to the *Apollo 8* astronauts at the White House, saying "there has been a great deal of interest as to who would be the new head of NASA. I will admit right now that we have searched the country to find a man who could take this program now and give it the leadership that it needs, as we move from one phase to another. This is an exciting period, and it requires the new leadership that a new man can provide." He added "but after searching the whole country for somebody, perhaps outside the program, we found, as is often the case, that the best man in the country was in the program, and that is why I am announcing today that Dr. Paine, who is now the Acting Director of NASA, will be appointed the Director of NASA."[43]

With Nixon's choice of Paine, NASA got a leader who over the next 18 months would be an unceasing advocate for a space program more ambitious than Richard Nixon felt he could afford or that the U.S. public and the Congress would support. The gap between what Paine thought was desirable for the nation to undertake in space and what the Nixon administration decided was fiscally and programmatically possible frustrated Tom Paine, but he never lost his enthusiasm.

That NASA would indeed be frustrated in its ambitions was not clear in early 1969 as the review of options for the future in space got underway and as NASA readied itself to send Americans to the Moon. In the enthusiasm surrounding the lunar landing, it was not unreasonable for NASA to expect that the White House would want to continue the kind of ambitious space effort that had led to that remarkable achievement. With Tom Paine leading the charge, NASA set as its top priority making sure that the Space Task Group would recommend an ambitious post-Apollo effort aimed at landing on another celestial body. Having reached the Moon, the space agency now would set its sights on voyages to Mars.

Chapter 3

After the Moon, Mars?

NASA Acting Administrator Thomas Paine told a reporter a few days after the November 1968 presidential election that he intended to present the incoming Nixon administration with an ambitious proposal for future human space flight. He was true to his word. In his first communication to President Nixon, on February 4, 1969, Paine urged the new president to "give early personal attention to the question of the future direction and pace of the nation's space program." He noted, in words he and his advisers thought would appeal to the new people in the White House, that "the future position in space of the United States relative to the USSR is at stake" and that "significant opportunities exist now for new leadership and initiatives." Casting space choices in terms of U.S.-Soviet competition was rather tone deaf on Paine's part, a characteristic that was to persist through his time at NASA. Richard Nixon during his campaign and then in his inaugural address had made it clear that he was seeking areas of cooperation, not competition, with the Soviet Union.[1]

Later in February, Paine followed this plea with proposals to increase the NASA budget for the coming fiscal year in ways that would preserve the ability to produce more Saturn V launch vehicles, allow a second, more scientifically rewarding, phase of lunar exploration, and accelerate the pace of space station development; these were the items that Lyndon Johnson had refused to approve in his final space budget decisions. Paine also sent to the president on February 26 a lengthy and impassioned argument for an immediate commitment to a large space station as the first major post-Apollo space goal.

The creation of the Space Task Group (STG) was a blow to NASA's hopes to get early approval of a major new space initiative; the president not surprisingly took the position that he would wait until he received the STG recommendations before making any commitment to new space ventures. Thus influencing the STG to take a position supportive of NASA's aspirations became a very high-stakes objective for the space agency, and particularly Tom Paine.

There were good reasons for Paine's attempts to get an early decision on a new program to follow Apollo. If no major new start were approved in the first year of the Nixon administration, NASA was facing both a hiatus in developing new capabilities for human space flight and a shutdown of the production lines for existing capabilities. Subsequent missions to the Moon after the first lunar landing would be based on already developed and purchased Apollo/Saturn equipment, as would the orbital workshop that was the only approved post-lunar landing human space flight project. The workshop and however many lunar landings would be attempted would be completed by 1975 at the latest, and more likely by the end of 1973. After then, there was a real chance that the U.S. program of human space flight would come to at least a temporary end. Paine and his associates were convinced that no U.S. president would accept such a situation, and wanted to press their case for quick approval of new human space flight efforts to avoid a lengthy hiatus. They also wanted to preserve NASA's identity as an engineering and systems development organization, not just as an operator of existing space capabilities, and to maintain as much as possible of the large personnel and facility base developed for Apollo. They thought it self-evident that the nation should continue an ambitious program of human space flight; according to NASA senior strategist Willis Shapley, "it was really a cultural shock, not really realized for many years [after 1969], that you did have to justify" the human space flight program.[2]

The Space Task Group—Getting Started

The first meeting of the STG was set for March 7. It was a "principals only" gathering. Attending as the Department of Defense (DOD) member was Secretary of the Air Force Robert C. Seamans, who had been assigned by Secretary of Defense Melvin Laird to be his surrogate on the STG. In formal organizational terms, this role might more appropriately have been filled by Director of Defense Research and Engineering Johnny Foster as DOD's senior science and technology official, but Seamans had been a top official in NASA from 1960 to 1968 and the Air Force also managed the bulk of DOD's space activities. This made Seamans's assignment logical. Others attending were Vice President Agnew, science adviser DuBridge, and NASA Acting Administrator Paine, whose nomination for the permanent position had been announced the previous day.

The principals agreed to appoint a senior staff representative from each of their organizations "to lead and coordinate the necessary studies." This "Staff Director's Committee" was to carry out the bulk of the STG work. Staff representatives included Homer Newell, seconded by Milt Rosen, from NASA; Russell Drew from the Office of Science and Technology (OST); Jerome Wolff from the vice-president's office; and Nevin Palley from DOD. Palley worked for Foster, not Seamans. The group also agreed to include as high-level STG "observers" Robert Mayo, director of the Bureau of the Budget (BOB), who was already at the meeting; Glenn Seaborg, chairman of the Atomic Energy Commission (AEC); and Undersecretary of State for Political Affairs U. Alexis Johnson. Reflecting on the meeting, Paine felt that it had gotten "the new administration's review of the U.S. space effort off to an excellent start: the right problems were addressed, the urgency of timely decisions recognized, and a reasonable process for reaching wise conclusions organized."[3]

Initial NASA Proposals

Paine on February 24 had responded to a January 23 letter from BOB Director Mayo asking NASA to identify areas for budget reductions. Rather than offer such reductions, Paine requested an additional $189 million for Fiscal Year 1970. The proposed budget additions were:

- $70 million for increasing the stay time on the Moon of the lunar module, developing a lunar rover vehicle, and other enhancements to allow the six additional Apollo missions (*Apollo 15–20*) then planned after the first four landings to carry out more intensive scientific activities;
- $52.2 million to preserve the option of continuing to produce Saturn V boosters; without additional large rockets, NASA would not be able to launch the large space station that was central to its post-Apollo planning and to carry out other large-scale future missions;
- $66.6 million for accelerating the pace of space station and space shuttle definition studies.[4]

Two days later, Paine sent directly to President Nixon a nine-page memorandum on "Problems and Opportunities in Manned Space Flight." The memorandum made NASA's case both for the additions to the FY1970 budget and for an early presidential commitment to a large space station. Paine organized his justification for the space station in several steps. First was accepting "as a matter of policy [that] the nation must and will continue in manned space flight," adding that "no responsible and thoughtful person, to my knowledge, advocates or is prepared to accept the prospect of the United States abandoning manned space flight to the Soviets to develop and exploit as they see fit." Paine then characterized a space station as "a central point for many activities in space," but added that "we believe strongly that the justification for proceeding now with this major project as a national goal does not, and should not be made to depend on the specific contributions that can be foreseen today... Rather, the justification for the space station is that it is clearly the next major evolutionary step in man's experimentation, conquest, and use of space."[5]

This justification met a critical response. DuBridge asked the Space Science and Technology Panel of the President's Science Advisory Committee (PSAC), his elite external group of science and technology advisers, to assess Paine's February 26 memorandum. That panel was chaired by Lewis Branscomb, a physicist and director of the Joint Institute for Laboratory Astrophysics in Boulder, Colorado. During the presidential transition, the panel had prepared an assessment of NASA's status that was a significant input into the Townes transition task force on space. The panel was "not reassured by the characterization of the space station's justification as a technological end in itself, accompanied by a reluctance to discuss the station in terms of its potential contribution to science, applications, and defense."[6]

Early STG Decisions

The second meeting of the STG principals took place on March 22, 1969. With respect to NASA's request for additional funding in the FY1970 budget, the STG principals accepted the advice of the Staff Directors Committee, which recommended:

- that high priority be accorded to funding for preserving the option of continued Saturn V production, but that the production rate be subject to review;
- that augmented lunar exploration capability be provided, with the pace of future lunar missions also to be subject to further study;
- that the amount of FY1970 funding for these two purposes be a matter of negotiation between NASA and BOB;
- that no additional funding for space station and space shuttle studies should be approved; and
- that no immediate presidential statement on the future of human space flight was desirable, although a broad policy statement by the president as astronauts returned from the first lunar landing might be worth considering.

The STG members also agreed that NASA and DOD should study their separate requirements for a new space transportation system, and then jointly determine whether a single system could satisfy those requirements. This set in motion a process that three years later would result in the decision to make a large space shuttle the central initiative of the post-Apollo space program.[7]

The STG principals discussed whether their deliberations should be constrained by any a priori limits on funds available for future space budgets. DuBridge noted that "he and many others would indeed want to have a vigorous program of five or six or even seven billion dollars annually. But realities must be kept in mind." Vice President Agnew stated "very strongly" that he opposed such constraints, and BOB Director Mayo, reserving for his organization the initiative with respect to budget decisions, agreed, saying that "it would be bad to constrain the planning by imposing funding restrictions at the outset. These would have to be introduced later."[8] This decision not to set in principle an upper limit on the post-Apollo NASA budget would allow NASA later in the STG process to come forward with the totally unrealistic proposal for early missions to Mars, an undertaking that would require during the 1970s a NASA budget well above that which had enabled Project Apollo.

NASA's Paine was not happy with the tone of this meeting. During the session, he distributed a three-page plea saying "to put it bluntly, the U.S.

manned flight program is going to go out of business, unless some decisions and steps are taken to keep it going." Paine told the other STG members that the "dichotomy" between "science and the practical applications of space" and "manned space flight...makes no sense to me," since both were only "means for accomplishing various goals." He found it "ironic" that "at the moment of its greatest public triumph, our manned flight program is declining and in need of help." Paine argued in support of his request for immediate presidential endorsement of a space station that "continued development of manned space flight capability is essential to maintaining a national position of power in space."[9]

Paine also suggested that there was a need for "a new banner to be hoisted" around which the NASA human space flight team could rally. He was joined in this call by Vice President Agnew. NASA had been courting Agnew since his STG role had been announced, and had invited him to the *Apollo 9* launch on March 3, making him the guest of honor at a luncheon following the lift off. During his time with the vice president at the launch, Paine was at his enthusiastic best. This experience convinced Agnew, if he needed convincing, of the importance of a vigorous space effort. At the March 22 STG meeting, Agnew argued that, "in his very strong opinion," the United States needed "an antidote to earth-based problems," and that dramatic space accomplishments could provide such a counterbalance. He raised the question that would permeate much of the STG's deliberations. "Where was the Apollo of the 1970s?" he asked. Could it be that the United States should undertake a manned expedition to Mars?[10]

Marking Time

After this March 22 meeting, the STG principals would not gather again to discuss the substance of their report for over four months; the next meeting took place only on August 4. In the meantime, the STG-related staffs of NASA, DOD, and OST engaged in discussions without reaching a consensus. According to Paine, "everybody put forth his own view and listened somewhat impatiently to the other people's view and the discussions were fairly general and hadn't really arrived at much of anywhere."[11]

Congress and the "Public" Consulted

The STG did organize a session to inform interested members of Congress about STG activities. That meeting produced little of substance. James Schlesinger, deputy director of BOB, attended as an alternate for Mayo, and reported that there was talk of "technology, pride, scientific knowledge, and spiritual uplift" and that a "promotional motive" ran "virtually unchecked throughout the meeting."[12]

The STG also organized two sessions with a group of "Invited Contributors" to get some sense of public attitudes with respect to the future in space. Science adviser DuBridge in April had suggested that "a detached and unbiased group of well-informed people could cast a considerable amount of light" on what kind of space program the nation should undertake. He proposed that a group "that represents the general public" be formed under the auspices of Vice President Agnew. The vice president approved this proposal and told his assistant Jerome Wolff "Let's go!"[13]

Of the 31 invitees, 18 attended the first meeting on July 7. One of them was former child movie star Shirley Temple Black; Agnew's assistant Wolff assured the vice president that "as you suggested, the little girl who sang 'On the Good Ship Lollypop' will be with us."[14] Agnew opened the meeting, telling the group "it would be ludicrous to say that you are the man in the street and that this is participatory democracy. Your profile is clearly that of America's intellectual, industrial, civic, and political leadership. But it is accurate to say that you are here to represent the man in the street and your participation reflects the finest tradition of participatory democracy. We are asking you to advise us on policy decisions that we hope the man in the street will be happy to live with for the next decade." There was a second meeting of the invited contributors on August 1, this time to hear briefings on the potential for enhanced international space cooperation and on Russian space plans. Many of the invited contributors submitted thoughtful letters after these meetings, but there is no evidence that their views had any direct influence on the content of the STG report or its recommendations.[15]

Additional Inputs

In addition to the thoughts of the invited contributors, there were several other inputs to the STG, none of which had much direct impact on the group's final report except perhaps to provide background context. One contribution came from the American Institute of Aeronautics and Astronautics (AIAA), the leading aerospace professional society. The AIAA report noted that the society's members "have discussed at length the appropriateness of establishing a single national space objective for the next ten years, comparable with the lunar objective of the sixties." It said that "such a course is not recommended," because "the proliferation of useful space applications which is foreseen during the next few years is so great that a single objective would be over-restrictive." With respect to human space flight, the report gave higher priority to activities in Earth orbit rather than continuing lunar exploration, and concluded that "it would not be reasonable" to commit to developing the capability for human missions to the planets. The AIAA report gave higher priority to developing "a partially reusable space transportation system" to deliver medium and large unmanned payloads to orbit than to a commitment to an "entirely new space station." This cautious approach was somewhat surprising, coming from the organization representing aerospace professionals who stood to benefit from an ambitious post-Apollo space effort.[16]

The Space Science Board (SSB) of the National Academy of Sciences, the country's top nongovernment space science advisory body, also submitted an input to the STG process. As had been its traditional position, the SSB remained skeptical of the value of human space flight. Its report suggested that robotic exploration of the solar system and the use of space-based observations to follow up on the rapid pace of discoveries in astronomy and astrophysics should have high priority, as should "the development of applications of space technology to the economic and social uses of mankind." The SSB added "in the future, we can foresee possible roles for man," but "we do not believe the country is, at the moment, ready to decide as to the nature and extent of the long-term manned program."[17]

NASA Planning in Disarray

Tom Paine's intent was to have NASA's input into the STG deliberations emerge from the planning process initiated under Homer Newell's direction in 1968. Newell made an initial presentation of the proposed NASA submission to the STG to Paine on May 27. Paine was not impressed "with the level of imagination and the level of innovation and the level of forward thrust" of Newell's proposals; he characterized the product as "good, workmanlike, but sturdy and unimaginative." He directed Newell to work on developing a more exciting prospectus.[18] During June, a strategic focus began to emerge in Newell's plan—exploration of the solar system with both robotic and human missions. This was perhaps the first time that *exploration*—going to new places to learn about them—was put forward as a justification for moving forward in space, distinct from scientific discovery. A Newell position paper suggested "a commitment to the principle of manned planetary exploration would give focus to the exploration theme, and would guide related program activities of the agency." By late June, Newell had a revised NASA "core plan" ready. It called for

- a 12-person space station by 1975
- a space shuttle by 1977
- a space station in polar orbit by 1977
- a space station in synchronous orbit by 1978
- beginning a build up to a 50-person space base in 1977, when the space shuttle would be available
- a small lunar base by 1976
- a lunar orbit station by 1977.

Newell suggested that a program of this scope could be accomplished for a NASA budget of $70 billion over a ten-year period, with budgets starting at $4 billion per year and increasing to $8 billion per year later in the 1970s. By comparison, NASA at that point was citing the cost of the Apollo program as $25 billion over eight years, so that the plan Newell was proposing was almost three times as expensive as the lunar landing effort. This proposal was totally disconnected from political realities, and was typical of NASA's misreading of its likely post-Apollo environment. Newell also suggested that "the United States begin preparing for a manned expedition to Mars at an early date," arguing that "the question for us to ponder is not whether man will go to the planets, for surely he will, but when this will take place and whether America will take the lead."[19]

The plan developed by Newell and his associates formed the body of the July 9 NASA submission to the STG, titled "America's Next Decade in Space."* Included as an appendix was "a summary of one of the many studies produced in NASA's planning effort." The report cautioned that "since the programs outlined [in the appendix]...are not official NASA proposals," their "cost and schedule estimates must be used with care since in many cases they are quite preliminary." These caveats were quickly rendered inoperative. By the end of July, what had been an appendix to the official NASA plan became its core.

What was contained in the appendix was "one way in which a versatile low-cost earth orbital space capability [i.e., the space shuttle] may be used as the basis of an integrated total space program." This "integrated plan" was the brainchild of NASA's Associate Administrator for Manned Space Flight George Mueller. It had been developed in relative secrecy without consulting other elements of NASA, not as part of Homer Newell's planning process. In this, it was typical of Mueller's style, which was highly individualistic and control oriented. Mueller had become convinced that Newell's effort was not likely to produce the kind of approach to the future that could gain political and public support, and viewed himself as a "white knight, saving the agency from itself."[20]

Mueller had earlier come to the conclusion that high priority should be given to lowering the costs of space operations by developing not only a space shuttle but also reusable space "tugs" to move payloads from low Earth orbit to other destinations between the Earth and the Moon; he characterized the combination of the shuttle and tugs a "Space Transportation System." His plan stressed three characteristics:

- *commonality*: the use of a few major systems for a wide variety of missions;
- *reusability*: the use of the same system over a long period for a number of missions; and
- *economy*: the reduction of "throw away" elements in any mission.

As Mueller had previewed his planning effort to associates in the human space flight community, there was considerable skepticism that his proposed development schedules and cost targets were realistic. Mueller paid little attention to such doubters. He "forced people to give him numbers that were a lot lower in many areas than people wanted to give him," resulting

---

* The Department of Defense also prepared an extensive report on its proposed plans for the 1970s and submitted it to the Space Task Group. That submission, and DOD-specific space issues, will not be discussed in this study.

George Mueller at the *Apollo 11* launch. (NASA photograph)

in costs that were "vastly underestimated." Eventually Mueller's colleagues gave his scheme their support, recognizing that "the integrated plan was successful at telling the story, even if it was a fairy tale."[21]

The integrated plan retained the Saturn V to launch its heavy hardware elements. Other components of Mueller's plan were:

- a 33-foot diameter "core module" capable of operating as a 12-person space station in Earth orbit by 1975 and in lunar orbit by 1976. The same module could also be used to develop a larger space base through in-orbit assembly and by 1980 could be used to create a geosynchronous station;
- a space shuttle as a fully reusable Earth-to-orbit transportation system, available to support the initial space station in 1975 and fully operational by 1977;
- a reusable, chemically fueled space tug capable of moving crew, spacecraft, and equipment throughout cislunar space, the area between the Earth and the Moon;
- a reusable nuclear-powered tug, to be operational by 1979 and capable of operating in cislunar space and beyond;
- human-tended and fully robotic spacecraft for science and application missions.[22]

As he became aware of Mueller's integrated plan in its fully developed form, Tom Paine decided that it should be central to what NASA would propose to the STG. In doing so, he was accepting what was in essence a very clever repackaging of the hardware proposals identified by Newell's planning process, but with more optimistic estimates of NASA's being able to overcome technological challenges and to meet ambitious, likely unrealistic, schedule and budget targets.

After hearing what his organization was preparing to propose, Paine also concluded that what was still missing was a truly bold goal. The objective of the integrated plan was developing capabilities that would allow the United States to carry out whatever activities it decided to pursue in the Earth–Moon region. But it lacked a unifying focus for the use of that capability. Vice President Agnew, with Paine listening carefully, had told the meeting of Invited Contributors on July 7 "when I consider the potential of a manned mission to Mars—and I recognize many cogent arguments counter it—I conceive of it as the possible overture to a new era of civilization." Comparing a human mission to Mars to the exploratory sea voyages of the fifteenth and sixteenth centuries, Agnew asked "would we want to answer through eternity for turning back a Columbus or Magellan?...Would we be denying the people of the world the enlightenment and evolution which accompany every great age of discovery?" On July 16, in the hours preceding the launch of the historic *Apollo 11* mission to the Moon, Agnew went public, telling reporters at the launch site that it was his "individual feeling that we should articulate a simple, ambitious, optimistic goal of a manned flight to Mars by the end of the century." After the launch, Agnew told the launch team that he had "bit the bullet...as far as Mars is concerned." Agnew's statement at the launch was not spontaneous; it had been planned in advance, and Tom Paine was likely in on the planning.[23]

Spurred on by Agnew's statement and by his own sense that there was a need for a dramatic goal for the 1980s to focus NASA's activities in the 1970s, Tom Paine in July 1969 also "bit the bullet"; he decided in the excitement of *Apollo 11* that it was time for NASA to propose sending Americans to Mars, not by the end of the twentieth century, three decades away, but as soon as possible.

A Mission to Mars?

The vision of human missions to Mars had long been central to those dreaming about exploratory voyages into the solar system. The prospect of former or even current life in some form on Mars had for many years intrigued scientists and explorers. Even if there were no life to be found there, Mars seemed a much more interesting celestial body to explore than was the Moon.[24]

The notion of getting ready for Mars missions in the 1980s as the rationale for developing a space station, a space shuttle, nuclear propulsion, and other new capabilities in the 1970s, while retaining the Saturn V for heavy lift assignments, had been in the background of Newell's planning for some months. However, it did not figure prominently in Mueller's integrated plan, which was focused on operations in Earth–Moon space. What Paine decided to do in July 1969 was bring the "Mars in the 1980s" goal to the forefront, to see if the nation and the White House were ready to take on another Apollo-like challenge in space.

Adding "Humans to Mars" as a Space Task Group Goal

From Paine's perspective, there were several reasons for adopting a Mars goal. NASA's continuing attempts to gain support for a space station program on the basis of its being the next logical step in developing human space flight capability or its use as a scientific laboratory had gotten little support from other STG members. By picturing it as a necessary precursor to a human mission to Mars, Paine hoped to present a convincing rationale for early station development. Not only the space station but also development of the space shuttle, space tug, and nuclear rocket stages and continued production of the Saturn V had to happen in the 1970s if a Mars landing in the 1980s were to be adopted as a national goal. Emphasis on Mars was also based on a rationale for the U.S. space program that went beyond advancing technological capability and applying that capability to provide tangible benefits on Earth. The Mars emphasis recognized exploration for its own sake as a legitimate goal of space activity.

Paine's own personality was such as to find the Mars focus attractive. His basic strategy during the STG deliberations had been to "err on the bold, bold, bold side." He thought that in the wake of the successful launch of *Apollo 11* chances for approval of a major new space goal were as great as they were ever likely to be. Paine saw the Mars mission as an "offer" that NASA should make to the country, an offer to undertake another tremendously challenging but very exciting national enterprise like Apollo. Paine judged that it was "worth the effort to at least hoist the banner and see if anybody would rally to it."[25]

Paine undoubtedly was influenced in his willingness to have NASA identified with the Mars focus by the repeated requests by Vice President Agnew during the STG deliberations and elsewhere for an "Apollo for the seventies" and by Agnew's now public support for Mars as that goal. Paine may also have thought that the vice president's support would have significant influence on President Nixon's space decisions. That judgment turned out to be deeply flawed. Spiro Agnew had even less influence on White House policy choices than most vice presidents.

Paine had by the day after the *Apollo 11* launch, July 17, decided to develop a proposal for an early human mission to Mars for presentation to the STG. He ordered his planners to come up with a "very strong, very far out, but down-to-earth technical presentation" which would "substantially shake up" the STG. Such a presentation would necessarily minimize the many technological uncertainties associated with sending astronauts on the months-long Martian journey. The decision to add the Mars focus to Mueller's already ambitious integrated plan was essentially Paine's; Mueller "would have been more conservative." It is unlikely, given the tone of STG deliberations to date and the signals regarding budget constraints that NASA was already getting from the White House, that Paine believed that a crash program to send humans to Mars as soon as technically feasible would actually gain political support. Rather, by presenting an accelerated Mars effort as doable, Paine hoped that a program leading to a Mars landing later in the 1980s might not seem too ambitious, and thus be acceptable to the other STG members and ultimately to the president.[26]

The possibility that NASA would propose an early Mars mission to the STG evoked early skepticism of the mission's technical feasibility. At a July 15 meeting of the STG Staff Directors Committee, Russ Drew of OST indicated that he thought that sending people to Mars "was not technically feasible in this century, let alone in the 1980s." One of the NASA representatives at the meeting, Milton Rosen, found Drew's perspective "incredulous." Rosen pointed out "that NASA was within days of putting men on the moon after eight years' work starting from scratch." He added that "the preliminary design of a Manned Mars launch vehicle, based on nuclear propulsion, was completed" and that "program plans were well advanced for putting men on Mars in the 1980s."[27] Rosen's views were typical of the technological hubris of the NASA leadership as *Apollo 11* sat on the launch pad.

Wheeling Up the "Big Gun"

Although Mueller was not ready to suggest sending humans to Mars, the team that had developed the integrated plan under his guidance had also prepared a scenario in which the hardware systems developed through the integrated plan could be used for a Mars landing in 1986. Paine heard a briefing on this scenario on July 19, as he waited in Houston for the next day's landing on the Moon. This briefing likely solidified Paine's decision to confront the STG with a technically plausible approach to a human Mars mission, one that would build upon the plan he had already selected as NASA's preference for the 1970s.

Then, on July 23, Paine decided to "wheel up NASA's big gun," the charismatic director of its Marshall Space Flight Center, Wernher von Braun, to take the lead in preparing the STG presentation. Von Braun was a well-known spokesman for pushing the frontiers of space exploration. After being brought to the United States from Germany at the end of World War II, he had readily adapted to his new country and had become widely known as a space visionary through his appearances on television, magazine articles, and in numerous talks around the country. Von Braun had long been thinking about the technical requirements for sending humans to Mars, and after being exposed to Mueller's thinking in May 1969 had also directed his center's Future Projects Office to develop an approach to using the integrated plan hardware for a two-year mission to Mars. He was thus well prepared to respond to Paine's request that he prepare a presentation based on the earliest feasible date for an initial Mars mission.[28]

Wernher von Braun with a Saturn 1B booster on its launch pad in the background. (NASA photograph)

Von Braun was later to raise some reservations about his role in presenting the Mars mission proposal to the STG. In a 1970 interview, he suggested that "I have never in the last two or three years strongly promoted a manned Mars project. I have supplied some data on how one would mount a Mars project, a manned Mars visit with today's technology, but I in fact have always actively advocated not to pursue such a thing at this point in time. People...have tried to cast me in the image in the last few years as the Mars or bust guy in this agency, which I am definitely not." He continued, in a not very veiled rebuke to Tom Paine, that "I, for one, have always felt that it would be a good idea to read the signs of the times and respond to what the country really wants, rather than try to cram a bill of goodies down somebody's throat for which the time is not ripe or ready." He wondered "how bullish you can get in a bear market," adding that "there may be too many people in NASA who at the moment are waiting for a miracle, just waiting for another man on a white horse to come and offer us another planet." But the political environment "is more difficult and more demanding than it was with that *carte blanche* from Kennedy," since "we have turned from a visionary society to an introspective society in the last ten years." As his biographer notes, while von Braun may have had serious reservations about being used to present an ambitious Mars plan to the STG, "he certainly kept quiet about them in 1969."[29]

Agnew on July 25 sent a memorandum to the STG members and observers announcing an August 4 meeting of the STG. Paine had decided to have the meeting at NASA so von Braun could use the space agency's elaborate three-screen projection system for his presentation. Agnew's memo said that as an important item of business "the recommendations of the Staff Directors for the Principals will be discussed." But first, NASA would make a presentation "on a proposed major new program goal which would focus United States space efforts during the coming decade."[30]

To Mars in 1981?

Paine led off the NASA presentation on August 4; he suggested that "Apollo 11 started a movement that will never end, a new outward movement in which man will go to the planets, first to explore, and then to occupy and utilize them." He then turned the meeting over to von Braun, who described a "typical manned Mars mission," which he claimed represented "no greater challenge than the commitment made in 1961 to land a man on the moon." This was a remarkable (and unrealistic) claim, given the myriad technological challenges associated with a two-year flight into deep space. Because the opportunities for Mars missions could be identified with high accuracy, von Braun was able to use precise dates in presenting his mission profile. The round trip to Mars would take 640 days, departing Earth orbit on November 12, 1981, and returning on August 14, 1983. The mission would be carried out by two spacecraft, each carrying six astronauts (all male). After arriving at Mars, the spacecraft would remain in Martian orbit for 80 days. First making sure the Martian surface was safe for human presence, three crew members from each spacecraft would land for 30- to 60-day exploratory sorties. The trip back to Earth would take 290 days and would include a swing by of Venus. After arrival back in Earth orbit, the crew and Martian samples would transfer to the space station, then be returned to Earth using space shuttles. Von Braun told the STG members that the plan he had outlined could be carried out with a NASA budget peaking at $7 billion in 1975 and then leveling at $5 billion/year in the 1980s.[31]

Paine closed the presentation by saying "with the successful Apollo landing on the Moon, we know that man can lay claim to the planets for his use. We know further that man will do this; the question is, which nations and when?" He was less optimistic than von Braun about the costs of the program, suggesting that it would require "a budget rising to $9 to $10 billion" in the second half of the 1970s. He suggested that "a commitment in principle to these achievements must be made now."[32]

Negative Reactions to the "Humans to Mars" Goal

Even before this presentation to the STG, Agnew's call at the *Apollo 11* launch for sending Americans to Mars had quickly produced a variety of negative reactions. Senate Majority Leader Mike Mansfield (D-MT) said that he would rule out any such venture "until problems here on earth are solved." He was joined in his criticism by Senator Edward Kennedy (D-MA).

Both Mansfield and especially Kennedy were already on record as opposing a high priority for post-Apollo space efforts. Even more telling was the skepticism of NASA's traditional supporters. Senator Clinton Anderson (D-NM), chair of the Senate's Committee on Aeronautical and Space Sciences, on July 29 said "now is not the time to commit ourselves to the goal of a manned mission to Mars." On August 11, Anderson's counterpart in the House of Representatives, George Miller (D-CA), chairman of the House Committee on Science and Astronautics, called the setting of a Mars goal "premature," suggesting that "five, perhaps ten years from now we may decide that it would be in the national interest to begin a carefully planned program extending over several years to send men to Mars." The members of Congress were joined in their criticism by *The New York Times*, which as the *Apollo 11* spacecraft was on its way to the Moon called discussion of a Mars mission "scientifically and technically...premature" and warned with some degree of hyperbole that "any forced-draft Martian analogue of the Apollo project would divert hundreds of billions of dollars that are more urgently required to meet the needs of men and women on earth." The general public also was skeptical. In a nationwide poll taken just after the *Apollo 11* mission, respondents were asked: "There has been much discussion about attempting to land a man on the planet Mars. How would you feel about such an attempt—Would you favor or oppose the United States setting aside money for such a project?" Of those queried, 53 percent opposed a Mars mission; only 39 percent supported it. President Nixon was an avid consumer of poll data; this kind of response is likely to have caught his attention as he weighed his decisions on future space efforts.[33]

Even Paine, while still pushing for the kind of vigorous program he thought NASA should undertake, was by the time of the August 4 STG meeting sensing that commitment to an early mission to Mars was not in the cards. Using von Braun's presentation material, he had made two speeches in the first days of August about a Mars mission. He described the speeches as "trial ballooning a little bit to see what kind of comment there would be to discussions of how a Mars mission could be carried out." From these speeches "came the first rumblings of a public reaction, which was that those trial balloons were going to be shot down, and that Mars was not going to be the thing we were going to hang the program on, that the idea 'after the Moon, Mars' was too simplistic a view. We have to come up with a better program rationale than Jack Kennedy sent us to the Moon, Dick Nixon sent us to Mars."[34] Even so, Paine continued to push hard for a STG report that would recommend setting Mars missions during the 1980s as a national goal, primarily as a way for gaining support for NASA's ambitious plans in the 1970s.

Space Task Group Debates Alternatives

If Paine's faint hope was that the August 4 presentation, which he had intended to "substantially shake up" the STG, would lead to a decision to recommend the program he and von Braun had outlined, he was quickly disappointed. Immediately following the presentation, the STG principals began to discuss the content of their report, and it was soon clear that they were not in agreement with the NASA proposal.

Speaking after von Braun and Paine, Secretary of the Air Force Robert Seamans indicated that he was not prepared to endorse the humans to Mars goal, and in fact thought that the focus of NASA's activities during the 1970s should be on space applications of direct service to mankind rather than on creating the capabilities needed for human exploration. Seamans had been a member of the transition task force on space headed by Charles Townes, and his comments to the STG on August 4 echoed many of the themes of that transition task force report. Before the meeting he had prepared a letter to the vice president outlining his views, and he used that as the basis for his remarks. He supported continued missions to the Moon, but only on a "careful step-by-step basis reviewing scientific information from one flight before going on to the next." Seamans argued for the use of Apollo hardware for additional missions in Earth orbit, including investigations of the planet's environment, but he judged that it was premature to "commit ourselves to the development" of a large space station. Seamans, in contrast to the bullish assessment of the space shuttle recently completed by a DOD/NASA team (see chapter 9), suggested that "it is not yet clear that we have the technology" for a reusable space transportation system that would produce major reductions in the cost of transporting payloads into space, and suggested "a program to study by experimental means including orbital tests" the feasibility of such a system. With respect to human missions to Mars, Seamans did not think "we should commit this Nation to a manned planetary mission, at least until the feasibility and need are more firmly established." The funds needed for such a mission "would compete with the resources needed to provide immediate benefits from NASA's capabilities." Given the ambitious proposals that NASA had just presented, Seamans felt he was "sort of like a skunk at a garden party" for espousing such a "go slow" view. Agnew expressed his disappointment with Seamans's views, suggesting that while it was difficult to argue in terms of concrete payoffs for the ambitious NASA proposal, it represented "a new vista for mankind."[35]

Undersecretary of State Johnson indicated that he was sympathetic to Seamans's perspective, and science adviser DuBridge indicated that PSAC was thinking along similar lines. DuBridge suggested that a NASA program at the $4-$5 billion level for the next twenty years could achieve many of NASA's objectives, although on a stretched-out scale. Although he was an observer, not formally a member of the STG, budget director Robert Mayo spoke next, commenting that Seamans and DuBridge "had made his speech already." Mayo's comments carried particular weight, since it would be through budget decisions in the fall of 1969 that any recommendations that the STG might make would begin to be implemented. Mayo was quite cautious, arguing that pursuing the ambitious NASA program would make it impossible to meet the budget needs of such high priority issues as alleviating poverty and better control of the environment, in addition to avoiding a budget deficit. Glenn Seaborg, another STG observer as chairman of the Atomic Energy Commission, disagreed, saying that "the country can certainly afford the suggested space program and still take care of its domestic needs." In the subsequent discussion, Mayo indicated that while he recognized "some social dividends to space," he did not see "how we could announce an exciting new goal when we have these problems on earth that need to be solved." Agnew and Mayo engaged in a spirited debate over national priorities that ended with the vice president calling the budget director "nothing but a cheapskate." DuBridge suggested that the target date for an initial Mars landing be set at 1990 or even the end of the century. Paine objected, saying that such slow forward movement "would change the character of NASA." He continued to argue that NASA needed a definite goal and decisions by President Nixon on specific things that NASA should do next. Agnew closed the discussion by suggesting that perhaps the STG should suggest a first mission to Mars in the 1980s as the culmination of a broadly based space effort.[36]

The STG principals and observers, without their staff present, then discussed the actual content of their report, at that point due on September 1, less than a month away. They had before them a draft of the report's summary and recommendations section prepared by Russ Drew of DuBridge's staff. Drew had identified four "major issues...for which additional guidance is requested." These were:

1. Shall there be a single powerful theme or goal for the post-Apollo decade?
2. If so, what should that goal be, and how should it be presented?
3. Should there be a large space station program, and should it precede the availability of a low-cost transportation system?
4. Should a reusable space transportation capability be developed, and how should the program be managed?

Drew's draft noted that "there was complete agreement [among the staff directors] on the importance of programs that are directed toward the application of the nation's space capabilities to a wide range of problems." There was also "general agreement" that "exploration of the solar system and beyond" should be "an important continuing broad objective of the Nation's space program."[37]

At the suggestion of the vice president, the STG members agreed that rather than present a single recommended program of human space flight, the report would provide the president with three options:

- a "vigorous" program along the lines presented at the meeting by Paine and von Braun, with funding for NASA increasing to between $7 billion by the mid-1970s and $8–10 billion in the latter half of the decade;

- an "intermediate" program with a commitment to sending humans to Mars but with no fixed date for such an achievement, and with NASA's budget increasing to $5–6 billion by the mid-decade;
- an "austere" program with funding level at approximately $4 billion per year, with no commitment to a Mars mission, while retaining the option of such a commitment at a later date.[38]

Defining the STG Options

At the conclusion of the August 4 meeting, NASA was given the assignment of defining the programmatic content of these three options. This was the role that the agency had sought from the very beginning of the STG process, when Tom Paine had argued vigorously against the proposal that space program options should be defined by DuBridge and his external advisory panel. NASA took full advantage of this assignment, and by mid-August submitted to the Staff Directors Committee three options, each of which included the same hardware elements, derived from Mueller's integrated plan, and each of which included human missions to Mars; the difference among the plans was in their schedules and annual budget requirements, not in their content. Each included simultaneous development of the two new systems that were NASA's top priority objectives for the next few years—a large space station and a space shuttle. Although at the August 4 meeting the STG principals had suggested that NASA prepare a $4 billion/year "austere" option that included a continuing human space flight effort, NASA argued that such an option was not feasible, and thus refused to provide it. The NASA options were:

- *Program A*, described as "maximum progress technically feasible," and "comparable to the 1961 decision to go to the moon." This was essentially the program that had been presented to the STG on August 4;
- *Program B*, described as "maximum returns from an economical program"; and
- *Program C*, described as "minimum consistent with continuing technological advance."[39]

It was clear from the way that NASA presented its options that Program B was its preferred choice; if adopted that option would commit the Nixon administration during its second term to NASA budgets greater than those at the peak of the Apollo effort.

The White House Gets Involved

As the STG effort moved toward its conclusion, President Richard Nixon and his inner circle of advisors were focused on capitalizing on, for broader policy and political purposes, the excitement surrounding the successful *Apollo 11* mission. Nixon purposely avoided saying anything about future space efforts in the many remarks he made both in the United States and during his around-the-world trip following the *Apollo 11* splash down. However, Nixon could not help but be aware of Vice President Agnew's call for a human mission to Mars, given his regular reading of his daily news summaries. He had talked about the space program with both Tom Paine and, separately, with Frank Borman on the trip to the *Apollo 11* landing, and he had indicated his interest in foreign astronaut participation in U.S. space flights, an interest that Paine either misinterpreted or amplified without the president's approval to include non-U.S. hardware contributions to post-Apollo space system development. Nixon in his conversation with Paine did not share his broader views on the future in space, nor did he refer to the STG deliberations.

At lower levels in the White House hierarchy, however, there was growing attention being given to the debates within the STG and to what options would be presented for presidential decision. As noted in chapter 2, Assistant to the President Peter Flanigan had since April been assigned the space portfolio; following space issues for Flanigan on a day-by-day basis was his 30-year-old assistant Tom Whitehead, who had the technical background that most others on the White House staff lacked.[40] As he began to familiarize himself with NASA's planning for its future, Whitehead quickly had become concerned that the process was heading towards an outcome that was not in President Nixon's interests. On June 25, he alerted Flanigan to his "uneasiness" regarding the STG review. His main concern was that "NASA and others will use the enthusiasm generated by a success of Apollo 11 to create very strong pressures on the President to commit him[self] and the Nation prematurely to a large and continuing space budget." Whitehead suggested that "a strong case can be made for constraining the NASA budget to its present level or slightly lower, while at the same time permitting the United States to maintain a strong space program, including manned space flight." He looked to Fiscal Year (FY) 1971 budget deliberations later in 1969 as providing "an opportunity to review significantly different alternative levels of spending so that the President will have meaningful options to consider." In order to create such options, Whitehead suggested "Bob Mayo has to be reassured that the President's interests would be served and the President is personally interested in a serious evaluation of several alternative NASA budget levels including one in the vicinity of $2.5 to $3 billion"; such a budget level would reflect a significant reduction from NASA's FY1970 budget of almost $4 billion. Whitehead also suggested that "the President should be informed that NASA is making very strong public statements about future commitments," creating the possibility that he "may find himself in a very difficult situation in the next few months" unless he insisted on such budget options as a way of countering "pressure being generated by NASA in the press and on the Hill." Whitehead was "not arguing here for a reduced NASA budget," but rather suggesting that there should be "a serious analysis of a $2.5 to $3 billion level in space programs, including its costs and potential accomplishments." In his judgment, there were "significant budgetary, scientific, and political factors that suggest that this could be a desirable alternative for the President." Whitehead also suggested that either he or Flanigan "call Bob Mayo to emphasize the importance of including at least three major options in the fiscal year 1971 budget review process." He also suggested that Flanigan write a memorandum to the president "suggesting that NASA be calmed down during the enthusiasm of Apollo 11."[41]

Whitehead's views obviously ran very counter to what NASA was hoping to achieve by having its future plans evaluated in the context of *Apollo 11* excitement. Had they become known to NASA, they might have raised a warning flag about the path that NASA was pursuing, but apparently they were not communicated except to the BOB, and then not until late August. On August 20, Whitehead discussed budget options with Schlesinger, the BOB deputy director. Whitehead told Schlesinger that "the President is not eager to proceed with an expanded space program, and in fact would like to see it significantly reduced in the near future." Whitehead also claimed that he had discussed such a posture with "other White House people" and found "none who indicated any real problem with significant reductions in the space program." He asked Schlesinger to make sure that a $2.5 billion option was included in both the STG report and the guidance being given to NASA as it prepared its FY1971 budget proposal. The head of the BOB unit in charge of the NASA budget, Don Crabill, who was also part of the STG Staff Directors Committee, asked whether Whitehead had spoken directly with the president; Schlesinger "thought not." Thus there is no evidence one way or the other regarding whether Whitehead was representing Richard Nixon's actual views, or rather using the president's name as a justification for his own skeptical perspective, a frequent practice among the Nixon White House staff. Crabill told Schlesinger that he and other NASA budget examiners thought that a $2.5 billion NASA budget for FY1971 was "equivalent to a no-manned-space-flight position."[42]

In an August 22 conversation with Schlesinger, Crabill learned that Flanigan, likely in response to Whitehead's suggestion, had at some point "telephoned Dr. Paine and instructed him to stop public advocacy of early manned Mars activity because it was causing trouble in Congress and restricting Presidential options." Flanigan, saying that he had discussed the issue with the president, had suggested to Scheslinger that Nixon "would like options even lower than $2.5 billion." Following this guidance, Schlesinger asked Crabill to prepare an additional budget option to "define a $1.5 billion per year space program."[43]

These White House conversations were taking place as NASA was pushing the STG to recommend its Program B, which called for a 1983 launch of a mission to Mars and a NASA budget during the later 1970s of almost $8 billion per year. NASA was insisting that at an annual budget of $4 billion it could not carry out a viable program of human space flight during the 1970s. Even NASA's Program C had a budget increasing to almost $6 billion by the mid-1970s. The alliance between Vice President Agnew and Tom Paine was plowing ahead toward a sure confrontation with the Nixon White House, with the content of the STG report the immediate focus of that confrontation.

Finalizing the STG Report

Although the target date for submitting the STG report to the president had been set in February as September 1, it became increasingly clear during August that more time would be needed to reconcile the differences among the STG principals. Rather than strongly advocate the views of the President's Science Advisory Committee contained in its report to the STG, which had endorsed the space shuttle but not the space station, DuBridge in these final weeks gave priority to his role as STG staff director in trying to find a way to bridge the differing views among his colleagues on the Staff Directors Committee. DuBridge's assistant Russ Drew took the lead in drafting the report, but DOD's Nevin Palley, Agnew's assistant Wolff, and NASA's Newell were also deeply involved in that effort. By the end of August, a draft report had been produced that in Newell's view represented "a consensus, one that could be accepted by all members" of the Staff Directors Committee and forwarded to the STG principals. Newell suggested that the goals and objectives of the draft report were those that NASA "probably would have chosen by ourselves."[44]

Penultimate STG Meeting

Because Vice President Agnew had to be at the Western White House in San Clemente, California for a September 4 cabinet meeting, he scheduled a STG meeting on September 3 in nearby Newport Beach.[45] Both Newell and Milt Rosen of NASA were unable to attend, and so the senior NASA staff person present was DeMarquis Wyatt, a top agency planner; Wyatt was to play a key role in finalizing the STG report over the next ten days.

The meeting was rather contentious, as the STG principals for the first time learned of Whitehead's and Flanigan's insistence that the STG report include an option with the NASA budget for the 1970s at the $2.5 to $3.0 billion level. By this time the draft report included four program options, A through D, each still including the same program elements in the 1970s, with even option D requiring a peak budget of almost $6 billion per year even though it included deferring a decision to send astronauts to Mars. In option C, that decision would be made in the late 1970s and the initial Mars mission would leave Earth in 1986. Drew of OST and Mayo of BOB proposed, in accordance with White House demands, to add a Program E that would reflect a hiatus in manned space flight after the end of the Apollo program, with no new starts on a space station or space shuttle. An angry Paine said that unless the implications of such an option were spelled out in detail, which would take some time, he would not sign the STG report. Seamans introduced into the discussion a totally new program plan that he and the DOD staff had developed as an alternative to NASA's Programs C and D. Seamans's alternative plan put more short-term emphasis on space applications and robotic exploration and maintained a human space flight program by extended use of Apollo-derived spacecraft and launch vehicles through most of the 1970s. This would be followed by sequential development, first of a space shuttle and space tug, then in the 1980s a space station, with a decision whether to send people to Mars made in the mid-1980s. Seamans argued that such a human space flight program could be carried out for $2 billion a year, thereby keeping NASA's budget in the $4 to $4.5 billion a year range for the next two decades.[46] Vice President Agnew suggested including the Seamans plan in the report rather than a Program E without human space flight; Mayo responded that this alternative would not satisfy the White House directive. Seaborg commented that the draft report before the principals was "very thoughtful," and that it made little sense at this late date to add a new option such as the one Seamans was suggesting. There was agreement with this position, and the Seamans proposal was tabled as far as the STG report was concerned (although it was embraced by the BOB staff preparing for the FY1971 budget review). Finally, the principals agreed that a Program E would be added to the report, but it would be added "to show a kind of limit that no one will want to adopt," giving the president "a better possibility of choosing one of the higher level options."

During the meeting, it became even clearer than it had been in August that the STG principals were not going to agree on a single program option to recommend to Richard Nixon. Paine suggested that all options be presented to the president without a STG recommendation, and then Nixon could consult with individual members of the STG and others to get their recommendations. Agnew agreed with this idea, saying that it allowed the inclusion of a Program E option even though none of the STG members agreed with it. The STG members decided that they would meet one more time to review the final draft of their report, revised to reflect the decisions and comments of this meeting. That meeting was set for September 11.

A revised draft of the STG report, now including Options A through E, was ready for review on September 8. The report noted that the STG had not attempted "to classify the space program in a hierarchy of national priorities." Rather, the STG had "concentrated on identifying major technical and scientific challenges in space in the belief that returns will accrue to the society that takes up those challenges." The draft recommended a "balanced program" aimed at

- "application of space technology to the direct benefit of mankind";
- "operation of space systems to enhance national security";
- "exploration of the solar system and beyond";
- "development of new capabilities for operating in space"; and
- "international participation and cooperation."

The draft noted that if there were significantly lower budget levels in the future, it would not be possible to develop new space capabilities and that at lower budget levels "if important increases in science and application programs were to be pursued, no manned space flight program would be possible." In its concluding section, the draft said that the STG had concluded "as a focus for the development of new capability," the United States should "accept the long-range option or goal of manned planetary exploration with a manned Mars mission before the end of the century."[47]

A High-Level White House Intervention

The text of the September 8 draft of the STG report appeared to make presidential choice of either Option A or Option B the best course forward. Selecting Option A would have required the White House to commit to simultaneous development of a space station and a space shuttle in its upcoming decisions on the Fiscal Year 1971 budget; selecting Option B meant that this commitment could be made a year later. It was already clear to the president's policy and budget advisors that, given the high priority President Nixon had assigned to avoiding running a deficit in government spending, the budget could not accommodate such a commitment in either year. The president, if the report was not changed, could be placed in the position of rejecting the recommendations of the group he had chartered to define the post-Apollo program.

Flanigan brought this situation to the attention of John Ehrlichman, who had emerged during the year as Richard Nixon's most trusted adviser on domestic policy. Ehrlichman in his 1982 book *Witness to Power* provides a vivid account of what followed. On the morning of September 11, just before the final STG meeting, Ehrlichman, Flanigan, and DuBridge met with Vice President Agnew. Ehrlichman told Agnew that the STG "owed it to the President not to include a proposal our budget couldn't pay for." Since an early Mars mission would be very popular, "if the committee proposed it and Nixon had to say no, he would be criticized as the President who kept us from finding life on Mars." Agnew argued that a mission to Mars was "a reasonable, feasible option." Ehrlichman "saw no excuse for Agnew's insistence" and was "surprised at his obtuseness." He "took off the kid gloves" and told Agnew "Look, Mr. Vice President, we have to be practical. There is no money for a Mars trip. The President has already decided that." He told Agnew "it is your job, with Lee DuBridge's help, to make absolutely certain that the Mars trip is not in" the report. Agnew, doubting that Ehrlichman was actually speaking for Richard Nixon, "demanded a personal meeting with the President." Ehrlichman's response was "I'll arrange it at once." Upon leaving Agnew's office, Ehrlichman asked Dwight Chapin to set up the meeting with Richard Nixon that the vice president had requested.[48]

Final Space Task Group Meeting

As the STG assembled for its final meeting, Agnew reported on his just-concluded confrontation with Ehrlichman, telling the group that the White House wanted to eliminate Option A from the STG report. Paine opposed such a step; DuBridge and BOB Director Mayo supported it. Seaborg suggested a compromise, in which Options A and E would be changed from potential choices to "dotted line" possibilities, meaning that the STG judged neither as viable alternatives. Agnew embraced this option and quickly checked with Ehrlichman, who found it acceptable. The group then decided to re-label Options B, C, and D as Options I, II, and III and to move the section containing the report's recommendations and conclusions from the back to the front of the document "to make them more noticeable and acceptable." The change in labeling the options made what was now Option II the choice that was most likely to be recommended to the president; this was the Program C option NASA had described in mid-August as requiring the "minimum investment consistent with continuing advance." The principals agreed that one person—OST staffer Russ Drew—should present the report to President Nixon, and that it was up to the White House to decide when and how to make the report public.[49]

Following the "truce" between Ehrlichman and Agnew, the request for an Agnew–Nixon meeting was quickly withdrawn. Instead Dwight Chapin asked President Nixon to approve a one-hour meeting in the Cabinet Room for the "Space Task Group to present [its] completed report and discuss its recommendations." Chapin wanted to schedule the meeting within the next week, since science adviser DuBridge was leaving on an extended trip on September 18. Originally set for September 16, the meeting was soon moved to the afternoon of Monday, September 15, to better fit President Nixon's schedule.

Final STG Report Prepared

The STG decisions to re-label the program options and restructure the report text led to a hurried effort over the next several days to reflect these decisions in the printed text of the STG report in time for it to be presented to the president four days later. NASA's Wyatt was the key NASA actor in this final revision. What had been Option A was relabeled "Maximum Pace." The report said that because that option represented "an initial rate of growth of resources which cannot be realized because such budgetary requirements would substantially exceed predicted funding capabilities," it had "been rejected by the Space Task Group." What had been Option E was relabeled "Low Level." The report noted that "the Space Task Group is convinced that a decision to phase out manned space flight operations, although painful, is the only way to achieve significant reductions in NASA budgets over the long term."

The "Conclusions and Recommendations" section of the draft report was moved to the front of the text and set in a different type face than the rest of the 29-page report. The basic recommendation was "*that this Nation accept the basic goal of a balanced manned and unmanned space program conducted for the benefit of all mankind.*" The group noted its conclusion that "*a forward-looking space program for the future of this Nation should include continuation of manned space flight activity.*" The STG recommended "*as a focus for the development of new capability,*" that "*the United States accept the long-range option or goal of manned planetary exploration with a manned*

Mars mission before the end of this century as the first target." This was a much softer goal than was contained in the program options presented later in the report, and in effect removed issues associated with a decision to send humans to Mars from consideration during the Nixon administration. The rewritten text noted that "schedule and budgetary implications...are subject to Presidential choice" and that decisions on what systems to develop and on what schedule would be determined "in a normal annual budget and program review process." The report proposed that NASA should "develop new systems and technology for space operations with emphasis on the critical factors of: (1) commonality, (2) reusability, and (3) economy, through a program directed initially toward development of a new space transportation capability and space station modules that utilize this capability." In particular, "should it be decided to develop concurrently the space transportation system and the modular space station, a rise of annual expenditures to approximately $6 billion in 1976 is required." However, "if the space station and the transportation system were developed in series...a lower level of approximately $4–5 billion could be met."[50]

Space Task Group Reports to the President

At 3:00 p.m. on September 15, President Richard Nixon met in the White House Cabinet Room with the members of the STG (with the exception of Glenn Seaborg, who was out of Washington). In transmitting the STG report to the president, Vice President Spiro Agnew commented that "the three options presented in the report provide properly balanced space programs, and that the range of choice provides flexibility in meeting budgetary constraints." Agnew suggested that Nixon choose Option II of the STG report, noting that "the cornerstones for any of the program options are two projects—the space station and the space transportation system."[51]

As planned, Russ Drew summarized the report and its recommendations. President Nixon responded that "he felt strongly that the Nation should move forward in space," and that "while the present financial burdens of the country may limit how fast we were able to move at this time, he wanted to be in a position to move faster in the future if circumstances permit." Nixon "tended to focus on the manned planetary mission" and welcomed the flexibility in the STG options to decide "in a couple of years" whether to undertake a mission to Mars in 1983. The president "liked the approach of the report. He was pleased that it rejected any substantial reduction in space activities and, at the other extreme, did not propose a crash program for a manned Mars landing." At the conclusion of the meeting President Nixon "stated a very positive personal view with respect to moving ahead" with U.S. space activities.[52]

The STG report and the NASA input to the STG, "America's Next Decades in Space," were released at a September 17 White House press conference attended by Agnew, DuBridge, Seamans, and Paine. Agnew made public his transmittal letter to President Nixon in which he had recommended

The Space Task Group presents its report to President Nixon on September 15, 1969. Clockwise from top right: Russell Drew, Office of Science and Technology; Thomas Paine, NASA; the President; Science Adviser Lee DuBridge; Budget Director Robert Mayo; Presidential Counselor Arthur Burns; (with back to camera) Milton Klein, Atomic Energy Commission; Bill Anders, National Aeronautics and Space Council; Robert Seamans, Secretary of the Air Force; Vice-President Spiro Agnew; Undersecretary of State U. Alexis Johnson; Jerome Wolff, Office of the Vice President; Frank Pagnotta, Office of Science and Technology. (Photograph WHPO 1962–4, courtesy of the Richard Nixon Presidential Library & Museum)

Option II. Seamans and DuBridge chose not to go on the public record with respect to their recommendation to the president, and Paine said he had not yet made his recommendation; he did so in a letter to the president on September 19. Like the vice president, Paine in his letter recommended that Nixon select Option II, "a balanced and challenging program." Ever the optimist, Paine added, "as the nation progresses toward meeting its other needs during the next few years, I would hope that we might be able to reexamine this and move closer to Option I."[53]

First Reactions

Press reaction to the STG report was generally positive. *The Washington Post* commented that the STG report "brought some rationality back to the discussion of whither the space program," noting that acceptance by President Nixon of the long-range goal of Mars exploration "would eliminate talk of abandoning manned space flight, which would be a foolish course of action, or of proceeding toward Mars in a crash effort." *The New York Times* characterized the report as recommending a "soft deadline for [a] trip to

Space Task Group Report, "Schedule of Accomplishments"

COMPARATIVE PROGRAM ACCOMPLISHMENTS

MILESTONES	MAXIMUM PACE	PROGRAM I	II, III	LOW LEVEL
Manned Systems				
Space station (Earth Orbit)	1975	1976	1977	—
50-Man space base (Earth Orbit)	1980	1980	1984	—
100-Man space base (Earth Orbit)	1985	1985	1989	—
Lunar orbiting station	1976	1978	1981	—
Lunar surface base	1978	1980	1983	—
Initial Mars expedition	1981	1983	II—1986	—
			III – Open	
Space Transportation System				
Earth-to-orbit	1975	1976	1977	—
Nuclear orbit transfer stage	1978	1978	1981	—
Space tug	1976	1978	1981	—
Scientific				
Large orbiting observatory	1979	1979	1980	—
High-energy astron. Capability	1973	1973	1981	1973
Out-of-ecliptic survey	1975	1975	1978	1975
Mars—High-resolution mapping	1977	1977	1981	1977
Venus—Atmospheric probes	1976	1976	Mid-80's	1976
Multiple outer planet "tours"	1977–79	1977–79	1977–79	1977–79
Asteroid belt survey	1975	1975	1981	1975
Applications				
Earliest oper. earth resource system	1975	1975	1976	1975
Demonstration of direct broadcast	1978	1978	Mid-80's	1978
Demonstration of navigation/traffic control	1974	1974	1976	1974

Mars," noting the absence of "the ringing phrases that had launched the Apollo Project in 1961" and saying that the "sooner-or-later Mars goal was carefully phrased for reasons of politics, economics and technology," since "neither Congress nor the American public seems in any mood to pledge the money for another accelerated, Apollo-like space project." Less positively, *Science* magazine called the report a "blurred vision of the future" with a primary objective of justifying "a long term continuation of a manned space program."[54]

NASA's Milton Rosen had assisted Homer Newell in Newell's role on the STG Staff Directors Committee. Reflecting on the outcome of the STG process, he told Tom Paine that "considering the initial attitude of a number of Space Task Group participants," the final STG report should be seen as "a favorable result," since it

- recognized "the importance of the first manned lunar landing" and the significance "of a focusing goal such as Apollo";
- accepted "a strong manned-flight activity as part of any acceptable future space program"; and
- accepted "exploration, in particular manned exploration of the planets, as the principal focus of activity for the future."

Rosen also suggested that the report "does not give much to anybody. After the ring of the glorious words in the report has subsided and the press has had a chance to examine it critically, it will be apparent to them that no commitments are involved." Rosen thought that the attitude of the press, and ultimately that of the public, would be "so what?"[55]

The STG report certainly did not produce a "so what?" response from NASA Administrator Paine. The report over the next six months became Paine's touchstone as he argued within the Nixon administration for budget and policy decisions that would allow NASA to implement the report's recommendations and as he traveled to Europe, Australia, Canada, and Japan seeking international engagement in the programs outlined in the report. What he was to discover during that time was that this was not a productive strategy. As the Nixon administration faced decisions on the NASA budget for the Fiscal Year 1971 and developed its policy response to the report, NASA would find its budget tightly constrained and its ambitious plans for the future dashed.

Chapter 4

Space and National Priorities

The Space Task Group (STG) report can be seen as a marketing document. The report recommended as being in the national interest a course of action that could be followed at several levels of investment. Like any other sales prospectus, it made the most positive case possible for investing in its proposed activities, without comparing that investment to alternative uses of available funds. The issue facing the Nixon administration in fall 1969 was how to react to the report's recommendations. To make that judgment, the administration, and ultimately President Richard Nixon, would have to decide where the post-Apollo space program fit into overall national priorities.

As the Nixon administration in late 1969 and January 1970 formulated its overall budget proposal for Fiscal Year (FY)1971, which would begin on July 1, 1970, the inexperience of Richard Nixon and his top White House staff in actually managing the federal government became evident. There was continuing uncertainty regarding the overall economic and fiscal policy context within which the budget was being formulated. Communication between the president and his top policy advisers, on one hand, and the Bureau of the Budget (BOB), on the other, broke down. There were several errors made in forecasting federal revenues, confounding President Nixon's intent to submit a balanced budget and forcing a last-minute round of budget reductions to achieve that goal. The cumulative result was a great deal of confusion regarding final budget decisions.[1]

NASA found itself caught up in this breakdown of the budget process. Tom Paine had hoped that the recommendations of the STG report could provide the framework for FY1971 budget choices. There was a conviction on the part of Paine and others in NASA that in the wake of the successful *Apollo 11* mission, NASA merited continued high priority among government programs and thus that the agency should receive funding commensurate with the STG report's more ambitious options. Given the chaos of the budget process, coupled with the opposition to the STG recommendations from key White House advisors and from BOB Director Robert Mayo and his staff, this approach did not prove productive. The results of the

FY1971 budget decisions were deeply disappointing to Paine and his associates. NASA's 1969 series of achievements, including four successful Apollo missions and two flybys of Mars by robotic spacecraft, were not rewarded; rather, the space agency's future remained almost as uncertain in January 1970 as it had been as the Nixon administration took office a year earlier. According to Paine, NASA "fought a retreating action through the entire budget process, beaten back but fighting lustily at every turn of the road." However "lusty" NASA's resistance to budget reductions might have been, it was ultimately unsuccessful.[2]

Evaluating the Space Task Group Report

On September 19, 1969, Thomas Paine recommended to President Nixon that he endorse Option II of the STG report, which led to a first mission to Mars in 1986, and suggested to the president that he soon make a statement to that effect. Before he forwarded Paine's letter to the president, Assistant to the President Peter Flanigan, the senior Nixon staff person with space policy responsibility, asked Robert Mayo for his comments on Paine's recommendations, which Flanigan supported. Flanigan was planning on preparing a presidential statement on space, as Paine had suggested, "in the near future." As he considered recommending to the president an immediate commitment to an ambitious space effort including a 1980s mission to Mars, Flanigan was concerned about whether such a commitment was politically sustainable. He wrote David Derge at the University of Indiana, the Nixon White House's preferred pollster, asking him to make sure that "the next Republican National Committee survey of public opinion include a question as to whether the public prefers the space budget to stay at the current levels, go up or go down, recognizing that an increase means an earlier Mars landing at the cost of expenditures at home."[3]

Budget Director Mayo was also preparing a memorandum for the president commenting on the STG report; he was basing that memo on an in-depth and skeptical analysis of the STG report prepared by the BOB staff. According to NASA's Willis Shapley, who had spent over 20 years at BOB before joining the space agency, "the budget people were terrified at the possibility of the public enthusiasm" in the aftermath of *Apollo 11* resulting "in another major commitment of some sort...With all the enthusiasm, the parades and all that, and with Tom Paine trying to exploit that, very clearly the whole name of the game from the budget side and from the people who were just afraid of an irrational decision of some sort, was to contain NASA."[4] In his September 25 memo, Mayo recommended that Nixon "withhold announcement of your space program decision until after you have reviewed the report recommendations specifically in the context of the FY1971 budget problem."

It was Mayo's recommended course of action that Richard Nixon chose to follow. Announcement of the overall Nixon approach to the post-Apollo space program would have to wait until after the review of NASA's Fiscal

Year 1971 budget proposal was completed, then anticipated to be sometime in December. It would be during that budget review that the NASA program would be evaluated in the context of national priorities.[5]

George Low Becomes NASA's Deputy Administrator

As the budget review went forward, an important new player in future space decisions entered the stage. Since he had left his position as NASA's deputy administrator in October 1968 to become acting administrator, Tom Paine had been without a deputy. The White House was under pressure to appoint Republicans loyal to Richard Nixon to various NASA positions. For example, as early as March 1969 a young Texas Congressman (and future president), George H. W. Bush, had noted that "NASA is about the only agency that does not have a pro-Nixon, Administration-oriented contact man," and suggested "correcting this situation...so that we can be assured of getting qualified Republicans and Nixon supporters into jobs there." The White House personnel office was sympathetic to this and similar pleas and urged Flanigan to find a qualified Republican for the deputy position. Flanigan suggested to Paine appointing Gordon McDonald, a California-based scientist; when Paine met with McDonald, he judged him not well qualified for the job. Instead, on September 19 Paine recommended the appointment as deputy administrator of George M. Low, a career NASA employee. Paine told the president that it had been "my hope initially to find a high-level candidate with qualifications similar to those of Mr. Low who wished to join the government from private life and, hopefully, with strong science, space engineering and Republican backgrounds," but that "my search for such an individual was unsuccessful." Paine characterized Low, then 43 years old, as "one of the country's most brilliant young technical managers." He pointed out that Low, who had served both as deputy director of the Manned Spacecraft Center in Houston and, after the 1967 Apollo 204 fire, as manager of the Apollo spacecraft program, had made essential contributions to Apollo's success.[6]

Low was an Austrian-born engineer whose family had immigrated to the United States in 1939, after the German takeover of Austria. He became a U.S. citizen in 1949 and received a M.S. in aeronautical engineering from Rensselaer Polytechnic Institute in 1950. He at that point was already working for the National Advisory Committee on Aeronautics (NACA), NASA's predecessor organization, and had risen steadily in responsibility within NACA and NASA during his 20-year career. Even at a relatively young age, he was widely known and respected within the aerospace profession.

As the White House considered whether to accept Paine's recommendation, Low traveled to Washington to meet with science adviser Lee DuBridge and Flanigan. DuBridge was not well briefed for the meeting; according to Low, he "was under the impression that I was already on the job" and wanted to discuss NASA's future. The meeting with Flanigan was "not quite so satisfactory." The meeting lasted only ten minutes, and Flanigan was "quite

Thomas Paine swears in George Low as NASA's Deputy Administrator, December 3, 1969. (NASA photograph)

provocative" in his questions. Low felt that he "was not communicating very well at all"; the meeting ended abruptly when Flanigan announced that he had an appointment with the president.[7]

Low apparently made a better impression on Flanigan than he thought. On October 21 Flanigan sent a memorandum to President Nixon recommending that Low's appointment be approved. He told the president "I have met Mr. Low and he is obviously a very capable individual." Flanigan noted that Frank Borman, the president's favorite astronaut, had characterized Low as "a man who has done a superior job. Perfectly capable of assuming utmost responsibility." After his meeting with Low, Flanigan checked again with Borman, who indicated "his complete support" of Low's appointment. Ehrlichman, likely after clearing the appointment with the president, initialed the "Approve" box on Flanigan's memorandum. Low's confirmation hearing was on November 25, and he was sworn into his position by Tom Paine on December 3.[8]

Low would become a central participant in 1970–1972 space policy and program debates and decisions. He had a low-key, steady personality that was an effective complement to Tom Paine's more ebullient style, but was also very tough-minded and more politically astute than Paine. Low was meticulous in style, and, like Bob Haldeman, Nixon's chief of staff, on an intermittent basis kept a detailed personal diary.

NASA Budget Review

The White House review of the budget NASA was requesting for Fiscal Year 1971 began in earnest on October 8, 1969, when NASA submitted a FY1971 budget request of $4.497 billion, an over $600 million increase from what President Nixon six months earlier had approved for FY1970. Thus began what the veteran official in charge of NASA's budget preparations, Bill Lilly, called "one of the most screwed-up operations anyone had ever seen in terms of how a budget was received and processed—the infighting between the White House staff and the Bureau of the Budget, [NASA] getting contrary directions from both sides, and it was a mixed up process all the way through." Tom Paine characterized the budget review as "byzantine." Decisions made during this budget review were of critical importance to NASA's future, not just the next fiscal year but also beyond, since they could either support or reject the path forward set out in the STG report.[9]

First Steps

The FY1971 budget process had actually started six months earlier, when BOB Director Mayo on April 4 had indicated to NASA areas of particular interest to BOB with respect to upcoming budget decisions. These included

- "Should the U.S. undertake the development of a long duration manned orbital space station in the FY1971–73 period?"
- "Should a grand tour mission to the outer planets be undertaken in the next decade?"*

On May 23, Mayo added to these two areas for intensive study the issue of the Apollo launch rate—whether there should be one, two, or three launches to the Moon a year after the first successful lunar landing. The question of Apollo launch rate was of particular interest to a young analyst in BOB's Office of Program Evaluation, Richard Speier; that office carried out special studies for BOB in support of its budgeting function. Speier was arguing within BOB that by limiting Apollo launches to one per year and by not only cancelling future production of the Saturn V launcher but also halting manufacture of the last two already approved Saturn Vs, there could be a budget savings of $1 billion in FY1971.[10]

During summer 1969, the budget process moved forward in its normal rhythm, independent of the activities of the STG. Mayo in a July 28 letter to Tom Paine gave NASA two budget targets for FY1971. One, the "official target," was the maximum amount that would be available for NASA under

---

* The "Grand Tour" mission would fly by Jupiter, Saturn, Uranus, Neptune, and perhaps even Pluto. This was possible because of a once-every-175-years alignment of the outer planets. Whether to undertake such a mission was a controversial issue in NASA-White House dealings between 1969 and 1973, but will not be discussed in this study, which focuses on issues related to post-Apollo human space flight.

the current fiscal outlook. This figure was $3.5 billion. In addition, NASA was told that in planning its future activities it should assume budgets of $3.5 billion per year for the next eight years, the anticipated tenure of the Nixon administration. This could hardly have been a welcome message for NASA, given that the agency at the same time was preparing to brief the STG on an ambitious program leading to an early Mars mission and requiring substantial budget increases in coming years.

NASA was also given an alternative target of $4.6 billion, with budget levels rising to $6 billion in subsequent years. This target was provided "as a means of indicating priorities at a higher resources level, in case subsequent events enable changes in current plans." The large difference in the two target figures was not all that unusual in the early stages of the budget process, since they bracketed what the BOB staff thought at the time was the most likely outcome, a NASA budget in the $3.7–$4.0 billion range.[11]

Even as the STG was finalizing its report, NASA budget examiners within BOB were preparing a lengthy critique of the report and an analysis of possible NASA programs at four different budget levels, ranging from one program at $1.5 billion/year, two options at $2.5 billion/year, and one at $3.5 billon/year. The BOB staff characterized the draft STG report as "inadequate as a basis for Presidential decision," noting that the report assumed "a Presidential posture favoring rapid deployment of new manned space flight systems," but that "the combination of Defense and domestic budget commitments with concomitant budget demands for the next 2 to 4 years may make such a space posture untenable." The staff paper suggested that "the crucial problem with manned space flight is that no one is really prepared to stop manned space flight activity, and yet no defined manned project can compete on a cost-return basis with unmanned space flight systems. In addition, missions that are designed around man's unique capabilities appear to have little demonstrable economic or social return to atone for their high costs. Their principle [*sic*] contribution is that each manned flight paves the way for more manned flight."[12]

NASA and BOB Clash

NASA Administrator Paine in August had told his NASA colleagues to prepare a budget reflecting what became Option I in the STG report; the resulting requests totaled $5.4 billion. Paine's reaction was that these requests "far exceed the dollar level that can be reasonably expected." At this point, NASA's internal budget process was in "disarray," with "Apollo euphoria" prevalent and Paine and other senior NASA officials concentrating on the STG process.[13]

In submitting the NASA budget request of $4.5 billion, Paine characterized it as consistent with Option II of the STG report, the choice he had recommended to the president. Paine also reminded Mayo that the official $3.5 billion target had been "issued prior to the Task Group's report and recommendations." If that budget level were forced on NASA, said Paine, "major program decisions totally inconsistent with the Task Group's recommendations" would be needed, including "immediate decisions on terminating manned flight operations."[14] This was the first of several times during the budget review when NASA claimed that there would be drastic consequences with respect to human space flight if its budget was reduced, only to find ways of avoiding those consequences when it was forced to accept a lower budget.

Over the next month, the BOB space budget examiners reviewed the NASA request. They initially thought that NASA was likely to end up with a budget at the $3.7–$3.9 billion level, but they were directed by Robert Mayo in "strong words" to keep the budget at $3.5 billion. Driving Mayo's action was Richard Nixon's focus on balancing the budget in the face of continuing inflation and the end of the tax surcharge that had been in place to help pay for the costs of the Vietnam war, even as the conflict continued. The fiscal outlook was much less optimistic in October than it had been only a few months earlier; the president's policies were not producing the desired results in terms of controlling inflation and stimulating economic growth and increased federal revenues.

To reduce the NASA request by $1 billion, the BOB staff made "meat-axe cuts." There was no coherent rationale behind these cuts, but even so the staff composed a paper, delivered to NASA on November 13, that attempted to explain the reasoning behind the BOB's "tentative allowance" of $3.5 billion for FY1971. At this budget level, there would be only one Apollo launch per year. Saturn V production would be "suspended"; production capabilities would be mothballed, to be restarted if additional launch vehicles subsequently were needed. Additional research on space shuttle technologies would be required before detailed design and development of the vehicle would be approved. Space station development was deferred.[15]

In a strongly worded November 18 letter, Paine told Mayo that "the allowance and rationale are both unacceptable," since they failed to support "even the minimal requirements of a balanced forward-looking U.S. space program." He added that "the proposed rationale ignores and runs counter to the conclusions reached by the Space Task Group...By refusing to recognize the need for a planning rationale and by undercutting existing commitments, the BOB staff proposals would force the President to reject the space program as an important continuing element of his Administration's total program." Paine reiterated his argument that a NASA budget of less than $4 billion/year "would require decisions to suspend manned flights." He closed his missive by expressing "his disappointment that at this point in the budget process so much effort has been expended and so little accomplished."[16]

Paine and Mayo and their relevant staffs met on November 21 to discuss their differences, but according to one of those present "it was a fairly short meeting and quite—you would not say bitter—but it broke fairly quickly because we couldn't accommodate anything"; according to another participant, "Paine went away angry." Paine and Mayo did agree that being so far apart so late in the budget process was not a good situation, and directed their staff members to work together to try to narrow the differences.[17]

There was some movement over the next few days. NASA developed four new budget alternatives, ranging from $4.4 to $3.9 billion, but Paine insisted that in order to make "meaningful forward progress on the key space station and space shuttle programs without sacrificing key elements of the balanced STG program," a budget of $4.25 billion was the lowest that he and Mayo should "responsibly recommend to the President." Paine continued to use the STG report as his basis for the president's budget decisions; he suggested to Mayo that "your job and my job" was to help Nixon "redirect America's space efforts into the forward looking course charted by the Space Task Group." NASA's consistent strategy, whatever budget level was finally approved, was to keep in the budget some meaningful funding for the station/shuttle combination that was key to post-Apollo human flights. To do this, said Paine, "if we must sacrifice current important programs—like Saturn V production—so be it."[18]

At this point in the budget process, normal practice called for the BOB director to meet with the president to make his recommendations on budget level and associated issues and to explain to the president the areas where these recommendations were not accepted by the affected agency. The agency head was not to be present; he would be given a chance to appeal the president's tentative decisions once they were communicated to him. The Nixon–Mayo meeting took place on the afternoon of December 5. Also present at the meeting was John Ehrlichman and, for the portion of the meeting dealing with NASA, Peter Flanigan.

There was one problem lurking in the background of the meeting—by this time, Richard Nixon had discovered that he "just plain did not like Mayo" and did not relish dealing with the BOB director, whose "mannerisms and odd sense of humor thoroughly alienated the President." This dislike was shared by Ehrlichman and Flanigan, and colored the relations between Nixon's White House staff and the BOB through the remainder of the budget deliberations, with the two parties not communicating well and often working at cross purposes. By the time of his meeting with the president, Mayo had increased his recommended FY1971 budget for NASA to $3.7 billion; this figure included launching two Apollo missions a year and continuing Saturn V production.[19]

Richard Nixon Talks about the Future in Space

President Nixon traveled to the Kennedy Space Center to view the November 14 launch of *Apollo 12*, the second lunar landing mission; in doing so, he became the first sitting U.S. president to witness an astronaut launch. The weather for the launch was "dismal," but Nixon, his wife Pat, and his daughter Tricia sat under umbrellas as the Saturn V lifted off through rain and low clouds, generating a lightning strike that threatened to abort the mission. Nixon called the launch "spectacular."[20]

NASA Administrator Paine took the opportunity of Nixon's presence at the launch to press his case for a NASA budget at the level the agency had requested. Paine had received the BOB allowance the previous day, and made sure the president knew of his unhappiness with it. Speaking to NASA employees in the launch control center after the *Apollo 12* crew—Pete Conrad, Alan Bean, and Richard Gordon—were safely in orbit, Nixon commented on his reaction to seeing the launch in person. He compared it to seeing a football game live rather than on television, because "it is a sense of not just the sight and the picture but of feeling it—feeling the great experience of all that is happening." Then, in his first public comments on the space program in the two months since the STG report was submitted, Nixon told the crowd

> You can be assured that in Dr. Paine and his colleagues you have men who are dedicated to this program, who are making the case for it, making the case for it as against other national priorities and making it very effectively.
>
> I leaned in the direction of the program before. After hearing what they have to say with regard to our future plans, I must say that I lean even more in that direction.
>
> I realize that within those of the program, between scientists and engineers and others, there are different attitudes as to what the emphasis should be, whether we should emphasize more far exploration or more in taking the knowledge we have already acquired in making practical applications of it.
>
> All of these matters have been brought to my attention. I can assure you that every side is getting a hearing. We want to have a balanced program, but, most important, we are going forward. America, the United States, is first in space. We are proud to be first in space. We don't say that in any jingoistic way. We say it because, as Americans, we want to give the people of this country, in particular our young people, the feeling that here is an area that we can concentrate on a positive goal.[21]

That the president was so aware of the arguments about the future direction of the NASA program may have come as a surprise to Paine; the NASA chief must have been heartened by Nixon's words. But those words turned out to be much more rhetoric designed to reassure the NASA workforce than a reflection of Nixon's actual attitude toward future space efforts. That attitude was soon to be reflected in Nixon's budget decisions.

"Final" Budget Decisions Are Not Final

In the meeting with Mayo, Ehrlichman, and Flanigan on December 5, President Nixon decided to give tentative approval to the BOB recommendation of a NASA budget for FY1971 of $3.7 billion, but also decided to suspend production of additional Saturn V boosters. It is likely that Flanigan had significant influence on the president's views. By the time of the budget meeting, he had become much more cautious with respect to NASA's future plans than had been the case in the immediate aftermath of the STG report.

He also had become attuned to the reality that there was limited public support for ambitious post-Apollo space activities. On December 6, he sent a memorandum to the president reporting that "the October 6 issue of *Newsweek* took a poll of 1,321 Americans with household incomes ranging from $5,000 to $15,000 a year. This represents 61% of the white population of the United States and is obviously the heart of your constituency." Of this group, Flanigan reported, "56% think the government should be spending less money on space exploration, and only 10% think the government should be spending more money."[22]

Nixon's budget decisions were communicated to Paine by Flanigan, not Mayo as would normally have been the case. Flanigan told Paine that "the President says that he doesn't have enough money within the next couple of years and must accept limitation of activity," that "the President will agree that at some time we will go to Mars," that Nixon "did not see the need to go to the moon six more times," and that "the President was alarmed [in the sense of being concerned about their future costs] about the space station and shuttle."[23] Nixon's skepticism regarding the value of additional lunar landing missions was to be a recurrent theme during the next two years.

In a December 17 letter to Nixon appealing the tentative budget decisions, Paine once again gave priority to getting started on the station and the shuttle, saying "if, because of today's severe fiscal constraints we must sacrifice some current operations...so be it. The important thing is to press forward now with our new program." Closing his five-page letter, Paine told the president "I believe I would be remiss and do you and your Administration a disservice if I did not place before you as you reach these important decisions on America's future in space the relevant facts, consequences, and potentialities." He requested a meeting with Nixon to discuss his appeal.[24]

An indication of the context in which President Nixon would evaluate that appeal came soon after the December 5 Nixon–Mayo meeting. One influence was Flanigan's December 6 memorandum reporting on the negative public attitude toward increases in space spending. In addition, an entry in the president's carefully read daily news summary discussed the Hunger Conference taking place in Washington that week. It noted that "constant references were made to space" as an example of spending that "could have been far better spent on hunger." After reading this report, Nixon asked his advisers Ehrlichman and Daniel Patrick Moynihan "whether you agree that some of our money would be better spent on hunger."[25]

Another signal that NASA was not going to succeed in its budget appeal came as the *Apollo 12* crew visited President Nixon in the White House on Saturday, December 20. The crew and their wives (except for Alan Bean's wife, who was ill) had dinner with President and Mrs. Nixon in the White House family quarters, then watched the movie *Marooned*, a story about three astronauts stranded in orbit. This was a rather odd choice for the occasion, given that all three of the *Apollo 12* crew hoped to fly in space again, but the movie had just been released to critical acclaim. Like the *Apollo 11* crew, the astronaut families stayed overnight at the White House and joined the Nixons the next morning for coffee, then attended a White House worship service. The *Apollo 11* visit to the White House the previous month had been a warm and relaxed affair, but Pete Conrad sensed the president's "apparent lack of interest in the space program." Conrad was "disappointed and disillusioned" after his White House visit. He suggested that "the President paid very little attention to any discussions on space and exhibited no technical interest. He also appeared to have very little knowledge of what had gone on in space and what was going on in the future." Conrad on several occasions "tried to bring up the future of space, the space station, the space shuttle, Mars missions, and was very quickly turned around and the subjects went back to small talk."[26]

Tom Paine had a 20-minute meeting with President Nixon on the afternoon of December 23 to make his case for a higher NASA budget. In advance of that meeting, Flanigan made his recommendations to the president on dealing with NASA. He suggested that Saturn V production should be suspended, that study funds for the space station and shuttle should be reduced, that the frequency of Apollo launches to the Moon should be reduced to "an average of 1-½ per year...thereby extending the period of manned space flight beyond the presently planned date of 1974," that university research funds should be eliminated "as requested by the President," and that the newly opened NASA Electronics Research Center be closed. Paine in his December 17 appeal letter had once again claimed that the steps NASA would have to take to accommodate a NASA budget of $3.7 billion would mean that "U.S. manned flight activity would end in 1972 with an uncertain date for resumption many years in the future." Flanigan called this claim "unacceptable," since it would place the "onus" for terminating the current human space flight "on the President," while NASA would "create commitments for very expensive programs that will require excessive outlays in the next few years." Flanigan was quite aware of NASA's "crying wolf" strategy in the budget negotiations, and by this point had become extremely skeptical of its validity.[27]

Notes taken by Ehrlichman at the December 23 meeting dealt with only two issues—whether to continue production of the Saturn V and, if the decision on that issue was to suspend production, whether to "close Kennedy [Space Center] in '72." Nixon did not respond to Paine's arguments at their meeting; rather, the president made what he thought was his final decision on the NASA budget on December 26, approving a $3.735 billion NASA budget that confirmed the suspension of Saturn V production and the closure of the Electronics Research Center. NASA was told that it should launch Apollo missions no more than twice a year in order to extend the time the Saturn V would be in service. Only a low level of study funds for the space station and shuttle was approved. The budget decisions were accompanied by the message that "the President was quite favorably inclined to the NASA program but that he just did not have the money to spend on it."[28]

NASA Budget: Ratchet One

In a normal "budget season" President Nixon's December 26 decisions regarding the NASA FY1971 budget would have been the end of the process until the budget was made public a month or so later. But this was not normal year in budget-making. Nixon's December 26 budget choices had a lifetime of only a few days. The increasingly detailed involvement of Flanigan and his assistant Tom Whitehead during the preceding month had convinced them that additional reductions to the NASA budget could be made without undercutting the president's space priorities. Flanigan had not been present at the December 26 meeting when Nixon had approved the $3.7 billion NASA budget, and in its aftermath suggested to Ehrlichman that a lower NASA budget was both desirable and feasible. In addition, Bryce Harlow, Nixon's top assistant for Congressional relations, advised the president that a NASA budget at a $3.7 billion level was likely to run into opposition in the Congress. Based on this counsel, the issue of the NASA budget level was reopened at the end of December; within the first few days of January, the NASA budget was "ratcheted" down to a lower level.

The involvement of Flanigan and especially Whitehead in the budget process had begun in late November and intensified throughout December. There was little precedent for such intense White House policy staff involvement; this was traditionally seen as the role of the BOB. But Richard Nixon, with his desire to control major decisions from the White House and his distrust of the Washington "permanent government" epitomized by the career staff of BOB, supported involving his White House staff in budget decisions with major policy implications. The result was a significant level of tension between the White House staff and the BOB staff, with neither side helping the other and very little communication between the two. Personal antagonism between Nixon, Ehrlichman, and Flanigan on one hand and Mayo on the other only exacerbated the situation.

As BOB was preparing its recommendations on the NASA budget in November, Flanigan and Whitehead had been monitoring the wide differences between NASA and BOB on the budget's level and content. They judged that neither NASA nor BOB was likely to develop budget choices that met the president's rather unclear priorities. Flanigan had communicated this perspective to Nixon and got clearance to begin developing alternate options. Given this guidance, Whitehead "turned with a vengeance" toward that task.[29] In a December 2 white paper, he observed that decisions with respect to the FY1971 were "particularly important," since "deceptively small budget issues for FY71 entail enormous (up to $100 billion) budget commitments for future years." Even so, he thought "the issues and options that have been defined for the President and the information to support them are scarcely up to the quality appropriate for a Presidential decision." He summarized the situation as he perceived it:

- "Low- cost opportunities for Presidential initiatives have been suppressed."
  Those opportunities included the "prosaic-sounding Apollo Applications

program" and robotic planetary exploration mission such as the Grand Tour.
- "Manned lunar landings have been scheduled at the rate of three per year at a cost of almost $1 billion per year over a rate of one per year, without this issue ever being presented for Presidential consideration."
- "The Budget Bureau has consistently been uncooperative in White House staff efforts to produce information on lower-cost options for Presidential consideration." In Whitehead's view, the BOB career staff seemed "to suffer from an institutional tendency to save the President and his staff from hard decisions, to compromise with agencies as far as possible, then to defend the agency base."[30]

It was quite unusual for White House policy staff to be delving into the technical details needed to craft and then cost out alternative programs in an executive agency. Whitehead peppered NASA with questions with respect to various "building blocks" for alternative programs. A veteran NASA official, skeptical of this activity, noted that the White House people "came up with impossible alternatives…They couldn't understand why…even though it would take you less than four months to check out and launch a vehicle, why you basically couldn't launch it [only] once a year."[31]

In his analysis of the FY1971 budget situation, Whitehead made three additional observations:

- "While the space program is interesting to most of the public, it ranks very low in their priorities for increased Federal spending." Whitehead suggested that "there is no space program or mission on the horizon that offers popular appeal comparable to the first lunar landing, so that space is not likely to climb in the public eye as a desirable use of Federal funds."
- Whitehead was skeptical of the political arguments in support of a high Apollo launch rate, noting that "it is unclear how much domestic and international political benefit accrues to the President and the Nation at the higher launch rates…A major consideration is avoidance of another Sputnik-like event, but we now appear far ahead of the Soviets." He added "the existing supply of 8 Saturn 5 vehicles potentially could be stretched to cover 9 years of manned activities."
- Finally, Whitehead observed that "there is no need now to make program commitments in order to preserve the 1986 Mars landing option." Richard Nixon in the aftermath of receiving the STG report and again as he discussed the NASA FY1971 budget had indicated that he wanted to preserve that option. Whitehead added "the President can at any time make a forward-looking statement on the future of the space program without any large funding commitments."[32]

Flanigan's late December intervention in the NASA budget process had an immediate effect. BOB Deputy Director Schlesinger on December 29 informed his NASA unit that it had to find a way to cut the agency's budget by $1 billion, likely as a reaction to the intervention by Flanigan and Whitehead. Working overnight, the unit was able to come up with $800 million in possible cuts. These cuts were apparently too draconic. Meeting with Nixon on the morning of December 30, Mayo and Ehrlichman decided that the NASA budget would be cut by "only" $225 million. Nixon agreed, saying that it should be made known that he was ordering these budget cuts to "slow down and stretch out" the post-Apollo space effort, reflecting a re-ordering of the priority of space compared to other national efforts, and that he had rejected the recommendation of the STG for a "crash program to Mars," even though sending people to Mars remained the "long-range goal."[33] Paine was called to the White House on the afternoon of December 30 to get the news of additional budget cuts, not from Mayo but from Flanigan and Bryce Harlow.

Paine and his associates spent New Year's weekend revising the NASA budget to meet the new expenditure limit. Paine wrote Mayo on January 2, 1970, telling the budget director that Flanigan had "made it clear that the controlling decision was the necessity to hold NASA FY1971 outlays to $3,600 million." Paine informed Mayo that he and Flanigan had agreed that NASA would be free to revise its plans as it chose, as long as the result was $3.6 billion in outlays (the funds actually spent during the year). Paine told Mayo "that I would, of course, accept and meet this expenditure limitation like a good soldier... *provided* that I have the flexibility to adjust program details and budget authority." Still pushing for approval of the STG recommended program, Paine added "this is the year, and the FY1971 budget is the instrument, in which President Nixon's initiatives in space will go on the record books." Paine's letter was apparently the first time Mayo had heard of the agreement that Flanigan had made with NASA; he felt "double-crossed."[34]

Then Flanigan wrote Paine and Mayo on January 6, laying down several conditions that NASA had to meet:

1. "The Manned Space Flight Program will be carried out on the previously agreed-upon schedule" of two launches per year.
2. "There is no commitment, implied or otherwise, for development starts for either the space station or the shuttle in FY72."
3. "The President's option with regard to the final Saturn 5 launch, as to whether it will be a lunar mission or a second Experimental Space Station is still open."[35]

These supposedly final decisions on the NASA budget soon became known to the Washington space community. *The Washington Post* headlined a front page story on January 11 "Nixon Rejects Big Outlay for Space in the '70s." Paine felt that it was important in terms of the morale of the NASA and contractor workforce to provide some insight into what was going on, and on January 9 and again on January 12 urged Flanigan to allow him to make a statement "explaining the actions we're taking in the most positive way." Paine on January 12 sent a draft of the statement he proposed to make the next day to the White House for approval. The statement was heavily edited to remove any indication that the statement was being made at the president's request and to delete sentences such as "the President accepts the recommendations of the Space Task Group as our basic space plan for the 1970's." Indeed, there was no mention of the STG in Paine's statement as issued. According to George Low, there were times in the days just before January 13 when the White House vacillated regarding the wisdom of making the statement at all, and White House edits "were in part substantive (e.g. don't talk about manned Mars landings or the grand tour) and in part were more or less nit-picking." Final approval of the statement came only 30 minutes before Paine's 2:00 p.m. January 13 press conference at which it was to be released. At the press conference Paine tried to put a positive spin on the impact of what he termed an "austere" NASA budget, but the headline the next day in *The New York Times* said "50,000 NASA Jobs to Be Eliminated." (The 50,000 number included both NASA civil servants and contractor employees.)[36]

With the agreement with Flanigan on budget levels and constraints and with the January 13 press conference, NASA had good reason to believe that its FY1971 budget had at last been finalized. That turned out not to be the case.

NASA Budget: Ratchet Two

On December 30, President Nixon signed a tax reform bill that he characterized as both "good and bad." One of the negative effects of the bill was that it would make it more difficult to balance the FY 1971 budget. Even so, as he signed the bill the president repeated his frequent pledge to present a balanced budget, saying that failing to do so would be "irresponsible and intolerable." This pledge flew in the face of warnings he had been getting from BOB's Mayo as final budget decisions were being made that it would be impossible to achieve a balanced budget without increased government revenues. The Treasury Department and BOB had discovered at the end of December that their revenue estimates, taking into account the impact of the tax bill, were wrong, and that there was an almost $4 billion gap between the proposed FY1971 budget of $205 billion and projected revenues. The issue facing the president was how to close that gap in order to achieve a balanced budget. He could either agree to a tax increase of some kind or further cut the budget.[37]

The Treasury Department quickly came up with a "painless" tax increase package as a means of rapidly generating additional revenue; it involved speeding up collecting estate and gift taxes and levying higher excise taxes on liquor, tobacco, and gasoline. That package would produce a revenue increase in FY1971 of $4.5 billion, more than enough to cover the projected gap. There was one catch to this approach; it depended on the willingness of the Congress to quickly pass another bill incorporating the new tax increases.

On January 3, Nixon approved this approach to achieving a balanced budget; he then called Arthur Burns, his conservative economist counselor, to tell him that news. Burns was scheduled to become chairman of the Federal Reserve Board at the end of January. Although he had lost standing vis-à-vis overall domestic policy within the White House, in his new position his agreement on the path Nixon was taking to achieve a balanced budget was essential. Burns did not agree; he insisted on a properly balanced budget, not one balanced through tax "gimmicks." This meant, Burns argued, additional budget cuts. Nixon had little choice but to agree.

The president announced his decision to seek additional budget reductions at a January 13 meeting of the cabinet, begun just as the NASA press conference announcing the first round of additional budget cuts was winding up. The meeting lasted over three hours. Mayo, present even though he was not a cabinet member, argued that further budget cuts were not possible. Burns's position was argued by Secretary of Housing and Urban Development George Romney, who "exhorted his colleagues to cut even deeper into their own budgets and capped his plea by an astonishing sermon calling on all members of the Cabinet *and* the President, to take a 25 per cent pay cut." Following the meeting, President Nixon ordered "anguished department heads to make still greater cuts to achieve a Burns-style balance." The budget-reduction exercise was dubbed "Operation Paring Knife."[38] It ended up resulting in nearly $4 billion in additional budget reductions, so that the budget proposal President Nixon sent to Congress on February 2 requested $201 billion in expenditures for FY1971, with revenues estimated at $202 billion.

NASA was not represented at the January 13 cabinet meeting, but the next day Paine was advised by Ehrlichman and Mayo that NASA's share of the overall budget reduction would be a reduction of an additional $200 million. This amount had been decided by, or at least cleared with, Nixon. (Mayo later suggested that Nixon had decided on the $200 million NASA reduction even before the January 13 cabinet meeting and thus it was not integral to the "Paring Knife" process.[39]) The NASA leadership quickly identified $51 million in cuts that could be made through a series of small reductions in science and applications programs, but to reach the $200 million reduction, they thought, Apollo missions 17, 18, and 19 would have to be canceled. (Apollo 20 had been canceled in May 1969 so that the upper stage of its Saturn V booster could be used as the basis for the planned orbital workshop, later named Skylab.) Paine wrote the president another strongly worded letter on January 15, informing him of the $51 million reduction but saying that additional reductions to reach the $200 million figure "would require actions which you have specifically instructed me you do not wish to take—actions which would cripple the space goals of your administration and dissipate the Apollo team." These actions included canceling the final three Apollo missions and reducing funding for the space station and shuttle. The job loss accompanying this action, said Paine, would be an additional

15,000 positions in addition to the 50,000 person job reduction he had just announced in his January 13 press conference. Paine said that if NASA were forced to take the whole reduction "I must discuss the problems involved with you personally."[40]

Reacting to Paine's letter, on January 16 there were a series of conversations between NASA and BOB. By late afternoon, Mayo phoned Paine and told him that BOB would accept the $51 million reduction and that no additional cuts would be needed. Paine phoned Flanigan with this news, recognizing the breakdown in communication between BOB and Flanigan's office likely meant that Flanigan was not party to the BOB decision. He was correct. Flanigan's reaction was anger; he said "Do you mean Mayo capitulated?" Flanigan informed Ehrlichman of the agreement, who in turn relayed the news to Nixon, who was at Camp David. The word quickly came back that the agreement was not acceptable; NASA would have to accept the full $200 million reduction. This message was communicated to Paine as he was enjoying a dinner at a Washington hotel in honor of Charles Stark Draper, the head of the MIT Instrumentation Laboratory. A loudspeaker announcement asked Paine to call the White House; Paine made the call "knowing damn well that they were not calling me to say we had more money."[41]

NASA was able to achieve the additional budget reduction by stretching out the schedule for Apollo launches and the launch of the orbital workshop and reducing funds for space station and shuttle studies. No Apollo missions were canceled; the White House had once again called NASA's bluff with respect to saying a reduced budget would mean the early end of human space flight. The final NASA budget was $3.3 billion, $400 million less than Nixon had approved in early December, 25 percent less than NASA's budget request of the preceding October and 15 percent less than NASA's FY1970 budget. New NASA Deputy Administrator George Low noted that "the whole budget situation has been tremendously confused... The series of consecutive cuts, each one of which was defined as being the last cut, is quite hard to understand." Low thought that Richard Nixon was "assessing as we go along the mood of the country." Low referred to a January 17 editorial in the *Washington Star* newspaper bemoaning the NASA budget cuts but saying "cutting the space program is exactly the right thing to do in this period of fiscal restraints." Low judged that "the President feels that he would be severely criticized if he did not make a major cut in the space program," given all the other budget reductions he was proposing.[42] NASA had been caught up in a chaotic confrontation between budget choices and broader fiscal considerations, reinforced by a breakdown in the White House policy-making process. That chaos obscured a stark reality—that through its decisions on the FY1971 NASA budget, the Nixon White House and ultimately the president himself had significantly reduced the priority of the space program among the whole range of government activities. In the form of modest funds for continued study of the space station and space shuttle, NASA's hopes for the future were still alive, but just barely.

President Nixon Explains His NASA Budget Decisions

The meeting with the president that Administrator Paine had requested in his January 15 letter was set for 4:00 p.m. on January 22. Earlier that afternoon, the president had delivered his first State of the Union message to a joint session of the Congress. He had said "the Seventies will be a time of new beginnings, a time of exploring both on the earth and in the heavens," but otherwise made no mention of the space program. As was standard practice in preparing Nixon for a meeting, Flanigan composed a briefing memorandum. He told Nixon that the purpose of the meeting was to allow Paine "to express his convictions regarding the importance of the Space Program as it relates to your Administration." He added that Paine had taken the first two cuts in the NASA budget "in a spirit of complete cooperation." But with regard to the final cut, "he did resist as he believed NASA was bearing a disproportionate share of the reduction." Flanigan characterized Paine as "consistently loyal and cooperative." He suggested that "no doubt you will wish to assure Dr. Paine of your personal interest in and support for the Space Program in the long run."[43]

The Nixon–Paine meeting went off as scheduled; Ehrlichman as well as Flanigan were present. Nixon began the meeting by saying "how much he regretted having to make the last additional cuts in NASA's '71 budget. He understood these were very severe and he had done it most reluctantly," but had no choice given the overall budget situation. He worried that "NASA might find it difficult to defend even this low space budget" against charges it represented misplaced priorities. The president said that "the polls and the people to whom he talked indicated to him that the mood of the people was for cuts in space and defense." Nixon also said that the people of the country seem to think all they want is a nice environment and a turning-away from challenge and sacrifice. Even so, thought Nixon, there were areas like "science, space, and the SST [supersonic transport] the nation must put money into."

Paine asked Nixon what he should tell the NASA workforce about the thinking behind the budget cuts. Nixon responded that the FY1971 NASA budget should be "rock bottom" and that he was "committed to the space program for the long-term future," adding "we should have a strong space program and it should be on an increasing [budget] curve." Paine's conclusion after the meeting was that Nixon "honestly would like to support a more vigorous space program if he felt that the national mood favored it." This seems to have been a valid reading of Nixon's position; in the hours following his meeting with Paine, Nixon called Bob Haldeman, directing him to make sure that the message accompanying the release of the FY1971 budget would include "the flat statement 'We shall plan to go to Mars.'"[44]

Conspicuously absent from discussions in the preceding weeks on the space budget was Tom Paine's putative White House ally, Vice President Spiro Agnew. Paine had thought as the STG process went forward that Agnew's recommendations would carry weight within the Nixon administration, and that Agnew as chair of the Space Council could play an ongoing role in space policy and budget decisions. By the end of the budget process, Paine certainly recognized that these assumptions were not valid. Agnew had become marginalized in administration policy discussions, and the Space Council had not carved out a useful role. Thus it was of limited consolation for Paine to receive a January 30 memorandum from Agnew, saying that while the vice president could not fault the "decision to reduce all budgets in a fashion commensurate to absolute national requirements," he was "concerned about our ability to maintain the high quality of performance that NASA enjoys." Agnew told Paine "you may be assured that I will do whatever I can to persuade the President to move the space program back to a more ambitious level at the earliest possible moment." There was little to no chance that Agnew could be successful in such an undertaking.[45]

For 11 months, Thomas Paine had been depending on the work of the Agnew-led STG and the recommendations in its report to provide the charter for the bold space program he thought was in the nation's, and NASA's, interest during the post-Apollo period. He had consistently tried to use the report as a basis for arguing against cuts in the NASA budget. With the continued reduction in that budget, Paine's aspirations were close to being dashed. In an almost plaintive sentence in his record of the meeting with President Nixon, Paine lamented "the President didn't mention the Space Task Group Report."[46]

Chapter 5

The Nixon Space Doctrine

The decisions about the NASA budget for Fiscal Year (FY) 1971 that emerged from the chaotic budget process were a result of two general influences. One was the need to fit spending on space within the very tight constraints on discretionary government spending if the overall federal budget were to be in balance with expected revenues. This meant determining how the civilian space program would fit within the Nixon administration's overall priorities. In developing the FY1971 budget, the Bureau of the Budget (BOB) had identified the administration's highest priority domestic goals: implementing revenue sharing between the federal and state governments, reducing the crime rate, expanding family and food assistance, increasing manpower training, environmental protection, and improving surface and air transportation.[1] Space did not make this list of top priorities, and that had been reflected in the FY1971 budget decisions. The other influence was the rather ad hoc policy framework President Richard Nixon and his policy advisers used to evaluate the recommendations of the Space Task Group (STG) and the NASA budget proposal based on those recommendations. The White House had not articulated a strategic perspective on the space program to guide it as it evaluated the STG's proposed initiatives. The Nixon administration, by treating space as a domestic rather than foreign policy issue, did not feel compelled to evaluate future space activities in the context of broader geopolitical goals beyond the general thought that there should be increased emphasis on cooperation rather than competition.

The FY1971 budget decisions reduced the priority of space spending within the overall federal budget to a ranking significantly lower than it had held at the peak of the Apollo program in 1966, when the space agency commanded 4.4 percent of total government spending and 19 percent of nondefense discretionary spending. By the time Congress approved NASA's budget for FY1971 in mid-1970, NASA's share of federal spending had shrunk by almost two-thirds, to 1.6 percent of the total and 7 percent of discretionary spending. This was certainly not a budget allocation that could support the kind of program NASA was advocating for the 1970s.

Crafting a Presidential Space Statement

Almost from the start of his administration, there had been suggestions that President Nixon spell out his views on the future in space in a formal statement or speech. As the STG report was submitted, one of the NASA senior staff who had been working with the group, Milt Rosen, observed that "the President is going to make...a policy statement on space, and presumably one comparable in importance with the statement of President Kennedy in 1961."[2]

It was the need to respond to the STG report that ultimately led to the formal statement of Richard Nixon's post-Apollo space policy; that pronouncement is called here the "Nixon space doctrine." But the presidential statement, issued only on March 7, 1970, six months after the STG report had been submitted, was hardly the kind of clarion call to leadership in space that President John F. Kennedy had proposed to the Congress and the nation on May 25, 1961.[3] Rather, it was carefully balanced between providing a positive but very general vision for future space development and making it clear that the space program would no longer be treated as it had been during Apollo, as "special," operating outside the normal process for setting national priorities and based on highly mobilized efforts to achieve challenging goals. In a rather negative way, the Nixon space doctrine was indeed comparable in importance to Kennedy's 1961 setting of a lunar landing as a national goal, for it set out a framework for making space decisions that not only Richard Nixon, but most subsequent occupants of the White House, have used over the past 40 plus years. The framework put in place by Richard Nixon on March 7, 1970, has thus had a far more lasting impact on national space policy than John Kennedy's 1961 decision to go to the Moon.

Drafting a Nixon Space Statement

In recommending that President Nixon endorse Option II of the STG report, NASA Administrator Tom Paine on September 19, 1969, had also suggested that the president quickly issue a statement announcing that endorsement. Peter Flanigan, the assistant to the president with oversight responsibility for the space program, agreed with Paine, and intended to take the lead in preparing such a statement. Although an immediate declaration was opposed by BOB Director Robert Mayo, Flanigan persisted in his effort, asking his assistant Tom Whitehead on October 6 to "draft a statement that the President might use, picking Option 2 but providing his flexibility along the lines suggested in my memorandum of October 4." In that memorandum, Flanigan had argued that he did not "believe that the President can delay until the budget review to respond to the Space Task Group report to him" and had proposed a presidential statement saying "that after a review of the Space Task Group's report...we should plan on a Mars landing in the mid-1980s," without also endorsing the STG recommendation that NASA should first develop a space station and space shuttle during the 1970s. Science adviser Lee DuBridge joined Flanigan in arguing for an earlier statement, saying that "many thousands of people employed in the Space Program, as well as many millions of citizens, are anxiously awaiting an indication of the President's proposals for the future."[4]

Despite the urgings of Flanigan and DuBridge, the White House decided that no immediate presidential space statement was desirable; Mayo's position that such a statement should follow and reflect, not guide, FY1971 budget decisions prevailed. Given the lack of time pressure, Whitehead did not complete an initial outline of a possible statement until mid-November. In transmitting his draft to Flanigan, Whitehead noted that it was "a compromise between strong positive words and the restraint necessary to maintain the President's flexibility in budgeting." He alerted Flanigan to the fact that he had "not specifically referred to Option II of the STG," since "to do so would have the effect of locking us into the spending stream projected for that option as a floor on NASA expectations." Whitehead suggested that "a draft outline should be sent to the President along with a memo showing what we are and are not letting Paine commit us now to begin spending on."[5]

Many of the features of the eventual presidential statement issued in March 1970 were already present in Whitehead's November 17, 1969, draft, which listed three goals for the nation's space efforts—exploration, science, and Earth applications. Notable was that exploration was separated from science as an activity "worthwhile in and of itself." The outline suggested a policy shift "to a continuing program of exploration and application" which would be "a continuing process rather than a series of crash timetables." Listed among "major program goals and initiatives for the next decade in space" were continued lunar landings "paced at a rate to maximize scientific returns"; a "newly designed Experimental Space Station" (This was the orbital workshop soon to be named Skylab); and a "longer lived Space Station Module that will serve both as a near-earth space station and a building block for manned interplanetary travel." A Mars landing, "perhaps as early as 1986," would follow. The outline called for efforts to "lower the costs of space launches," but did not mention the space shuttle. Rather, it suggested that "our recently developed rocket technology will provide a reliable launch capability through the next decade," with continuing research "to make possible even lower costs for launching space payloads in the future." A final initiative was to "expand international cooperation." With respect to funding, the outline suggested that the president should say "we will seek to provide a stable level of expenditures to enable steady progress consistent with other pressing national priorities," but also hold out the hope "to be able to expand our effort in some years and move some accomplishments nearer in time."[6]

Accelerating the Schedule

One way that Richard Nixon got information was through his daily news summary. After reading a November 26 column titled "Future of Space

Program is Reaching a Critical Point," Nixon asked Flanigan to accelerate the public release of his statement on the post-Apollo space program. The column had claimed that the "space program was sinking into some kind of political swamp," absent presidential guidance, with "confusion among scientists and technological communities about the future of the program, and much more dangerous confusion in the government." It noted that "the political climate was not favorable to any decision" and warned that "the White House had best be prepared for a political hurricane when President Nixon finally decides what to do next." Nixon on December 2, as he prepared for his December 5 meeting on the NASA budget, suggested to Flanigan that "the week after next might be an appropriate time" to issue the space statement.[7]

The president's request set the White House machinery in rapid motion. On December 9, Flanigan told staff secretary Ken Cole, the coordinator of White House activities, that "we are currently preparing an outline of a speech or statement for the President regarding the future space program. It is thought that this will be delivered or released in approximately 10 days." In turn, Cole suggested to Jeb Magruder, deputy director of the White House Office of Communications, that "it's not too early to begin drawing up a game plan" for the announcement of the presidential decision. Cole added that "whatever the decision, there will be something there for somebody to stand up and say hurrah for the President."[8]

Whitehead sent a revised version of his outline for the space statement to the White House speechwriting office on December 12; that office, headed by Ray Price, would turn the outline into presidential prose. Whitehead had made a few significant changes from the November version of the outline, reflecting comments made by the Office of Science and Technology (OST) and BOB staffs. The pace of lunar exploration would not only be designed to maximize scientific return but also to be "consistent with the minimum launch rate for safety and reliability." This addition reflected an ongoing debate between those advocating only one Apollo launch per year and NASA, which thought launches every four or at most every six months were needed to maintain the performance of the launch team. The 1986 date for a human Mars landing was deleted, and not replaced by any target date for when such a mission might occur. In the launch vehicle section of the outline, the just-made budget decision to suspend production of the Saturn V booster was noted, with the comment that production could "be resumed at any time in the future as the need arises." A sentence was added saying "we will begin to design a space shuttle that will be re-usable to provide frequent, reliable, and low-cost launches for a wide range of payloads." This was a significant step in decoupling the shuttle from its NASA-advocated role as a logistics vehicle for the space station, and reflected the views of OST and its external advisers of the shuttle's importance as a lower-cost launch vehicle for all U.S. space missions.[9]

To this point in time, the White House had not shared the outline of the space statement with NASA. On December 16, as the text of the space statement was being prepared in the White House for a planned December 18 release, Whitehead sent the outline to Paine, promising to send him a full draft of the proposed statement "as soon as it is available." On the same day, Whitehead shared with Paine the high-profile plan for public release of the statement that had been developed by the White House Office of Communications. That plan called for a short speech by the president before the statement was released. Nixon would be accompanied by Vice President Agnew, science adviser DuBridge, and NASA Administrator Paine. As follow-up to the release of the statement, a variety of activities were planned, including obtaining "strong endorsement" from the aerospace industry, preparing statements for astronauts to use in public appearances, placing astronauts and Paine on various news shows, giving advance briefings for Congressional space committees, scheduling NASA briefings in both Washington and Houston for space reporters, preparing short speeches for use by supportive members of Congress, and preparing an information packet for wide distribution "on the application of space technology to earth technology."[10]

A first draft of the space statement did not emerge from the speechwriting office until December 17. Given the delay in preparing the statement, its release was postponed until December 23. The draft was distributed for comment on December 18 to Agnew, Flanigan, Paine, Whitehead, DuBridge, Mayo, director of the Office of Communications Herb Klein, and National Aeronautics and Space Council Executive Secretary Bill Anders; comments were due on Monday morning, December 22.[11]

Release of Statement Postponed

The review of the draft statement went forward over the next few days in parallel with final decisions on the NASA FY1971 budget. DuBridge submitted a "revised re-draft" that had been "reviewed with representatives of the Vice President's office, BoB, DoD, and Mr. Flanigan's office." As Agnew's office reviewed the statement, it had proposed that the vice president as well as the president make remarks to the press as the statement was released. Agnew was seeking a meaningful post-STG role in space policy, and the Space Council staff had not been able to insert itself into the ongoing debates on reacting to the STG report and shaping the NASA FY1971 budget. Tom Paine sent to the White House not only comments on the draft statement but also a totally rewritten draft reflecting NASA's hope to get President Nixon on the record as formally endorsing the recommendations of the STG, especially by initiating space station and space shuttle development immediately and setting a mission to Mars before the end of the century as a long-term goal.[12]

The choreographed release of the Nixon space statement was abruptly postponed just before Christmas. It is likely that the decision to take a re-look at the NASA budget in late December was the proximate cause. It would have been embarrassing for the White House to release a space statement just as the NASA budget was undergoing additional reductions, and that possibility may well have become evident to the White House policy and communications staff in the December 20–22 period.

As the release of the statement was postponed, the hope was to have it available soon after the president made what were anticipated to be his final decisions on the NASA budget. After a few days' hiatus for the Christmas holiday, the release was rescheduled for January 3, 1970. But this release was also postponed, as "Rachet 1" of the NASA budget took place over the turn of the year and "Rachet 2" seemed a possibility. On January 5, 1970, staff secretary Cole told speechwriter Keogh that "public release of this statement has been delayed indefinitely."[13]

When to Release the Space Statement?

As they met on January 22, 1970, after the final budget decisions, President Nixon, Paine, and Flanigan agreed that the statement should be issued before the *Apollo 13* launch in April. Nixon stressed that the statement should be written in a way to avoid opponents of the space program being able "invidiously" to compare "his positive statements on space to problems in poverty and social problems here on earth." He did not want to be put in a position of appearing as if "he is taking money away from social programs and the needs of the people here to fund spectacular crash programs out in space."[14] This was another example of the impact of treating space as a domestic issue, competing for funding with other domestic programs.

Following the presidential meeting, Flanigan reported that "Dr. Paine sees no necessity for the President's Space Statement being made in the very near future. In fact, he believes the ideal time would be between the last week in February and the middle of March." The release was then scheduled Saturday, February 28, in time for it to be reported in that Sunday's newspapers. Flanigan told Paine of the date, suggesting that if it was "not appropriate would you please let me know" and asking Paine to be sure that any changes in the early January draft of the statement "are discussed with us early enough so that we can staff them through the speechwriting office." Paine had suggested that a delay in releasing the statement would allow NASA to insert in the draft "some additional information...to give it more sex appeal."[15]

Paine reminded Flanigan that he would be out of the country beginning February 22 on a two-week trip to Australia and Japan "to develop possibilities for further space cooperation." This would mean that Paine would not be in Washington if the statement were released on February 28. Additional discussions between NASA and Flanigan's office led to a decision to delay the statement's release by one week, until Paine had returned from his overseas trip; the release was then set for Saturday, March 7. At the January 22 meeting among Nixon, Paine, and Flanigan, the desirability of increased attention to international partnerships was discussed, and Paine had suggested that the statement should be revised to "put somewhat more emphasis on international cooperation."[16]

Richard Nixon's "Pet Idea"

That suggestion set off a brief and ultimately unsuccessful attempt to add an attention-getting angle to the space statement. The idea was attractive to White House speechwriters; lead writer James Keogh "was enthusiastic about casting the message as a call for more international cooperation," since "if this were the central theme, the message would take on a novel and exciting quality which the present draft is lacking."[17]

Early Interest in Increased International Cooperation

There was substantial background to White House interest in international space cooperation. As the Nixon administration entered office in January 1969, Arthur Burns had identified three themes from the space transition task force deserving of detailed attention. Two of these themes were a need for an overall review of the space program and the possibility of significant reductions in launch costs. These items had been combined in the decision to create the Space Task Group. The third theme was increasing the amount of and broadening the character of international cooperation in space. President Nixon had asked Secretary of State William Rogers to assess ways of achieving these objectives. Rogers responded to the president on March 14, 1969, saying that "we are interested in space cooperation, not only for its intrinsic scientific merits, but also to further specific foreign policy objectives." Rogers identified "major new opportunities for international cooperation." These included "foreign participation in the U.S. manned flight program, including foreign scientist-astronauts." He told the president that he was examining the benefits of Nixon making at the successful climax of the first lunar landing mission "a major public statement on the international values of our ongoing space program."[18]

Such a statement was not issued. Although he said nothing specific about increased international cooperation at the end of the *Apollo 11* mission, President Nixon did address space cooperation as he spoke before the General Assembly of the United Nations on September 18, 1969. Nixon told delegates from around the world "I feel it is only right that we should share both the adventures and the benefits of space." He said that the United States would take "positive, concrete steps...toward internationalizing man's epic venture into space—an adventure that belongs not to one nation but to all mankind, and one that should be marked not by rivalry but by the same spirit of fraternal cooperation that so long has been the hallmark of the international community of science."[19]

Flying Foreign Astronauts?

The possibility of having non-U.S. astronauts go into space on U.S. spacecraft had interested Richard Nixon from the start of his presidency. He asked Henry Kissinger soon after his inauguration to explore broadening international space cooperation, and especially "participation of foreign astronauts in the US program." Nixon may have mentioned this idea to Frank Borman when the *Apollo 8* crew visited the White House on January 30, 1969. At any rate, as Borman returned from touring Western Europe, he recommended that President Nixon invite the European Space Research Organization, the intergovernmental agency created to pool resources for Europe's space science efforts, to nominate two European scientists to train at NASA as astronauts. Borman followed his phone call with a letter to Secretary of State Rogers proposing that the United States "immediately request an international agency to select a certain number of qualified scientists from different nations of the earth to join our program to participate as scientists/astronauts in future earth-orbital space stations."[20]

The subject of non-U.S. astronauts came up again on the July 23 flight across the Pacific Ocean aboard Air Force One to meet the returning *Apollo 11* astronauts, as Borman discussed the idea with the president and Henry Kissinger. Nixon remained intrigued, and asked Borman to follow up with Kissinger. Borman laid out his thinking in an August 5 memo. He proposed that the United States immediately begin discussions with Europe and Japan to nominate scientist-astronauts who could "participate in the earth orbital flights…in the mid-1970's." He also proposed "a rather dramatic call for Japanese-European experiments to be flown on the space station." He suggested that "the appropriate time to undertake negotiations" leading to foreign participation was "the immediate future."[21]

NASA Administrator Paine was also on the flight across the Pacific, and he met separately with Nixon and Kissinger. Nixon authorized Paine to begin discussions with potential international partners, particularly in Europe, with respect to their possible participation in the post-Apollo program. Soon after returning to Washington after the *Apollo 11* landing, Paine met with the head of the European Space Research Organization to brief him on U.S. post-Apollo planning. He stressed that the opportunity "to associate their own astronauts with us in future programs" had to be considered "in the context of substantive joint contributions" to those programs. Linking flight opportunities to sharing the costs of hardware development was to remain central to Paine's thinking on international cooperation.[22] Paine had either misread or misinterpreted Richard Nixon's interest in enhanced cooperation, which was focused on flying non-U.S. astronauts, not on joint development of or major foreign hardware contributions to post-Apollo space systems. What the president had in mind *was* clear to Flanigan, who told Nixon of Paine's initial conversations with European representatives, saying that based on these discussions, Paine "would prepare a plan for the inclusion of foreign nationalists [sic] in future U.S. space activities."[23]

Between October 1969 and March 1970, Paine traveled to Europe, Canada, Japan, and Australia, promoting the STG report as reflecting what the United States was very likely to do in space in the coming years, even as he knew full well that the Nixon administration was resisting approval of the major programs the STG had recommended. His rather paternalistic goal in Europe was "to stimulate Europeans to rethink their present limited space objectives" and "to help them avoid wasting resources on obsolescent developments [such as their own launch vehicle]." Paine also sent the STG report to the Soviet Union in the hopes of promoting "complementary or cooperative space programs." These efforts to create substantial international involvement in the U.S. post-Apollo space program will not be discussed in detail in this study, other than to note that they became controversial within the upper reaches of the Nixon administration.[24]

By late November Richard Nixon was becoming impatient with the lack of any action with respect to flying non-U.S. astronauts. He asked "is there still no way to get multi-national participation in some of our future space flights? I have raised this with Paine and Borman and I know there are some technical problems but it is a pet idea of mine and I would like to press it." He asked Peter Flanigan to "jog the bureaucracy" on the issue.[25]

Flanigan did discuss the issue with Borman, and Borman responded in a December 2 memorandum, saying "it was perfectly feasible and desirable to invite foreign participation in the space program at the present time." He equated "foreign participation" with flying foreign astronauts, saying that "the inclusion of foreign astronauts in our programs would lead to further cooperation at the engineering level and hopefully to more direct financial participation" on the part of other countries. While NASA's Paine believed that financial contributions were a necessary prerequisite to flight opportunities for foreign astronauts, Borman (and seemingly Richard Nixon) thought the flight opportunities should precede, and perhaps lead to, financial involvement. Borman noted that in principle a foreign astronaut could be part of an Apollo lunar landing mission, but he recommended against such a step, saying that "the Apollo hardware is extremely complicated and requires long training periods for proper utilization." In addition, there were already a number of U.S. astronauts who had been training for a long time and who "would quite properly wonder at the sudden inclusion of a foreign crew member." As he had suggested in August, Borman repeated "the time to take the initiative in this field is ripe."[26]

Paine also responded to Flanigan's query about flying foreign astronauts by lobbying for approval of a NASA FY1971 budget that allowed rapid progress on the space station and space shuttle. Paine told Flanigan "obviously, we can't fly foreign astronauts if we are not going to have anything to fly them in—a Space Shuttle, or anything to fly them to—a Space Station." Flanigan responded in a manner suggesting either a slip in attention or that he was still not fully familiar with NASA's programs, saying "how about flying them in the Apollo obligations [*sic*—should be applications] program?" While Borman had suggested that foreign astronauts could fly on the orbital workshop, which is what Flanigan was referring to, Paine did not offer that possibility, saying that there were too many American scientist-astronauts hoping to be on one of the planned three flights to the workshop to open up a slot for a foreign participant.[27]

With the decision to postpone the release of the space statement until March, the urgency of responding to President Nixon's query about flying foreign astronauts diminished. But Nixon did not forget his "pet idea." On February 12, after reading a report regarding Paine's international activities, the president, clearly impatient, tried to force the issue. He informed the National Security Council that "he would like to have a program which could be announced as soon as possible for German, Japanese, British and French astronauts to participate in our space program." Nixon wanted "to have this program initiated in the earliest possible year."[28]

It may have been the possibility of announcing an invitation for foreign astronaut participation to which Paine was referring in January when he said that the delay in releasing the space statement would allow NASA to "add more sex appeal" to the draft. The president's persistent raising of this issue appears to have catalyzed action on this concept. On February 26, NASA proposed a modification of the January draft of the space statement that would include

> the first official announcement on foreign astronauts. Foreign astronaut participation is linked to space shuttle-space station projects as the first practical opportunity for foreign astronauts in the current U.S. program. Foreign astronaut participation is also tied to "broad involvement" and "contribution" by the foreign nations to the space shuttle-space station programs so as to be consistent with our attempt to secure meaningful participation by the other countries.[29]

NASA's change was not accepted; there was opposition to such a step coming from the president's staff. In his February 10 memorandum discussing the possibility of making international cooperation a central theme of the presidential space statement, Lee Huebner of the speechwriting office had added a "caution," saying that Tom Whitehead was "*very* skeptical about over-selling internationalization," since "there has been little substantive progress" and the issue "is wrought with pitfalls." Given this, "the President could easily overpromise without being able to deliver." Whitehead perceived NASA as "engaging in some wishful thinking, trying to create new realities through public relations even though the tough questions in the area have not yet been hammered out." In addition, NASA was trying in its suggested language to link Nixon's interest in flying foreign astronauts to getting the sought-after presidential commitment to the space station and space shuttle.[30]

Whitehead's position, seconded by Flanigan, carried the day within White House policy circles, even in the face of the president's explicit request for a plan for foreign astronaut participation. This was not an isolated incident. Nixon's senior staff not infrequently ignored or countermanded his directives, especially those issued in a fit of anger, when they judged them not to be in the country's or the president's interests. In this case, Nixon had persisted in pushing his "pet idea," but either explicitly or by not being offered the option of adopting it as his space statement finally reached him for approval, his wish was overruled.

Announcing the Nixon Space Doctrine

In the week before the March 7 release of the space statement, there were some final edits to the draft that had been ready on January 3. The plan was to have NASA Administrator Paine return directly from Japan, where he had been discussing post-Apollo cooperation, arriving in Florida in time to meet with the president at his Key Biscayne retreat, then to be available to answer press questions after the statement's release.[31]

NASA was given one more chance to comment on the draft. The space agency suggested two substantive modifications and a few word changes. Instead of just a passing mention of the space shuttle, NASA suggested adding two sentences saying "we are currently examining the design of a reusable space shuttle that could evolve into a new space capability. With this capability, we could fully exploit and use space for the benefit of all mankind and at the same time substantially reduce the cost of space operations." This was another attempt to get the president on the record as supporting the shuttle. It was rejected. The other suggested addition reflected a vague mention of the intention to fly foreign astronauts: "Unmanned scientific payloads from other nations already make use of our space capabilities on a cost-shared basis; we look forward to the day when these arrangements can be extended to larger application satellites and astronaut crews." This suggested change was tentatively accepted by Whitehead; he told Flanigan's office that, if Flanigan "has any troubles" with the mention of foreign astronauts, "blow the whistle fast!!!"[32] Flanigan did not object, and the NASA change was incorporated into the statement.

On March 5, the statement went to John Ehrlichman for final review before being sent to the president. Ehrlichman recommended to Nixon "that you approve the Space Statement...for release this Saturday." After getting the president's verbal approval, Ehrlichman on March 6 checked the "Approve" option on the memo. This was the climax of the elaborate staff process that had begun exactly five months earlier with Flanigan's October 6 charge to Whitehead to begin drafting the space statement.[33]

There were at this midweek point still plans for President Nixon to meet with Tom Paine on Saturday in Key Biscayne before the statement was released. Flanigan prepared a briefing memorandum in anticipation of the meeting. Recognizing that NASA was not happy with the cautious tone of the statement and that Nixon was more positively inclined toward the space program than most of his advisors, Flanigan told Ehrlichman that, while he believed that "it would be desirable for the President to meet with Paine for a short time, I would urge that this not be an occasion for Paine to attempt to talk the President into reinterpretations of the Message, since we are not yet ready to make any further commitments on NASA programs." Flanigan told Nixon that the space statement "was designed primarily to put space in perspective vis-à-vis our other priorities and to set forth a rationale for planning the future direction of the space program." Flanigan reminded the president that the "thrust" of the statement was "more explanatory of a rationale than a listing of program initiatives," and recommended that Nixon suggest to Paine that he "address the rationale as well as program initiatives in his press briefing." With respect to international cooperation, Flanigan told the president "this area turns out to be more difficult than might be expected." Flanigan counseled Nixon, if Paine were to raise the question of the level of presidential commitment to the space station and the space shuttle, to "stress the need to consider a full range of options and make design and development decisions only after more technological and cost unknowns are resolved."[34]

As it turned out, Paine and Nixon did not meet on the morning of March 7; the president took most of the morning off from official duties. Nor did any of the activities that had been planned in December to accompany the release of the statement take place; by this time, the statement was modest enough in aspiration to convince the White House it did not merit high visibility. Flanigan had suggested in early February that "much of the interest in the future of the space program has been dissipated"; the White House press and communications staffs apparently agreed.[35] In May 1961, John Kennedy had announced his decision to go to the Moon in a nationally televised address before a joint session of Congress. In 1970, Richard Nixon's space policy was announced in the form of a statement issued by the White House press office; Nixon himself was nowhere to be seen.

The final version of the space statement differed little from the draft that had been ready for release in January, with the exception of incorporating some, but not all, of NASA's suggested changes and linking the rationale put forth in the statement to the administration's FY1971 budget decisions. The document was released as a "Statement by the President." The statement noted that "over the last decade, the principal goal of our nation's space program has been the moon" and that it was now time to "define new goals that make sense for the Seventies." Those goals had to be chosen while recognizing "that many critical problems here on this planet make high priority demands on our attention and resources. By no means should we allow our space program to stagnate. But—with the entire future and entire universe before us—we should not try to do everything at once." It mentioned the STG report and said that "after reviewing that report and considering our national priorities," Nixon had "reached a number of conclusions concerning the future pace and direction of the nation's space effort."

Having said that there was a need to "define new goals that make sense for the Seventies," the statement did not spell out such goals, at least in a way similar to President Kennedy in 1961. Rather, it called for an approach to space that was both "bold" and "balanced." It identified "three general purposes" to "guide our space program": exploration, scientific knowledge, and practical applications. Six "specific objectives" were identified:

- "We should continue to explore the Moon."
- "We should move ahead with bold exploration of the planets and the universe." The statement identified as a "major but longer range goal...we will eventually send men to explore the planet Mars."

- "We should work to reduce substantially the cost of space operations." The statement noted the need in the "longer-range future" for a means of transporting payloads into space that would be "less costly and less complicated" and said "we are currently examining...the feasibility of reusable space shuttles as one way of achieving this objective."
- "We should seek to extend man's capability to live and work in space." The statement discussed the "Experimental Space Station (XSS)." (NASA by this time had christened the orbital workshop as Skylab, but had not convinced the White House to use the new name in the statement.) It said that "on the basis of our experience with the XSS, we will decide when and how to develop longer-lived space stations."
- "We should hasten and expand the practical applications of space technology."
- "We should encourage greater international cooperation in space."

The core policy element of the statement set out the approach to treating space as "an investment in the future." The final version of this policy declaration differed little from what had been in the January draft:

> We must realize that space activities will be part of our lives for the rest of time. We must think of them as part of a continuing process—one which will go on day in and day out, year in and year out—and not as a series of separate leaps, each requiring a massive concentration of energy and will and accomplished on a crash timetable. Our space program should not be planned in a rigid manner, decade by decade, but on a continuing flexible basis, one which takes into account our changing needs and our expanding knowledge.
>
> We must also recognize that space expenditures must take their proper place within a rigorous system of national priorities. What we do in space from here on in must become a normal and regular part of our national life and must therefore be planned in conjunction with all of the other undertakings which are also important to us.[36]

The overall message of the president's space statement was that NASA's days of operating outside of the continuing competition for government resources were over. The Apollo program in 1962 had been formally assigned the government's highest national security priority, giving it preferred access to scarce resources, and it was difficult for the NASA leadership, indeed for most of the space community that had grown up alongside Apollo, to accept a future in which that priority was drastically reduced, with space becoming just one among many areas of government activity. Yet a realistic reading of the Nixon space statement in the context of the overall policies of his administration should have made clear that this was the space agency's most likely prospect.

Chapter 6

The End of the Apollo Era

In his press conference after the March 7 release of the presidential space statement, NASA Administrator Paine tried to put a positive spin on the document, calling the program that the president had announced "bold, diversified, very wide ranging." But Paine in a rare note of realism did recognize the challenge of reorienting NASA to new objectives, saying "what we are really faced here in this change as President Nixon's space program replaces the old space program of the 60's is we are essentially taking a $3.5 billion enterprise which has been going in one direction, a very single-minded purpose, and completely changing it around and moving in a new direction. That is a tough job."[1]

The reality—that a new direction was needed and that it was not going to be based on accepting the recommendations of the Space Task Group (STG)—sank in fairly quickly. As it defended its FY1971 budget request to the Congress in spring 1970 NASA was publicly persisting in its hope to develop simultaneously both the space shuttle and the space station, presenting them as a single, inseparable "station/shuttle" program. NASA also told the Congress that it intended to launch seven more Apollo lunar landing missions, Apollo 13–19. But even as these programs were being justified, to mixed Congressional reaction, behind the scenes the NASA leadership was beginning to recognize that there was essentially no possibility of getting the budget allocations over the next several years needed to support the agency's ambitions. Something would have to give, and over the summer of 1970, that "something" became both abandoning plans to develop the space station and the space shuttle in parallel and canceling two of the six Apollo missions remaining after *Apollo 13* was launched on its fateful flight in April 1970. By the time NASA submitted its budget request for Fiscal Year 1972 in September 1970, the only major new program for which the space agency was seeking approval was the space shuttle. In a little over 12 months, the shuttle had transitioned from a necessary complement to the top-priority space station to the single large program on which NASA was staking its future. The totality of the changes in the NASA program made during the first nine months of 1970 added up to the end of the Apollo era in NASA's history, even though four more Apollo launches to the Moon would take place in 1971 and 1972, a Skylab orbital workshop based on Saturn V hardware would be launched in 1973 and visited by three astronaut crews using Apollo spacecraft, and an Apollo spacecraft would rendezvous and dock with a Soviet spacecraft in 1975. After those missions, there would be no more use of the launchers and spacecraft developed for Apollo. Unless NASA could get presidential approval for the space shuttle, the U.S. human space flight program would come to an end.

First Adjustments

All of these final Apollo missions used equipment already in production by 1970. The ability to produce more Apollo spacecraft and Saturn launchers would soon be abandoned.

No More Saturn V Launchers

NASA in July 1969 had awarded 11-month contracts to study the preliminary design of a Saturn V-launched space station to leading aerospace companies North American Rockwell and McDonnell Douglas. The space agency had set the parameters for the studies based on George Mueller's integrated plan. The initial station module was to be 33 feet in diameter, the size of the first and second stages of the Saturn V booster that would be used to launch it. This "core module" would be able to support a 12-person crew and have a ten-year lifetime; it was to be the first step on a path to having an increasing number of humans living and working in space.

The FY1971 budget decision to suspend for an indefinite period production of the Saturn V cast an immediate pall over this plan. NASA would need one Saturn V to launch the initial module, and additional boosters if the subsequent low-Earth orbit infrastructure buildup contemplated in the STG report were to be pursued. However, the seven remaining Saturn V vehicles of the original 15 ordered at the start of Apollo were already committed to the six remaining Apollo missions after *Apollo 13* and to Skylab, and prospects for restarting Saturn V production in a few years appeared dim.

As noted in chapter 2, the process of shutting down the production line for the Saturn V had begun in 1968, even before Richard Nixon had arrived at the White House. Then-NASA Administrator James Webb had rejected a request to begin procuring long lead-time equipment for a next production run of the Saturn V on the grounds that there was no approved requirement for those additional launchers. The Saturn V had received a brief reprieve in early 1969 as the STG recommended adding the funding to NASA's FY1970 budget needed to keep the production line open in order to preserve President Nixon's option to approve an ambitious post-Apollo space program. That decision had been reversed in the December 1969 budget negotiations; Tom Paine had chosen to sacrifice funding for additional Saturn Vs in order to obtain White House approval for funds to study the space station and space shuttle. The FY1971 presidential budget proposed "suspending" Saturn V production, with the idea that production could be restarted if additional heavy-lift boosters were needed in the future.

By mid-June 1970 NASA Deputy Administrator George Low concluded that restarting Saturn V production was an unrealistic hope, given NASA's budget outlook. This meant that the only way to have the massive boosters available to launch the initial large space station module or a second Skylab mission was to cancel one or more Apollo missions and use the Saturn V boosters assigned to those missions for those launches. Low judged that NASA would "not get the amount of funding we anticipate in 1972 or 1973" and that "there seems to be a disenchantment in America and particularly in Congress with additional flights to the moon." Low discussed his ideas on canceling one or more Apollo missions with Tom Paine, who "originally was very negative," but upon reflection "talked about this in a much more positive vein."[2] The final decision that NASA would not retain the industrial capability required to restart Saturn V production was not made until 1972, but by mid-1970 it was virtually certain that there would be no more of the Moon rockets produced. With this decision, the United States gave up for decades to come its capability to launch astronauts for voyages beyond the immediate vicinity of Earth.

A Shuttle-Launched Space Station?

In the first half of 1970, an alternative approach to developing a space station emerged. The Aerospace Corporation, the national-security-oriented engineering and systems analysis organization that had done most of the work on a joint Department of Defense–NASA study of the space shuttle submitted to the STG in June 1969, had continued to examine possible uses of the space shuttle. One of those options was using the shuttle to launch a number of smaller modules that could be assembled in orbit to create a space station with capabilities similar to the Saturn V-launched version. Some in NASA found this approach intriguing, and by April were suggesting that NASA's space station study contractors begin to examine "Shuttle-sized modules" as the basis for a station. By mid-May, NASA at the engineering level had made its decision; a directive to the study contractors said that "additional work on the 33-ft. diameter space station will be deferred" and that further study effort would focus on "modular station concepts 15-ft. in diameter." (That diameter was based on the width of the payload bay of the shuttle design NASA was studying.) After some additional in-house study, this decision was formally announced on July 29, 1970; that was the day that the Congress passed the NASA appropriations bill, which included no funds for the Saturn V. (There had been some faint hope that the Congress would reverse the Nixon administration decision to suspend production.) Henceforth, NASA's industry partners would study only a shuttle-launched station.[3]

Space Station Exits the Stage

However, the shuttle-based approach to keeping space station development alive as an immediate post-Apollo prospect had a short lifetime. The NASA leadership in mid-July 1970 met to formulate the agency's program for the next five to ten years. They took into account the president's March space statement, the funding the agency would request in its FY1972 budget submission, due on September 30, and an estimate of the budget it could expect in the subsequent few years. A key result of these discussions was a decision to return the space station to preliminary study status rather than seek FY1972 approval to begin its detailed design and development. This decision effectively postponed the station for a number of years. Associate Administrator for Manned Space Flight Dale Myers, who had joined NASA in January 1970 as George Mueller's successor, told Low that he was "moving out to the shuttle first because...an interim space station, without a proper logistics system, would be dead-ended." Low agreed, recognizing that "a space station without a shuttle makes no sense at all...a shuttle without a space station does."[4]

This was a momentous choice. It meant that NASA would abandon its plan for simultaneous development of the station and shuttle that had been at the heart of its post-Apollo aspirations; rather, NASA would first seek approval to develop the space shuttle, postponing station development until after the shuttle began flying later in the 1970s. It also meant that the shuttle would have to be sold as a general-purpose, lower-cost launch system and as the way of keeping astronauts flying in space, not as a logistics vehicle for a space station, its original rationale.

Even with the decision to give shuttle schedule priority vis-a-vis the station, the link between the space shuttle and an eventual space station remained unbreakable; in NASA's view, one of the highest priority requirements driving space shuttle design would be its ability to launch modules large enough to be assembled into a viable space station. NASA told the White House as it submitted its budget request in September 1970 that "we have made a major decision to defer development of a space station...to a later time and to orient the space station studies we will continue in FY1972 toward modular systems that can be launched as well as serviced by the space shuttle."[5] The space station for the time being might be postponed, but it would not be forgotten.

Retreat from the Moon

The human space flight program that emerged from these July meetings also anticipated canceling two Apollo missions. Budget constraints were an important reason for NASA's willingness to forgo those trips to the Moon. But there was another factor in play. Some influential individuals within the NASA human space flight leadership had by the start of 1970 become skeptical of the wisdom of flying additional missions to the Moon after the 1969 successes of *Apollo 11* and *Apollo 12*. They argued that President Kennedy's end-of-the-decade goal had been met and there was no compelling reason to continue to accept the high risks associated with each lunar journey. According to one authoritative account, Robert Gilruth, the director of NASA's Manned Spacecraft Center in Houston, who some described as the "father of manned spaceflight," suggested that NASA should "stop now, before we lose someone." There is disagreement about whether these were actually Gilruth's views, but certainly the risk of each additional lunar mission was on the minds of NASA's leaders. The near-fatal accident during the April 1970 *Apollo 13* flight only reinforced their already-present hesitation to fly out the full Apollo schedule.[6]

However, NASA on its own was not free to finalize a decision to cancel an Apollo mission. The Apollo 16 through Apollo 19 missions would use an enhanced lunar module capable of longer stays on the Moon's surface and would carry a lunar rover able to carry the astronauts well beyond walking distance of the module. This combination would greatly increase the potential scientific yield from the lunar missions, and was eagerly anticipated by the segment of the scientific community interested in planetary science. Not flying latter Apollo missions would likely cause an uproar in that community.

Apollo Program Review

NASA thus decided to go through a formal consultation process before making a final decision on how to proceed. On August 5, Paine wrote John Findlay, chairman of the Lunar and Planetary Missions Board (a NASA-chartered advisory group) asking him to provide the board's views on the question "what additional values accrue to lunar science by retaining Apollo 15 and 19 in the lunar exploration program?" A similar letter was sent to Charles Townes, chair of the National Academy of Sciences Space Science Board, on August 13. NASA alerted the White House to what it was contemplating, saying that it was assessing two program alternatives. One would involve flying *Apollo 14–17*, then launching Skylab and the planned three astronaut visits to the workshop, and then launching *Apollo 18–19*; the other option was canceling *Apollo 15* (the last mission without the lunar roving capability) and *Apollo 19* and flying the four remaining Apollo missions before Skylab. The latter choice, which was preferred by NASA, would make two Saturn Vs available for future uses—"such as space station launches." NASA told the White House that it "would be in touch with you about September 1 to let you know the conclusions" of its review. Peter Flanigan responded quickly, saying that "it certainly seems to me that you are giving this problem the careful consideration it deserves" and asking whether someone from the White House "could profitably sit in on" the final review meeting "in order to hear the pros and cons of the arguments," rather than just having the White House be informed of NASA's conclusions after the review was completed.[7]

The review meeting was held on August 24. Myers presented a plan calling for the deletion of *Apollo 15* and *Apollo 19*, a step he estimated would save approximately $800 million over the next several years. Findlay reported that both the Lunar and Planetary Missions Board and the Space Science Board strongly preferred flying the remaining six lunar landing missions as "markedly superior from the point of view of scientific yield," but if a mission had to be canceled, "the loss of Apollo 15 from the program is serious, but the loss of Apollo 19 would be much more serious due to its capability for longer lunar surface EVA and its significant transverse capability." In response to Flanigan's suggestion, NASA had invited several White House representatives to the meeting. No one came from Flanigan's office, but Bill Anders from the Space Council and Russ Drew from the Office of Science and Technology attended. Anders was "extremely concerned" that, if *Apollo 15* and *19* were canceled, there could be a hiatus of up to four years in human space flights between the end of the Skylab program and the first flight of the space shuttle; he was later to suggest flying several Earth-orbiting missions using leftover Apollo spacecraft in this period.[8]

As NASA was preparing to make its decision, science adviser Lee DuBridge added his thoughts, writing Paine on August 28 to say that even if *Apollo 15* were canceled, he would "favor making every attempt to retain all of the other flights and I hope very much that it will not be decided to eliminate *Apollo 19*. This can cap the climax [*sic*] of all the others." DuBridge added "I understand the desire of some to keep Saturn V's in reserve. But they have been built for the Apollo purposes and there is no emerging purpose which seems clearly able to take precedence over the use of the Saturns for the additional Apollo missions. In addition, one must recognize that...there is a certain non-zero probability that one will be lost as in the case of Apollo 13."[9]

None of the arguments that NASA heard in August changed the agency's July's thinking—that the prudent course of action, given NASA's anticipated budgets for the next several years, its desire to get FY1972 approval to start developing the space shuttle, and the high risk associated with each Apollo mission, was to fly *Apollo 14* in January 1971, to cancel *Apollo 15* and *Apollo 19*, and to re-number *Apollo 16–18* as *Apollo 15*, *Apollo 16*, and *Apollo 17*, with *Apollo 17* being the final lunar landing mission. Paine informed President Nixon of this plan on September 1, saying that "the most compelling reason for the decision to delete these flights, which we have arrived at reluctantly but with overwhelming consensus, is the current and reasonably foreseeable austere funding situation for NASA." Paine told Nixon of the views of the scientific community in favor of not deleting the missions," but said that the scientific benefits of the two missions being canceled "do not, in our judgment, outweigh the benefits of other ongoing and future NASA programs and the risks involved in these difficult missions." Paine noted that "in view of Soviet progress on large launch vehicles, it is prudent to retain a modest Saturn V capability...Deleting the Apollo 15 and 19 missions provides a national reserve of two Saturn V's."[10]

Who Ended Apollo?

Richard Nixon has frequently been identified as the individual who decided to truncate the Apollo program. As the above account shows, this is not fully the case. Nixon's personal attitude toward the desirable number of Apollo flights was not consistent. In January 1970, Nixon and his advisors approved a NASA FY1971 budget that anticipated seven more Apollo flights, even though the president had in early December 1969 expressed skepticism regarding "the need to go to the Moon six more times" and "didn't care about building more [Apollo] hardware." After the April 1970 *Apollo 13* accident, which had a strong emotional impact on Nixon, the president indicated that the Apollo program would continue as planned. It was a Nixon decision to hold NASA to the tightly constrained budget that forced a choice between existing missions and getting started on future programs. But it was the NASA leadership that proposed not flying all remaining Apollo missions. In June, reflecting on NASA's future outlook, George Low had even contemplated canceling four, rather than just two, of the remaining six Apollo flights. He noted that "if we make a major program change like this, we will attribute it to the budgetary situation and to the manpower situation in NASA, and not to the fact that it may programmatically also make more sense."[11] The United States decided in 1970 to retreat from exploring the Moon; that decision had several parents, not just Richard Nixon.

Apollo, Kennedy, and Nixon

During the 1960s, the United States had spent close to $25 billion to develop the capability to launch large payloads into orbit and beyond and to land on another celestial surface. This capability included not only the production facilities and tooling for the Saturn V launch vehicle and the Apollo spacecraft but also the gigantic complex at the Kennedy Space Center required to launch the Apollo/Saturn combination to the Moon. To those such as James Webb who had fought for the political support and funding to create and use it, this capability represented an extremely valuable element of U.S. national power, not only in the context of the Cold War competition with the Soviet Union but also in terms of being a concrete and very visible symbol of U.S. ability to do in space whatever it decided was in its national interest. Sending astronauts to the Moon, Webb had argued throughout the 1960s, was only the first use of this capability. It could also enable a variety of other large-scale national security, exploratory, and scientific undertakings.

Richard Nixon and most of his policy and budget advisors did not share this concept of continued large-scale space undertakings as being important to U.S. power and pride. The March 1970 presidential statement on space had said that U.S. space activities should be viewed "as part of a continuing process—one which will go on day in and day out, year in and year out—and not as a series of separate leaps, each requiring a massive concentration of energy and will and accomplished on a crash timetable." Based on this perspective, through its post-Apollo budget and policy decisions the

Nixon administration made a conscious decision to abandon the capability that had been so expensive to develop and that had given the United States the possibility of an expansive future in space. John F. Kennedy in 1961 had characterized his decision to send Astronauts to the Moon as a "great new American enterprise...which in many ways may hold the key to our future on earth." A year later, Kennedy declared that the he had chosen "to go to the moon in this decade and do the other things not because they are easy, but because they are hard," and "because that goal will serve to organize and measure the best of our energies and skills." Richard Nixon did not share this view of the importance of space achievement; in sharp contrast to John Kennedy, Nixon in 1970 made the mundane proposal that "what we do in space from here on in must become a normal and regular part of our national life." Although Richard Nixon as he discussed the space program frequently linked "exploring the unknown" to continuing national vitality, there was little of such a grand vision in his actual approach to space decisions.

Intermission

With NASA's cancelation of two Apollo missions to the Moon, deferral of space station development, and the decision to make the space shuttle the centerpiece of its post-Apollo hopes, the curtain came down on the first act in the drama of setting the content and direction of the post-Apollo space program. NASA Administrator Tom Paine's hope of getting, in the months following the success of the *Apollo 11* lunar landing and the submission of the Space Task Group (STG) report, White House support for a fast-paced space effort in the 1970s had been decisively denied. The Nixon White House in shaping a post-Apollo space effort had decided not to build on the national investment in the capabilities that had made Apollo possible.

In February 1969 Richard Nixon had asked for a "definitive recommendation on the direction which the U.S. space program should take in the post-Apollo period." When seven months later he received that recommendation in the form of the STG report, he and especially his policy and budget advisors found it not at all to their liking. NASA, with the active assistance of Vice President Spiro Agnew, had in essence seized control of the STG; none of the other members had fought hard for a different recommendation than one centered on space station and space shuttle development during the 1970s, leading to missions to Mars in the 1980s. Secretary of the Air Force Robert Seamans in both August and September had presented alternatives to that approach, but had not been persistent in his advocacy. Science adviser Lee DuBridge, rather than act as an advocate for the views of his external advisory committee, which favored the space shuttle and was skeptical about the value of a space station, chose to be a mediator with respect to his fellow STG members, seeking an outcome that all could accept. Budget Director Robert Mayo decided to deal with space issues in the context of FY1971 budget decisions rather than argue within the STG for a program he thought the president could support. The STG participants from the Atomic Energy Commission and the Department of State had narrower interests that the totality of the post-Apollo program, and thus deferred to NASA's recommendations.

Even in the aftermath of the triumph of the *Apollo 11* lunar landing, the question of the content and pace of the post-Apollo space program had relatively low priority in the Nixon White House, as the president grappled with a recalcitrant economy and a looming budget deficit, not to mention various overseas involvements of higher interest to him. This lack of top-level interest in the future of the space program allowed a junior member of the White House staff, Clay Thomas "Tom" Whitehead, to exercise substantial influence on how the president and his senior advisers responded to the STG report. Although there was significant confusion and competition in roles between the White House policy staff, represented by Whitehead and his boss, Assistant to the President Peter Flanigan, and the Bureau of the Budget staff members dealing with space issues and their director, Robert Mayo, the two groups were united in their skepticism regarding the value of the kind of post-Apollo space program Paine was so insistently advocating. Their views carried the day with President Nixon, who by most indications was personally in favor of a more ambitious NASA program than his advisers favored. Nixon, apparently reluctantly, came to the conclusion that there was neither the public and political support nor the budget wherewithal to support such space ambitions. As Flanigan commented at the time, there was in the White House in 1969 and early 1970 "a feeling that the country had had enough excitement for now"; the result was "a series of negative decisions—no, we won't do this."[1]

The March 1970 presidential statement on space was deliberately noncommittal, seeming to echo the STG report by identifying the space station and space shuttle as desirable future developments, but also indicating that they and other NASA proposals would have to compete with other government programs for funding. To optimists like Paine, the statement seemed still to leave the door slightly open for future approval of some version of the STG program, but that was not a realistic reading of White House intent.

All of this was clear to Whitehead, who observed that "no compelling reason to push space was ever presented to the White House by

NASA or anyone else." Reflecting in 1971 on his space policy experiences, Whitehead suggested that

> this Administration has never really faced up to where we are going in Space. NASA, with some help from the Vice President, made a try in 1969 to get the President committed to an "ever-onward-and-upward" post-Apollo program with continued budget growth into the $6–10 billion range. We were successful in holding that off at least temporarily, but we have not developed any theme or consistency in policy. As a result, NASA is both drifting and lobbying for bigger things—without being forced to focus realistically on what it ought to be doing...We have cut the NASA budget, but they manage...to get a "compromise" of a few hundred million on their shuttle and space station plans. Is the President really going to ignore a billion or so of sunk costs when he gets hit for the really big money in a year or two?...There needs to be a sense of direction, both publicly and within NASA. The President's statement on the seventies in space laid the groundwork, but no one is following up.

Whitehead suggested that "we really ought to decide if we mean to muddle through on space policy for the rest of the President's term in office" and pointed out the need to answer a crucial question: "What do we expect of a space program?"[2] How that question was answered (or not) will be the central focus of the second act of this drama.

Act 2
What Next?

Chapter 7

A New Cast of Characters

As the curtain rose on the second act of the drama of post-Apollo decision making, there were a number of changes in its cast of characters, both at the White House and at NASA. The White House framework for making space policy decisions was changed by creating two new structures—the Domestic Council and the Office of Management and Budget—to oversee the development of policy and budget options for presidential decision. This meant that the heads of those new organizations would inescapably be involved in space-related deliberations. Science adviser Lee DuBridge left; he was replaced by a young engineer from the private sector, Edward E. David, Jr. Tom Whitehead, who as Peter Flanigan's assistant had been influential in shaping President Nixon's early space decisions, moved to a new position within the Executive Office of the President, but still stayed occasionally involved in NASA-related issues. There was a proposal to eliminate the National Aeronautics and Space Council and its staff; while this proposal was not acted on, the council staff were not able during 1970–1971 to become significant actors in the policymaking process, although the council's executive secretary, Bill Anders, became personally involved.

At NASA, Dale Myers, a senior executive from North American Rockwell, where he had been working on the Apollo spacecraft and then space shuttle studies, succeeded George Mueller as associate administrator for manned space flight on January 9, 1970. In that position Myers was in charge not only of the ongoing Apollo and Skylab efforts but also of studies of the space station and space shuttle. Wernher von Braun moved to the agency's Washington headquarters from his position as director of the Marshall Space Flight Center in Huntsville, Alabama. In Washington, he would lead the agency's planning effort; Tom Paine's hope was that he also could be a "super salesman" for NASA's ambitious post-Apollo aspirations. Then, after making one last attempt to gain support for such an undertaking, NASA Administrator Paine in August 1970 abruptly resigned to return to private industry. NASA was left with an acting administrator, George Low, as it fought in fall 1970 for approval of its proposals for future programs, particularly the space shuttle. In that struggle, NASA found itself dealing with a number of individuals new to the post-Apollo decision making process and skeptical of the value to the president and the country of a major commitment to developing a new capability for human space flight.

New White House Structures for Space Decisions

At the start of his administration, President Nixon had established an Advisory Council on Executive Management, headed by industrialist Roy Ash; it soon became known as the Ash Council. That council soon came up with recommendations for reorganizing the Executive Office of the President to better serve Nixon's interests. It would take until mid-1970 to turn the Ash Council's recommendations into reality; the dysfunction of the FY1971 budget process was an important influence on confirming to Nixon that a major change in White House organization was needed. Richard Nixon's goal was to centralize decision making in a few trusted individuals, with himself presiding as the final arbiter of his administration's actions without getting directly involved with his cabinet members or other top agency officials.

Domestic Council Created

The Ash Council recommendations for reorganizing the White House were unveiled on March 4, 1970, in a White House briefing to cabinet and sub-cabinet officials; the immediate reaction was concern, voiced most vocally by Secretary of Housing and Urban Development George Romney and Vice President Agnew, that such a structure would serve as a barrier to cabinet members being able to meet directly with the president. This in fact was precisely what Nixon had in mind. On March 12, the president sent a message to Congress announcing his intent to establish "a Domestic Council to coordinate policy formulation in the domestic area." This White House body would be provided with its own staff, and to a considerable degree would be a domestic counterpart to the National Security Council.[1]

John Ehrlichman was named the executive director of the Domestic Council. Ehrlichman during 1969 had steadily risen in influence among President Nixon's advisers. He had been named Nixon's top assistant for domestic affairs in November 1969; the creation of the Domestic Council, with Ehrlichman as its director, completed his ascendancy to Nixon's innermost circle of advisors. Creating the Domestic Council gave Ehrlichman a formal role in developing space policy, since NASA was considered a domestic agency. Even so, Assistant to the President Peter Flanigan, who during 1969 had had primary responsibility within the White House for overseeing NASA, continued with that role, operating outside the Domestic Council framework and retaining direct access to the president. This situation created some uncertainty with respect to space policy oversight, but Flanigan and his staff and Ehrlichman and his staff worked closely together on space issues in the ensuing months. In addition, Ehrlichman and the Domestic Council staff used the Office of Science and Technology (OST) for advice on technical issues, including space; later in the year Ehrlichman would ask new science adviser Ed David, "since policy, as opposed to programs, is so difficult to define," to list for him "those issues which could be considered domestic policy which are currently under study by OST. I have in mind matters such as our manned space program."[2]

A New Office of Management and Budget

Following up on another of the Ash Council's recommendations, the president also proposed to create within the Executive Office of the President an Office of Management and Budget (OMB) that "would be the President's principal arm for exercise of his managerial functions...The Domestic Council will be primarily concerned with what we do; the Office of Management and Budget will be primarily concerned with how we do it, and how well we do it." Although functions of the Bureau of the Budget (BOB) remained the core element of the new OMB, responsibilities such as overall management of the executive agencies and evaluating their performance were added to the organization's charter. In the BOB, only the director and deputy director were chosen by the president. In the new OMB, there would be in addition several presidentially selected associate and assistant directors; by placing political appointees in these positions, the intent was to more effectively link budget choices to Nixon's policy and political priorities.[3]

Chosen to be the first OMB director was George Shultz, at that point Nixon's secretary of labor. Shultz held a doctorate in economics and had come to the Nixon administration from the University of Chicago, where he had been dean of the business school. Shultz was a steady personality and was one of the few cabinet members who had established a good relationship with President Nixon during the administration's first year; in his new position, he soon became part of the president's inner circle of advisers. To clear the way for appointing Shultz, BOB Director Robert Mayo in June 1970 was named counselor to the president, a position with no substantive responsibility. Recognizing that he had been shunted aside, Mayo resigned in July to become the president of the Federal Reserve Bank of Chicago.

Selected as OMB deputy director with primary responsibility for budget issues was Caspar "Cap" Weinberger, who was chairman of the Federal Trade Commission, a regulatory agency. Weinberger had served as California governor Ronald Reagan's budget director before coming to Washington, and his budget-cutting fervor there had earned him the sobriquet "Cap the Knife." The OMB assistant director for energy, natural resources, and science, one of the new political appointees, was Donald Rice. He came to OMB from the Department of Defense, where he had been responsible for cost analysis, manpower and logistics requirements, and budget planning. Shultz, Weinberger, and Rice would from the time they took office in mid-1970 become key actors in the space policy process.

President Nixon with his new budget team: (l-r) George Shultz, President Nixon, Donald Rice, and Caspar Weinberger. (National Archives photo WHPO 8904–11)

Whitehead Switches Jobs

Although Tom Whitehead had been deeply involved as Peter Flanigan's assistant in developing the Nixon administration position on post-Apollo space efforts and had been the originator of the president's March 1970 space statement, NASA issues had in fact not been his primary concern in the first year of the Nixon administration. Rather, his major focus had been revising the policy and regulatory regime for telecommunications; it was Whitehead who was the moving force behind the Nixon "open skies" policy that permitted the domestic use of communications satellites. By early 1970, the White House decided that there were enough telecommunications-related issues on the policy agenda to merit a separate organization to deal with them; Richard Nixon on February 9, 1970, sent a message to Congress announcing his intention to establish an Office of Telecommunications Policy within the Executive Office of the President.[4] On September 22, Whitehead was named director of that office. Moving to head the new office meant that Whitehead would no longer serve as Flanigan's staff person for NASA issues; that responsibility was divided between Flanigan staffers Will Kriegsman and Jonathan Rose. Over the subsequent months, neither exercised the amount of influence on NASA issues that had characterized Whitehead's involvement. In addition, even as he directed the new office, Whitehead at critical moments would engage himself in decisions related to NASA's future.

Nixon's second science adviser, Edward E. David, Jr. (National Archives photo WHPO 7542–19)

A New Science Adviser

Science adviser Lee DuBridge decided in mid-1970 that it was time to leave Washington. DuBridge had not been able to exercise the influence he had anticipated in taking the science adviser's job, and was frustrated both by his lack of direct access to President Nixon and by cuts in science funding. A search for DuBridge's successor was initiated in early summer. It was soon successful. President Nixon's new science adviser would be Edward E. David, Jr., a 45-year-old engineer who had spent the prior 20 years of his career at Bell Laboratories, working in areas as diverse as computer science, undersea warfare technology, and developing an artificial larynx. David was the first presidential science adviser since the position was created in 1957 to come from an industrial rather than a university background. He was reported as being "very skeptical of the value of the man-in-space program," feeling that "we should push the space program but in a very studied fashion." David was sworn in as science adviser and director of the Office of Science and Technology on September 14, 1970. Russell Drew stayed on as David's top staff person on space issues.[5]

The Space Council Seeks a Role

Another of the early recommendations of the Ash Council was to abolish the National Aeronautics and Space Council (NASC), on the grounds that its policy coordination function could be performed by the combination of OST and OMB.[6] As discussed in chapter 2, the Space Council, composed of the head of NASA, the secretaries of state and defense, the chairman of the Atomic Energy Commission, the secretary of transportation (added by Congress in 1970), and chaired by the vice president, had seldom met at the principals level during the presidency of Lyndon Johnson, and its staff had had little influence on Johnson administration space policy decisions. Vice President Agnew in early 1969 had taken initial steps to revitalize the council, selecting *Apollo 8* astronaut Bill Anders as the council's executive secretary and trying to build up a high-quality professional staff under Anders' direction.

However, the Space Council staff did not play a significant role in the decisions with respect to the FY1971 budget or the content of the March 1970 presidential space statement. A key reason for the lack of influence on the part of Anders and his staff was that they were working for Vice President Spiro Agnew. Richard Nixon and his immediate advisors were disinclined to give Agnew any meaningful policy role, preferring to use him for political attacks on administration opponents and as a link to state and local officials. Agnew soon lost interest in space issues. Without the "top cover" of an influential vice president, Anders was largely left on his own to find ways to involve himself and his staff in ongoing policy debates. He had some success in this regard in areas such as space science and applications and aeronautics, and he got personally involved with Cap Weinberger with respect to the NASA program, but neither Vice President Agnew nor the Space Council as a body from 1970 on had any involvement in discussions related to the future of human space flight.[7]

As preparations for developing the FY1972 Nixon budget began, White House staff secretary Ken Cole on August 24 wrote the new director of OMB, George Shultz, reminding him of the Ash Council proposal to eliminate the Space Council and suggesting that "it seems appropriate to again consider" abolishing the council and that "perhaps this is a project that the Office of Management and Budget will want to undertake." The response to this suggestion took some time to develop. In September, OMB Assistant Director Dwight Ink commented that "the Space Council has not really played a significant policy role since its inception." He noted that Anders had "assembled a vigorous staff who want to exert more leadership, but the Space Council does not provide a viable base for their efforts." In October, OMB Assistant Director Don Rice indicated his "general feeling" that "organizations [such as the Space Council] spend money and make paperwork—both of which are bad until proven otherwise." OMB Associate Director Arnold Weber on October 29 suggested that "the Council should be abolished effective June 30, 1971." He added "the change in emphasis on space programs as we attempt to fit those programs into overall national priorities makes it unnecessary to retain" the council. The OMB recommendation recognized "some political and public relations problems," such as the appearance of "an insensitivity on the part of the Administration to the problems of the aerospace industry" and of "an attempt to reduce the stature of the Vice President."⁸

As it turned out, the White House in December 1970 decided to keep the Space Council. Vice President Agnew called Ehrlichman, inquiring about the fate of the council. Ehrlichman told him that "the President's State of the Union [speech] undoubtedly would involve changes in organizational structure which would contemplate elimination of the Space Council as a separate and independent entity." Agnew asked for a meeting to discuss the situation. Agnew persuaded Ehrlichman that the council's staff could be an asset in selling the administration's space and aeronautics programs to Congress and an effective liaison with the aerospace industry. These assignments would not involve the council staff in policy formulation, but rather use the staff as a "selling device." Ehrlichman agreed that it would be "bad politics to dismantle [the Council] now," since it could send a signal that such an action marked "the end of the space program." That was not a message that the Nixon White House wanted to send; there was already concern about the impact of aerospace unemployment on the 1972 presidential election. After lunching with Ehrlichman a few days later and learning that the council was not likely to be dissolved, Bill Anders told him "I believe the Council and its staff can fit into the reorganized White House team quite nicely and can provide valuable support to both domestic and national security interests across a broad front."⁹

Tom Paine Urges NASA to be "Swashbuckling"

Once *Apollo 11* had been successful in achieving the goal of a lunar landing before the end of the 1960s, Wernher von Braun had considered his work as director of the Marshall Space Flight Center completed, and during fall 1969 expressed to George Low "a strong interest" in moving to NASA Headquarters in Washington. Von Braun was burned out from his intensive efforts in getting the Saturn V ready for Apollo missions, and he and his wife, both raised as Prussian aristocrats, were ready to leave the rather provincial Huntsville, Alabama for life in Washington. Low and NASA Administrator Paine decided not to offer von Braun a headquarters line management position, but rather to invite him to become NASA's chief planner, supervising a "strong, but small staff," with the goal of "putting some imagination back into the future plans of the agency." In this role, von Braun would be both the "chief architect" of and "salesman" for the future NASA program. Von Braun indicated that he was "most interested in undertaking this assignment." He assumed his new position on March 1, 1970.¹⁰

An early von Braun project was to organize a long-range planning conference called by Paine. The purpose of the three-day conference was "to provide a long-term context against which current decisions can be tested" by expanding on the Space Task Group (STG) recommendations, which had focused on the 1970s and 1980s, to the year 2000. Paine invited visionary futurist Arthur C. Clarke to provide the keynote address for the get-together.

Paine's hope was that the combination of extending the time frame for consideration of space options and exposing his staff to Clarke's often far out thinking would result in a NASA long-range plan that could capture public and political imagination.[11]

The meeting took place on June 11–14. Paine's concluding remarks to the conference capture his exuberant personality, his fascination with things naval, and his lack of understanding, or perhaps acceptance, of the policy context in which NASA was operating in mid-1970. He urged his associates to adopt "a fighting ship analogy for the kind of society, the kind of rationale, actions, courage, and determination that we in NASA should have in the coming decades." Paine added "we need the discipline and determination and capability of a naval fighting ship," but that NASA should adopt a "swashbuckling, buccaneering, privateering kind of approach." He suggested that NASA should emulate "the concept of Admiral Nelson and his band of brothers, which certainly was one of the great management teams of all times." Paine added "we have got to enjoy the experience of living dangerously because that is really the only way to handle the kind of campaigns we are going to be waging."[12] This was certainly not the image of NASA that the White House had in mind as it tried to constrain the space agency's ambitions. Paine's exhortation to enjoy "living dangerously" was very likely to lead NASA, to continue the naval analogy, to crash on rocky shores.

In addition to his bullish long-range vision, Paine apparently had in his back pocket a short-term proposal for a major new initiative. Even as the STG was winding up its work the preceding September, NASA's Milt Rosen had suggested to Paine that he should seek "a commitment to have a permanent manned space-station in earth orbit in 1976" as a means of marking the two hundredth anniversary of the Declaration of Independence. This proposal was not mentioned during the FY1971 budget discussions, as NASA fought for the program laid out in the STG report, but it was also not forgotten. In mid-June 1970, as NASA planned its FY1972 budget request, Paine was arguing within NASA that "it is extremely important that in 1976 a major mission of new significance be considered." The leading possibility was a "first" space station that would be an advance beyond the Skylab orbital workshop, would have potential for up to ten years in orbit, and would make possible "participation by foreign astronauts or scientists." This "'76 spectacular" would be "a source of national pride."[13]

Apollo astronaut Jim Lovell called Peter Flanigan in July 1970, asking "if the Administration was looking for a space spectacular in 1976." Flanigan told Lovell that he had once suggested a change in the NASA schedule "in order to provide a meaningful launch just prior to the 1972 election," but that President Nixon had said that "he was not interested in this kind of grandstanding." Flanigan told Lovell "based on this...the Administration was not trying to design a space spectacular for 1976." This word may have gotten back to NASA planners; at any rate, the idea of a NASA mission tied to the country's bicentennial was not pursued.[14]

Paine in early July wrote the president, requesting an appointment to discuss the results of the long-range planning conference. Paine stressed that the purpose of the meeting "was not to discuss budgetary or detailed programming actions, or to review decisions," but rather "to give you a heretofore unavailable Presidential level long range view of man's future potential in space."[15] As the White House considered whether to schedule such a meeting, the first anniversary of the *Apollo 11* lunar landing on July 20 passed without any major celebration. One NASA idea had been a live television conference involving President Nixon and other heads of state, with Armstrong, Aldrin, and Collins standing by. The White House did issue a presidential statement, saying "this triumph of unique achievement, described by our first man on the moon, Neil Armstrong, as 'one small step for a man, one giant leap for mankind,' brought with it a moment of greatness in which we all shared, a priceless moment when the people of this earth became truly one in the joy and wonder of a dream realized."[16] But there was no White House desire to stage an event intended to recapture the excitement surrounding the first lunar landing or to encourage the agency to push for the kind of future Paine had in mind.

Paine Leaves NASA

On Saturday, July 25, Tom Paine called the Western White House in San Clemente requesting a ten-minute meeting on the following Monday or Tuesday to discuss a "personal decision." That decision, it turned out, was to leave his position at NASA to accept an unexpected and apparently unsolicited offer from his former employer, General Electric, to become its vice president in charge of the company's power generation group. This was a well-compensated position and Paine had the education of four children to pay for, but it is probable that he also was very frustrated by his inability to get the Nixon administration to accept his vision of the future in space. There is no evidence that the White House had encouraged Paine to resign; in fact, Peter Flanigan would later ask Paine to stay on until his successor was ready to take over.[17]

When George Low learned of Paine's resignation, he was surprised. In a July 25 telephone conversation, Paine had told Low that he would have "some important information" he would discuss once Low arrived in Washington; Low was in the process of moving his family from Houston. Low "momentarily thought that this information might concern Tom's resignation," but he "quickly discarded this idea" because Paine had "told me after Apollo 13 that he would not leave the agency until after we had flown a successful Apollo mission."[18]

Paine met with the president on the morning of Tuesday, July 28, to submit his letter of resignation, effective on September 15. Even after resigning, Paine continued his effort to convince Richard Nixon of the value of an ambitious U.S. space program. On August 10, Paine once again requested a 90-minute appointment with the president to present "NASA's projection of man's future in space to the year 2000." Although Ehrlichman and Flanigan recommended that the president schedule such a meeting, Nixon decided to "wait for [the] new man," that is, Paine's replacement. When the search for a new NASA administrator did not produce quick results, the meeting never occurred.[19]

In attempting to set NASA on an ambitious post-Apollo course, Tom Paine had reversed by almost 180 degrees the approach followed by his predecessor, James Webb. According to one of his closest associates, Paine from the start of his time as NASA administrator had "decided to be a promoter...a fighter for what he thought ought to be done. He always may have known that he wasn't going to get it all, but he would never admit it in advance." Where Webb had believed that NASA should create a broad basis of capability and allow the country's leaders to select specific missions to use that capability, Paine felt that NASA should take an "uninhibited look at what the program should consist of" and then ask "the public and the nation the biggest question that we could ask, namely, whether the United States was sufficiently wealthy and sufficiently adventurous to continue human exploration of the solar system." As he prepared to leave NASA, Paine continued to believe that NASA had asked "the right question, made the right offer," but that the country, including Richard Nixon and his associates, "may have made the wrong response."[20]

Paine's 23 months as the head of NASA left a mixed legacy. He brought to the fore those within NASA who had the most expansive view of the agency's objectives; by doing so, he tried to shake the agency out of what had been its rather cautious approach to the future. He adopted and expanded on George Mueller's ambitious integrated plan, giving priority to human space flight rather than robotic science and application missions and in the process perpetuating the split between NASA's human and robotic programs and antagonizing large elements of the external scientific community. Paine was willing to give up the repeated use of existing capabilities, particularly the Apollo/Saturn system, in order to get started on the next generation of human space flight projects. He took the lead in advocating international participation in NASA's post-Apollo human space flight efforts; that participation has been a hallmark of such efforts since.

Given the desire of those advising the president to avoid committing to major post-Apollo space projects, Paine's advocacy may have been a necessary counterbalance; he thought that "the responsibilities of leadership...required him to get approval for as large a space program as the traffic would bear." According to NASA's Homer Newell, there was "a difference of opinion as to whether Paine's attempts to force the space budget far above the levels the administration wanted to see kept it from falling lower than it did, or were counterproductive." One assessment noted that Paine's departure was "greeted with relief in the Bureau of the Budget and the White House staff"; another suggested that his resignation "came as a welcome relief to both the executive and legislative branches." A Bureau of the Budget veteran characterized Paine as a "glory hound" who was "unrealistic and unwilling to compromise." But to Flanigan, Paine's aggressiveness was not "counter-productive." Paine was a "good soldier" who accepted decisions after getting a full hearing. Ehrlichman compared Paine's bold proposals to a spring that "had to be stretched in order for it to come back to where it belonged."[21]

Who Would Replace Paine?

As he accepted Paine's resignation on July 28, President Nixon asked him to suggest potential successors. Paine replied quickly, telling Nixon that "it would be best to seek a replacement from outside" of NASA; this ruled out George Low and Wernher von Braun as candidates. Paine provided a list "of seven principal candidates of national stature." They were: James Fisk, president, Bell Telephone Laboratories; Thomas Jones, chairman, Northrop Corporation; Ruben Mettler, president, TRW Systems; Howard Johnson, president, MIT; Charles Townes, University of California, Berkeley; Frank Borman, who was in the process of leaving NASA; and George H. W. Bush, then a member of the House of Representatives and a candidate for the Senate from Texas (and a future president). Paine's personal recommendation was to select Borman, who was "the right age and temperament," would "add technical experience and charisma to your administration," could "deal effectively with the Congress," would "be received with enthusiasm by NASA and the press," and "can do an outstanding job maintaining the momentum in securing increased cooperation in space."[22]

Flanigan added several other names to Paine's list. One was Roger Lewis, chief executive officer of General Dynamics. He asked several people, including science adviser DuBridge, General Bernard Schriever, and Donald Kendall of Pepsi Cola, a Nixon confidant, to evaluate the various candidates. Flanigan tried to persuade Paine to remain in his job until a successor could be confirmed, but Paine said that this was not possible, and that in his judgment George Low was "entirely competent to manage the Agency for two months." Flanigan reported to the president that, after first being interested in the NASA position, Borman had "indicated a change of heart, saying that he had no great interest in the job." Even so, Flanigan was sure that "Borman would take the job if he knew you [Nixon] wanted him to have it." Flanigan added that "much as I would like to see the position held for George Bush should he not win in Texas, I have serious reservations about leaving it unfilled for two months," since this might be interpreted as indicating that "NASA and the Space Program were not important to the Administration. Given the current condition of the space industry, this would be an unfortunate inference." Donald Kendall and Nixon assistant Leonard Garment knew Roger Lewis and indicated that "he appears to be an exceedingly able individual and would make an excellent spokesman for NASA and the Administration." Based on this assessment, Flanigan recommended offering the NASA job to Lewis.[23]

It is not clear from the available record whether that recommendation was accepted and Roger Lewis rejected the offer, or whether action was deferred.

At any rate, Lewis was not nominated, and a month later, Flanigan was still seeking ideas for people to become NASA administrator.[24] Paine left NASA on September 15, 1970; the next day, George Low became NASA's acting administrator. Rather than being only a short-term replacement, Low would serve in that role for the next eight months. It fell to him to take the next steps in defining the program that NASA would pursue in the 1970s, particularly in terms of the negotiations with respect to NASA's FY1972 budget. In taking on that responsibility, Low would be dealing with a mix of new and continuing members of the Nixon White House. His style was very different than that of Tom Paine, but he had little more success than Paine in getting the kind of commitment to a major future program that NASA so badly wanted.

Chapter 8

The Space Shuttle Takes Center Stage

Based on the decisions made during the previous months, the human space flight program that NASA presented to the White House in September 1970 looked very different from the one put forward a year earlier. NASA hoped that this revised program, focused on beginning to develop the space shuttle, would be seen as sufficiently responsive to White House budgetary and program priorities to gain Richard Nixon's approval.

By shutting down the Saturn V and Apollo spacecraft production lines and by returning the space station to preliminary study status, NASA was in effect giving the Nixon administration only one alternative if there was to be a continuing U.S. human space flight program after the mid-1970s—to approve development of the NASA-designed space shuttle. This was a situation unacceptable to the new space actors in the Office of Management and Budget (OMB) and the Office of Science and Technology (OST); they would push NASA over the remainder of 1970 and particularly during 1971 to come up with alternative human space flight proposals or, at a minimum, alternatives to NASA's preferred shuttle design. These two organizations operated under the premise that President Nixon did not want to terminate U.S. human space flights, and thus pushed to find a way of continuing such flights that both made technical sense and also could be carried out in the context of a modest NASA budget, while also maintaining a balance between the human space flight effort and robotic science and application activities. Tensions between OMB and OST on one hand and NASA on the other would be the axis of space policy debates in coming months.

With White House failure to find a successor to Tom Paine, there was a de facto realization that George Low would serve as NASA's acting administrator as the NASA budget was being decided during the fall of 1970. Compared to Paine's call for NASA to be a "swashbuckling" organization, Low's thoughts as he became the agency's top official were much more somber.

> In the 1960's, the country was looking outward, and the national priorities included the Apollo goal, because this would establish clearly in our minds and in the minds of the world technological leadership by the United States...The

situation in the beginning of the 1970's is very different. We are now an introspective nation. We will do only those things that help ourselves and help ourselves at an early date.[1]

This rather dour perspective would color Low's actions as he sought a persuasive rationale to convince the White House to approve NASA's reduced post-Apollo ambitions.

Low's first responsibility as acting administrator was finalizing NASA's budget request for Fiscal Year (FY) 1972, due at OMB on September 30. The prospects for getting OMB approval to begin shuttle development in FY 1972, which would begin on July 1, 1971, were very much on Low's mind as the NASA budget request was prepared: "If we do not get a firm go-ahead for the shuttle this year, we will not have a viable space program in the middle 1970's... The question, then, is 'how do we approach OMB and the White House to get them to give us $500-$600 million more than they would like to approve?'"[2]

It would turn out that there was no positive answer to this question. Even though the process by which decisions were made on NASA's FY1972 budget was much more orderly than the chaotic approach of a year earlier, NASA did not get the definitive commitment to the shuttle it was seeking, In addition, there was some last-minute drama. There was serious thought given to canceling Skylab, NASA's experimental space station. A new consideration—the possibility that aerospace unemployment in areas that could affect President Nixon's reelection prospects in 1972—became part of the discussion about NASA's future, and was a major factor in the ultimate decision to proceed with Skylab. In addition, Nixon, shaken by the *Apollo 13* accident, personally tried to cancel the final lunar mission, *Apollo 17*, as excessively risky, but was persuaded not to follow through on that action. By the time final budget decisions were made in early January 1971, NASA's post-Apollo future remained uncertain, although there were some positive signs that a space shuttle would eventually gain White House endorsement.

New Actors and a New Issue

One impact of creating the Domestic Council as the structure for developing policy options for presidential choice was that NASA's FY1972 budget proposal was evaluated, as had been suggested in the March 1970 presidential space statement, in comparison to the budget proposals of other domestic agencies. The Domestic Council staff person assigned both to look for potential cuts in the overall budget and to track NASA issues was Ehrlichman's deputy Ed Harper, who held a doctorate in political science and who had worked in the Bureau of the Budget before joining the council staff. In mid-August, even before formal agency budget requests were submitted to OMB, Harper had provided John Ehrlichman with a list of potential budget cuts across the executive branch. Listed as among the "easier cuts to announce" were an "across the board" reduction of $40 million in the NASA budget;

Harper also identified the possibility of canceling Skylab, which would save $300 million. Another Ehrlichman assistant, John Whitaker, had provided a "political evaluation of cutback or elimination possibilities" related to the budget planning targets that OMB had provided to various agencies; with respect to NASA, Whitaker had suggested that "in principle for policy reasons, continue moon manned space flight on a stretched out basis, but cut out space shuttle and station. Real money ($2 billion) could be saved—[but] look at unemployment effect."[3]

Tom Paine, even as he was preparing to leave NASA, and Low met with George Shultz and Cap Weinberger, the new leaders of OMB, in early September. Low reported that "the meeting was fairly short but...fruitful. Shultz looks like the kind of person we could easily work with, if only he were going to be available to us. I'm not sure whether the same would be true of Weinberger." NASA was told that "the procedure that will be used by OMB this year is that they will try to delegate agency level discussions to one of the three political appointees at the Associate Director [actually Assistant Director] level." For NASA, that would be "a man by the name of Don Rice, whom we have not yet met." That would change quickly; Rice would establish himself as a formidable presence in NASA–OMB dealings over the 1970–1972 period. Also in early September, new science adviser Ed David came to NASA for a briefing on NASA programs. David "was attentive for about two hours while we ran through our entire program and commented very little," according to Low, who observed that it was "quite difficult, on the basis of this first meeting, to even form a first impression."[4]

Like most politicians, Richard Nixon throughout his first term as president worried about his prospects for reelection, and was concerned that job reductions in the aerospace sector caused by his cuts in the defense and space budgets could have negative political consequences in key electoral states, particularly California. Nixon and his long-time associates recognized that Nixon had won the presidency "by an eyelash in 1968, just as we lost by an eyelash in 1960, and thought during the first term we would likely win or lose by an eyelash in 1972."[5] Thus winning California loomed large in Nixon's reelection planning. Nixon was also interested in restoring the U.S. economy to a healthy condition, and believed that unemployment in high technology sectors ran counter to that objective. Nixon brought his long-time associate Robert Finch to the White House in June 1970 both because Finch was having problems handling the stress of his position as secretary of health, education, and welfare and because he wanted Finch's advice on strategy for the 1970 Congressional elections and the 1972 presidential campaign.

Harper from the Domestic Council staff wrote Finch on an "urgent" basis on September 23 about a "Key Election Issue: Federally Caused Unemployment." He reported that "cutbacks in Defense and NASA by 1972 will shrink by 30% in expenditures from 1968 levels, creating unemployment (850,000 workers)—especially among scientists and engineers (an additional 130,000)." He added that "the unemployment is very localized," with 43.5 percent concentrated in the Pacific region, with the Los Angeles area as the hardest hit.[6] The connection between aerospace employment and the space shuttle, already evident in 1970, was to prove an important factor in the final decision to approve the NASA-preferred shuttle at the end of 1971.

NASA Submits Its FY1972 Budget Request

In January 1970 Richard Nixon had approved a NASA FY1971 budget of $3.3 billion in outlays, the funds actually to be spent during the fiscal year. There had been attempts in both houses of Congress to make cuts in this request by eliminating funds for the space station and space shuttle, primarily on the grounds that they were the first steps toward missions to Mars, but these attempts were defeated. By mid-summer it was clear that Congress would approve a FY1971 NASA budget with only a slight reduction from the president's request. On the basis of Richard Nixon's comments at his January 22, 1970, meeting with Tom Paine that the FY71 budget level was the end of NASA budget reductions, NASA had hoped to get a budget target from the White House for FY1972 that was higher than its FY 1971 budget. But the poor economic outlook had persisted; NASA was disappointed when in August it received a budget target of $3.1 in new budget authority and $3.2 billion in FY1972 outlays, both reductions from the FY1971 figures. It was this highly constrained budget outlook and the anticipation that it was likely to continue in subsequent years that had colored the summer 1970 decisions to defer the space station and to cancel two Apollo missions.

The deadline for NASA to submit its budget request to OMB was midnight on September 30, and NASA went down almost to the last minute before deciding what to request and especially how best to justify its proposals. The budget requests from the various elements of NASA totaled over $4 billion, and it took some doing on the part of Low, his strategy adviser Willis Shapley, and his budget chief Bill Lilly to get the request down to $3.7 in new budget authority and $3.4 billion in outlays. This latter number was the one of most interest to the White House, given its short-term economic concerns with respect to limiting government expenditures; the NASA total was $200 million higher than the OMB outlays target. Low felt that "a budget at this level was the lowest level that I could submit in good conscience." On September 30, the budget submission letter was "written and rewritten, edited and re-edited, and finished typing by 8:30," reaching OMB "at 9:00 or three hours before the deadline."[7]

The budget letter spelled out the adjustments in its program that NASA had made in order to avoid "an unacceptable peaking of the NASA budget at over $5 billion in the middle 1970's," saying that the program laid out could be approved "without committing the nation to an annual budget level in excess of $4 billion." These adjustments represented a dramatic lowering of sights since the submission of the Space Task Group report a year earlier, which had forecast NASA budgets in the $8–10 billion range in the late

1970s. NASA argued that "the key element in our program for the 1970's is the space shuttle...We must start this development now to lay the foundations for the nation's future space program, and to bring about the major economies in later years." In justifying the shuttle, NASA said that "the space shuttle will be used for manned and man-tended experiments and to place unmanned scientific, weather, earth resources and other satellites in earth orbit and bring them back to earth for repair and reuse." Only in the future would the shuttle be used to "transport men, supplies, and scientific equipment to and from space stations." Deciding to characterize the space shuttle as an all-purpose launch and space operations vehicle was a major change, since it represented a claim that the shuttle could stand on its own merits, not primarily as an adjunct to the space station. NASA justified the shuttle as "cost-effective," a claim that was to become a controversial point in NASA–OMB interactions in the coming months.[8]

There was significant weakness in NASA's argument for approving shuttle development in FY1972; in essence, the shuttle concept was "not ready for prime time." NASA was focusing on a large, two-stage, fully reusable shuttle, but had not yet decided what version of such a system it wished to develop, whether it was technologically feasible, or how much it was likely to cost. Intensive contractor studies of fully reusable shuttle designs and alternate configurations were just starting. An independent study of shuttle economics requested by the Bureau of the Budget in early 1970 was also not complete. What NASA was asking OMB to approve was putting in the FY1972 budget a modest down payment of $190 million on shuttle development; more significant, that down payment was to represent a commitment that the shuttle had gained White House approval. The $190 million would allow NASA to award contracts soon after the start of FY 1972 on July 1, 1971, for detailed design and development of both an advanced technology rocket engine planned for the shuttle and the shuttle's "airframe," that is, the basic structures of the shuttle orbiter and booster. The results from the shuttle technical and economic studies were expected in the May–June 1971 time frame, and the proposition that NASA was asking OMB to approve in fall 1970 was that those results would justify an immediate start on shuttle development. This request—to approve in advance a multi-billion dollar, multi-year program to develop a not-yet-well-defined shuttle—was not a proposition OMB was likely to accept.

NASA Seeks Support

As OMB began its review of the NASA budget, Low set out on an intense effort through both face-to-face meetings and letters to communicate the NASA story, both inside the agency and to anyone outside the space agency who might offer support to NASA's plans. One of those targeted by Low was William Pickering, the long-time director of the NASA-affiliated Jet Propulsion Laboratory of the California Institute of Technology; Pickering had expressed some skepticism regarding whether NASA was indeed prepared to begin shuttle development. Low suggested to Pickering that "the *technology* for the shuttle appears to be as well or better in hand than the technology was for the Apollo lunar mission when that program got started." Low in 1961 had been in charge of human space flight at NASA headquarters and had prepared a key report saying that there were no technological barriers to a lunar landing mission.[9]

One of the meetings Low organized as he explained the NASA budget request was with science advisor Ed David and his space staff person Russ Drew. Low was quite surprised to discover that David and Drew were "very much opposed to Skylab." The two argued that the only reason for getting experience with long-duration space flight was preparing to send astronauts to Mars, and, since there was no intent in a relevant time frame of undertaking a Mars mission, there was no need for Skylab. Low found it "inconceivable" that "there would be serious consideration given to the cancellation of Skylab," given all the money that had already been spent on the program. Following this meeting, Low wrote a letter to David discussing the relative priority of Skylab and Apollo. With respect to Apollo, Low was rather guarded, reflecting his own concerns about additional Apollo missions, saying that although the final four Apollo missions would increase scientific understanding of the Earth–Moon system, the missions "would in another sense be dead-ended. No new capabilities or techniques would be explored that could be further exploited...no major new opportunities for leadership and prestige would likely accrue; and the potential of Apollo for international cooperation is limited." By contrast, with respect to Skylab "there has been no return from considerable investment to date...We simply have no data on man's ability to live and work in space for long periods of time." Low suggested that "on balance, the weight of evidence seems to favor Skylab over Apollo if a choice must be made."[10]

One of the other people to whom Low wrote in this period was national security advisor Henry Kissinger. Kissinger and his staff had not gotten deeply involved in NASA-related decisions, with the exception of monitoring the discussions in 1969 and 1970 between NASA and European space officials about possible European participation in the U.S. post-Apollo program. Low pointed out to Kissinger that, given the NASA decision to defer space station development, the space shuttle program provided the only opportunity for international participation in human space flight, something that the president wanted. He hoped that Kissinger would support a decision to begin shuttle development in FY1972, since without "forward motion on the space shuttle system...the prospects for the major advance in international cooperation that we have hoped for will dim to the vanishing point." The letter had little impact; Kissinger did not get involved in the budget process.[11]

Low also tried several times in October to set up a meeting with Peter Flanigan, but Flanigan "cancelled each time because of other commitments." In comparison to his active role in the deliberations that had led to the NASA budget decisions a year earlier, Flanigan was noticeably missing from the FY1972 discussions. The OMB was approaching its review of the NASA budget request in a much more orderly fashion than had been the case in late 1969 and trusted Nixon assistants were in charge of the budget process. In addition, the Domestic Council was monitoring space options. Flanigan may have felt no need to intervene in the budget process to make sure that the president's priorities were heeded.[12]

NASA's informal contacts with the OMB staff working under Don Rice had alerted it to the areas where OMB was considering NASA budget reductions. Trying to preempt such cuts, Low wrote Weinberger on October 28, saying that he wanted to make "especially sure" that several elements in the NASA budget request were "clearly understood and given careful consideration." Low gave particular emphasis to the reasons for going ahead with the space shuttle, saying that shuttle development "can be justified as a versatile and economical system for placing *unmanned* civil and military satellites in orbit, entirely apart from its role in conducting or supporting manned missions." This was the newly developed NASA argument as the agency recognized that the shuttle now had to be justified as a launch and orbital operations vehicle, absent a space station to service. Low added what would turn out to be a winning argument: "With the shuttle the U.S. can have a continuing program of manned space flight...without a commitment to a major new manned mission goal." Recognizing that the Nixon administration had no intention of setting out an Apollo-like goal for the post-Apollo space program, NASA was basically arguing that the country could have, almost "for free," a continued human space flight program by approving a system justified by reducing the costs of space launch and in-orbit operations, which incidentally happened to be operated by a human crew and could carry humans as passengers.[13]

A final NASA move in making the case for shuttle approval was to prepare for OMB Director George Shultz a paper "from a national—not just a NASA—standpoint of the need for and importance of a continuing program of manned space flight." Shultz was reputed to be skeptical about the value of humans in space, and the NASA paper was aimed at countering that skepticism. In his cover letter, Low emphasized "that manned flight to Mars is *not* a goal or justification of the program that NASA is recommending for the 1970's. Skylab and the space shuttle, for example, are necessary elements of the United States space program without a manned Mars mission." This statement was intended to rebut the claims of Congressional critics of the two programs such as Senators William Proxmire and Walter Mondale and Representative Joseph Karth, who had linked the station and shuttle in Congressional debates to preparing to send astronauts to Mars. Karth's attempt to cut station and shuttle funds from the NASA FY1971 budget had failed, but only on the basis of a 54–54 tie vote.

The 11-page NASA paper discussed both "the role of manned space flight as a *means* for accomplishing objectives in space" and "the importance of manned space flight to the United States as an *end* in itself." With respect to the former role, the paper stressed that the space shuttle was "*not a 'manned spacecraft'; it is a space transportation system*" that "would bring about a fundamental change in space operations and result in very substantial cost reductions." With respect to the latter role, the paper argued for "*acceptance of manned exploration of space as an important and continuing goal in its own right*," one which "the United States, as a great nation, should continue" and "take a leading role." It suggested that "manned space flight will continue to be the best and perhaps the only arena of worldwide interest where the United States can demonstrate at the same time technological strength, peaceful intentions, power without confrontation, and the openness of a free society."[14]

NASA Budget Review

As background for the OMB review of NASA's budget request, in late October Russ Drew of OST and Dan Taft, whom Don Rice had selected as head of OMB's NASA unit, collaborated in preparing a "space strategy paper." The paper noted that NASA "in the wake of its spectacular success with Apollo, has failed to generate a clear and substantial basis of public and Administration support for the decade ahead." This had led to setting NASA's budget "on an ad hoc basis, rather than as part of an overall coherent plan and in accordance with an accepted and continuing rationale to guide decisions." The paper postulated three possible strategies for NASA's future: (1) "an all unmanned science and applications program"; (2) "a mixed manned and unmanned science and applications program"; (3) "a mixed program with a strong manned Mars emphasis." Option 1, abandoning human space flight, was deemed infeasible because it was "not consistent with existing Presidential policy." The paper identified five NASA budget options, ranging from $2 billion per year to $6–8 billion annually. There was a focus on NASA's institutional base; at lower budget levels, up to six of the ten NASA field centers could be closed. The paper assessed likely public, industry, and political reaction to the various budget levels and subsequent programmatic and institutional actions.[15]

OMB Makes Its Recommendations

The strategy paper illuminated the consequences of various budget choices. It certainly influenced the OMB staff in its recommendations regarding the NASA FY1972 budget, with an OMB bias toward the lower budget options. The next step in the budget process was Don Rice's presentation to Cap Weinberger of his staff's recommendations with respect to the NASA budget. This "director's review" took place on November 3. Weinberger "tentatively decided" to accept the staff recommendation to terminate the Skylab program. Possible cancelation of *Apollo 17*, the final lunar landing mission, had been considered during the budget review, but the staff recommendation, which Weinberger accepted, was to continue with the mission.

Weinberger did approve a start on the new rocket engine intended for use in the space shuttle, but denied funding for moving forward on developing the shuttle's airframe. Approving engine development was an important first step in eventual White House approval of the space shuttle, and suggested that some version of the shuttle was likely to get approval in the months to come. These and other decisions, particularly the cancelation of Skylab, brought the NASA budget recommended by the OMB staff down to $2.8 billion in budget authority and $2.7 billion in FY1972 outlays. The latter figure was some $700 million below what NASA had requested and almost $500 million less than the budget target that had been provided to NASA a few months earlier. The NASA unit in OMB, led by Don Rice, was clearly setting itself up as a counterforce to NASA's already diminished post-Apollo aspirations. Although he tentatively approved the staff recommendations, Weinberger wanted a better sense of the context in which they were being made. He thus requested a more detailed analysis, taking into account "agency priorities, unemployment consequences and Soviet initiatives."[16]

The proceedings of the director's review were not supposed to be known outside the White House and Executive Office of the President. But the Space Council's Bill Anders attended the review meeting and on a very confidential basis called NASA's Low to communicate its results. In addition, Bill Lilly, NASA's budget chief, got feedback from some of the career budget staff. Based on this information, Low judged that while Rice and the OMB career staff were "quite negative to our programs," Weinberger had "carefully read our letters and is, in fact, trying to get a detailed understanding of the issues involved in the NASA budget." In addition to calling Low, Anders wrote a letter to Weinberger in support of the Skylab program. This may have been the beginning of Anders's relationship with Weinberger; over the following months Anders served as Weinberger's unofficial space advisor, providing an informed view independent of the information and recommendations the OMB deputy director was getting from his staff. This relationship gave Anders a way to have an impact on major space decisions. Anders found in Weinberger an individual who appreciated the value to the nation of a vigorous space effort; he took every opportunity to nurture that appreciation.[17]

Initial Presidential Decisions

Richard Nixon was scheduled to meet with John Ehrlichman and George Shultz on December 1 for an initial discussion of the NASA budget. The issues identified by OMB as requiring presidential decision were: whether to continue lunar exploration through *Apollo 17*; whether to start space shuttle development or just begin engine development and defer an airframe commitment; whether to cancel Skylab; and whether to cancel the NERVA (nuclear rocket engine) effort. Other NASA issues would be decided without direct presidential involvement.

Budget Options Assessed

As time for the presidential meeting approached, there were several new inputs into the decision process. One was the OMB paper that Weinberger had requested, putting the staff recommendations in a broader context. The paper compared the employment effects of canceling Skylab, *Apollo 17*, and NERVA. Job losses if NERVA were canceled were estimated to be 2,600; if *Apollo 17* were canceled; 6,000–7,000; if Skylab were canceled, 18,000–20,000, with 9,000 of those job losses coming in California. Science adviser David also weighed in, supporting retention of the *Apollo 17* mission. He said that *Apollo 17* "is of considerably higher priority" than either Skylab or NERVA and noted that canceling *Apollo 17* "would give rise to a considerable chorus of criticism among the scientific community. In my view, this is the wrong place to cut."[18]

Ehrlichman forwarded to President Nixon a memorandum on the employment impact of cuts in the NASA budget that had been prepared by Will Kriegsman of Flanigan's staff, who had taken over most of Whitehead's responsibilities vis-à-vis NASA. Kriegsman suggested, using the figures in the OMB staff paper, that Skylab not be canceled "because of the employment situation and because we have already invested $1B in the program." Instead, he proposed, "we should try to save some FY72 money by slipping Skylab's schedule 6 to 12 months," and that "we [should] defer the initiation of the Space Shuttle program." OMB had recommended $133 million to start shuttle engine development; Kriegsman suggested total deferral of this new start. He argued that "the problem with the shuttle is that it will cost $8-$10 B as a minimum over the next 10 years. Neither the economic nor the technical justifications are...sufficiently defined at this point for us to make such a commitment in the FY1972 budget." After reading Kriegsman's memo, Nixon, in a handwritten note on the document's final point regarding a shuttle commitment, commented "this is persuasive." That comment likely sealed the shuttle's fate for FY1972.[19]

Ed Harper also prepared several background memos to prepare Ehrlichman for his meeting on the NASA budget. Following up on Kriegsman's memo on aerospace unemployment, Harper told Ehrlichman "the employment factor in the NASA budget decisions is a significant but complicated phenomenon." He noted that, while the program that NASA had proposed would "result in a gradual increase in employment throughout 1971," the OMB recommendation "would result in a sharp decline continuing through calendar 1971 for a total cut of 20,000 aerospace employees." He also noted that while OMB and OST had given retaining the *Apollo 17* mission their highest priority and had given Skylab lower priority, NASA had ranked the lunar mission behind both retaining Skylab and starting the shuttle. His advice to Ehrlichman was "that the optimal budget decisions on the NASA options is to (1) continue Skylab, (2) slip the shuttle engine development, (3) continue with Apollo 17, and (4) cancel NERVA."[20]

Nixon's Decisions and NASA's Response

Shultz and Ehrlichman met with President Nixon on the afternoon of December 1. After hearing the OMB recommendation to cancel both Skylab and NERVA, Nixon indicated he was very reluctant to take those actions, with Skylab being a particularly "tough problem." Nixon suggested slipping the NERVA schedule by one year rather than canceling the program, and asked if there was also a way to stretch out the Skylab schedule to avoid terminating the program and thus causing immediate job losses. There is no record of the discussions regarding the space shuttle or *Apollo 17* during the meeting.[21]

Based on this presidential guidance, OMB developed a proposed NASA budget that included $3.3 billion in new budget authority (NASA had requested $3.7 billion) and $3.2 billion in FY1972 outlays (NASA had requested $3.4 billion). *Apollo 17*, Skylab, and a start on shuttle engine development remained in the budget, but NERVA was canceled and a start on developing the space shuttle airframe was not approved. Rice called Low on December 7 to communicate this result. Meeting with Low a few days later, Rice said that the major reason for retaining Skylab and thus approving the NASA budget at a higher level than the OMB staff had recommended "was the employment situation in the aerospace industry." *Apollo 17* had been approved "because of the inputs from the scientists."[22]

Low wrote President Nixon on December 14, requesting reconsideration of the NASA budget decisions. He offered two reasons for such action. One was "the grave *unemployment* situation in the aerospace industry." The other was that "the *Soviet challenge* in space science and technology threatens our hard-earned superiority." With respect to the former reason, Low argued that a "visible effect" in countering unemployment was possible by 1972 "by adding only the relatively small amounts needed to make a start on the *space shuttle airframe*." With respect to the Soviet challenge, adding funds for a start on the airframe would reduce the period during which the Soviet Union would be flying people to space while the United States was not "by a year and permit us to point clearly to the time when the US will again be first in space." Low was able to meet with OMB Deputy Director Weinberger as he hand-delivered the NASA appeal letter. The meeting "was not a very satisfactory one in that Weinberger received a half a dozen or so phone calls during the course of our discussions, and I was never really able to complete a point." Low left the meeting with the feeling that "our request for reconsideration on ... the shuttle would be denied."[23]

Richard Nixon, *Apollo 13*, and *Apollo 17*

What Low did not know as he met with Weinberger on December 14 was that Richard Nixon was having second thoughts about going ahead with the *Apollo 17* mission. The president had somehow gotten the impression that *Apollo 17* was even more risky than the three missions scheduled to precede it. Nixon did not want a repeat of the *Apollo 13* experience, particularly in mid-1972, when the *Apollo 17* launch was then scheduled, not least of all because it would come as he was campaigning for reelection. The near-tragedy of *Apollo 13* had made a strong impression on the president, and provided the background against which he decided that *Apollo 17* should be canceled.

"Houston, We've Had a Problem"

The *Apollo 13* mission was launched on the afternoon of April 11, 1970. Almost 56 hours later, with the spacecraft 200,000 miles from Earth, *Apollo 13* commander James Lovell reported to mission control in Houston that "we've had a problem here." Within a few minutes, NASA notified the White House situation room. National security adviser Henry Kissinger was informed at around 11:00 p.m. Kissinger called Nixon chief of staff Bob Haldeman, suggesting that President Nixon be awakened and informed of the situation, but Haldeman, in what Kissinger later characterized as "one of the mindless edicts by which Haldeman established his authority," refused to contact the president on the grounds that this was merely a "technical problem." At 4:00 a.m., Haldeman changed his mind and decided to inform the president; he also called Nixon press secretary Ron Ziegler, telling Ziegler to inform the press that the president was "in personal charge of the crisis." Kissinger describes Ziegler's interaction with the press as "verbal contortions to imply, without lying outright, that the President had been in command all night."[24]

The story of the herculean efforts undertaken by NASA and its industry colleagues to achieve the safe return of the *Apollo 13* crew—Jim Lovell, Jack Swigert, and Fred Haise—is well known and will not be repeated here. Once made aware of the risky situation, Richard Nixon became very emotionally involved in the crew's fate. There were at the time intense discussions within the White House on whether to send American troops into Cambodia to attack North Vietnamese sanctuaries. Even so, according to Henry Kissinger, "the rescue of the astronauts absorbed a great deal of Nixon's attention" and "took a heavy toll of Nixon's nervous energy."[25]

On the morning after the accident Ehrlichman suggested to Nixon that he might want to go to Houston to signal his personal concern about the fate of the crew; it took a call from Frank Borman to Haldeman to dissuade the president from making such a trip. Borman, who was in Houston, told the White House that Nixon's presence would be a distraction as the NASA mission managers struggled to find a way to get the crew safely back to Earth. Likely on the same call, Borman relayed to the White House the news that Vice President Agnew, who was in Iowa on a political trip, was intending to come to Houston "to take charge of the rescue efforts." The director of the Manned Spacecraft Center, Robert Gilruth, told Borman that "Agnew's interference was the last thing NASA needed or deserved," and asked "is there anything you can do to keep the Vice President away from here?"

In his call to the White House, Borman suggested that "Agnew's presence in Houston would be about as welcome as a Martian invasion." Haldeman kept an unhappy Agnew waiting for an hour at the end of an airport runway in Des Moines until he could consult with Nixon with respect to Agnew's plans. When he did reach Nixon, the president "fully agreed" that Agnew should not go to Houston. Haldeman relayed that order to Agnew, who was "mad as hell."[26]

The next day there were discussions among the president, Haldeman, and Borman on how to react to various outcomes of the *Apollo 13* crisis; the astronauts' survival was still very much in doubt. The three decided that if the crew returned safely, the president would go to Houston to congratulate the NASA flight control team, then fly to Hawaii with the astronauts' families to greet the crew as they returned to U.S. soil. If the crew did not survive, the president would go to Houston to "speak to the men of NASA and reaffirm his support of them and compliment them on their tremendous efforts to bring Apollo 13 home."[27]

Safe Return

Splashdown was set for just after 1:00 p.m. on the afternoon of April 17. Haldeman gives a vivid description of the events of the day:

> Apollo 13 day. They made it back and the P [President Nixon] was really elated! Started out in the morning with some general details, then into a lot of planning, etc., for his participation in the Apollo return. Had TV, squawk box, and [former astronauts] Collins and Anders set up in Alex's [Nixon aide Alexander Butterfield] office to keep him posted. Kind of anxious about results but basically confident that they'd make it, and all wrapped up on little specifics about the trip, which we have very well set up on contingency basis.
>
> [material deleted]
>
> For splashdown, P watched in Alex Butterfield's office with Alex, me, Anders, Collins and K [Henry Kissinger]. Was very cranked up. Ordered cigars for all on success when learned that was Chris Kraft tradition at NASA. Put through call to wives immediately, then waited to call astronauts till they were aboard *Iwo Jima* [the recovery aircraft carrier] and had called wives. Meanwhile P called all the Congressional leaders and George Meany, saying to all, "Isn't this a great day." He was really excited...Then talked to astronauts and told them of trip plans, then out to press to do likewise, then over to the EOB [Executive Office Building] at about 3:30, with no lunch. Took a nap.

Nixon biographer Richard Reeves adds an additional detail to the day's account. He suggests that as Nixon talked to the Congressional leaders after the splashdown, he was "having one drink after another," and that soon after he reached his hideaway office in the Executive Office Building, "the President was drunk, falling asleep on the couch." If that were indeed the

Apollo 11 astronaut Michael Collins (foreground) and Space Council Executive Secretary Bill Anders join President Nixon and Henry Kissinger to watch as the *Apollo 13* command module parachutes to a safe return. (National Archives photo WHPO 3359–7A)

case, Nixon recovered quickly; that evening he hosted a White House performance by country music singer Johnny Cash.[28]

On April 18, President Nixon flew to Houston. At the Manned Spacecraft Center, he presented the Medal of Freedom to the *Apollo 13* mission operations team. Then he flew to Hickham Air Force Base in Honolulu, Hawaii. There he presented the Medal of Freedom to Lovell, Haise, and Swigert. He told the crew that "this was a successful mission, a great mission on behalf of your country... You did not reach the moon, but you reached the hearts of millions of people on earth by what you did... We realize that greatness comes not simply in triumph, but in adversity."[29]

Richard Nixon's associates never passed up an opportunity to portray the president in a positive light. Even as they planned how the president would deal with the unfolding crisis, they made sure that his involvement would reflect well on Nixon as a national leader. In the days after the safe return of the *Apollo 13* crew, the White House approached *Life* magazine senior correspondent Hugh Sidey about "doing an inside story on the President's involvement in and the attitudes, etc. during the Apollo 13 crisis." It took several months for this suggestion to bear fruit, but eventually Sidey wrote a very positive account, saying that "the near tragedy of Apollo 13, a deeply emotional drama for all Americans, was even more so for the President." The Apollo astronauts, Sidey suggested, were an "obsession" for Nixon, who viewed them "as more than heroes." According to Sidey, Nixon, "in his single-minded manner... seems to be trying to assess and grasp the spirit of the astronauts."[30]

Richard Nixon's involvement with *Apollo 13* has been discussed in some detail because the episode reinforced to his associates the reality that Nixon would never accept a future U.S. space program not including human space flight as an important element. In addition, Nixon's concern for the astronauts' safety became linked to a political calculus in his mind regarding possible negative political fallout from a similar problem on a future Apollo flight. When he got the impression that *Apollo 17* was particularly risky, it is not surprising that Nixon's first instinct was to cancel the flight.

Final Budget Decisions

President Nixon made his decisions on various budget appeals in the days following Christmas. Included in Nixon's December 28 choices with respect to NASA were the decisions to slip the Skylab schedule, to restore NERVA to the budget at a low funding level, not to approve shuttle airframe development—and to cancel *Apollo 17*. Meeting with Ehrlichman and Shultz, the president first suggested shifting funds intended for *Apollo 17* to the Skylab and shuttle programs. As he discussed his options, he suggested that "politically" it was better not to launch the mission, or at least slip it, "at whatever cost," until after the November 1972 election. His final decision was to cancel the mission.[31]

These decisions were communicated by Weinberger to Low on December 31. To Low, the slip in the Skylab schedule and especially the cancelation of *Apollo 17* "were a complete surprise." Weinberger let Low know that these decisions were made "by the President himself, without any input from OMB." Meeting with his NASA colleagues to discuss how to respond, Low was told by Dale Myers that canceling *Apollo 17* so soon after two other Apollo missions had been eliminated would be "a devastating blow to morale." After phone calls to David and Rice to get more background on the budget decisions, Low decided to "do no more about this on New Year's Eve (By this time, it was 7 o'clock in the evening and we were in the midst of the biggest snowstorm in three years)."[32]

Low met with OMB Director Shultz and Rice on the afternoon of January 2, 1971, to get more information on the reasoning behind the budget decisions and to reemphasize NASA's perspectives regarding the relationship between the space program, Soviet competition, and aerospace unemployment. He also wanted to make a last effort to preserve the *Apollo 17* mission. He found that "Shultz was not all that interested in unemployment in the aerospace industry... He apparently still believes that the U.S. R&D capability can be maintained by retraining the aerospace scientists and engineers into other fields." Shultz asked Low whether there was a possibility "of using some of NASA's R&D capability to solve domestic problems." This was an idea that would rise to prominence in White House thinking during 1971.

In his apparent lack of concern about aerospace unemployment, Shultz was running counter to the president. Nixon had read a December 30 memorandum from the chairman of his Council of Economic Advisers, Paul

McCracken, which noted "unemployment among scientists and engineers in California increased from 0.9 percent to 2.4 percent in the past year as national priority changes reduced defense and aerospace spending." Nixon wrote a message to Ehrlichman and Weinberger on the memo: "As a matter of *top priority*, we must move with *maximum* publicity on all these fronts & any others which occur—get a *real plan* & act on it."[33]

Regarding the decision to cancel *Apollo 17*, Shultz reiterated to Low that "it was not a budgetary one, but was based on the fact that the President had been informed that Apollo 17, as the last Apollo mission, was of considerably higher risk than the previous one and that he [Nixon] did not want to undertake such a mission just before the elections." In response to Shultz's questions, Low said that that while the risk of flying *Apollo 17* was "substantial," it "may not be any higher than that for all other missions." Low told Shultz that it was possible, with "a good technical justification," to delay *Apollo 17* until December 1972, after the presidential election. Low recommended that a decision on whether or not to cancel *Apollo 17* be deferred for a year, but Shultz preferred the option of deciding immediately to slip the mission to December 1972, since such a decision "would save some money in Fiscal Year 1972," even though it would increase the overall cost of the mission. Keeping government spending down during the election year was an important objective to the Nixon White House. Shultz told Low that the president was aware of their meeting and that he would get in touch with Nixon "right away and let me know before the end of the day" whether he would reverse his cancelation decision if the *Apollo 17* flight were slipped until after the 1972 election. "About an hour later," Don Rice, rather than Shultz, called Low to say that "the president had accepted the delay in Apollo 17."[34]

As he met with Low, Shultz may have already known that the president had had second thoughts about canceling *Apollo 17*. The weekly magazine *Newsweek* in mid-December had noted the possibility of such a cancelation. This publicity had produced messages to Ed David from the scientific community opposing such a step. Writing the president on December 31, David argued that canceling the mission would "give the Administration an unfortunate image among opinion-makers in society" and was "likely to result in strong protests from responsible and influential people." David did not base his recommendation against canceling the mission on its scientific merit, an argument he knew carried little weight with Nixon. Rather he suggested that such a step would "make it much more difficult to rally the responsible elements to support the Administration's other forward-looking programs." Apparently independent of NASA's internal thinking, David suggested that "to counter many of the concerns that have been raised about the flight of the last Apollo mission in the few months before November 1972 [the election period]," the *Apollo 16* mission could be launched in February 1972 and the *Apollo 17* launch could be scheduled "in mid-November or December. This would have the double advantage of maintaining critical employment levels through this period and better phasing of launch and support personnel." Nixon in the margins of David's memo wrote "GS [George Shultz]—good.

Do." Nixon communicated this decision in a December 31 meeting with Ehrlichman, directing him to tell Shultz to take another look at the *Apollo 17* issue. Given this directive, it is not clear that Shultz actually called Nixon after his meeting with Low or had learned before the meeting that Nixon had decided to reverse the decision.[35]

There was one more contentious NASA–OMB interaction before the NASA FY1972 budget was made public on January 29, 1971. As the budget message was being finalized, there was a dispute between OMB and NASA about what it should say with respect to the space shuttle. After the initial budget decisions in early December, NASA suggested including language in the budget message indicating an administration commitment in principle, not just to the engine, but to the shuttle program overall. Low had suggested that the space shuttle "posture" should be that "the FY1972 budget provides for proceeding with the development of a space shuttle system," that "detailed design and development of the shuttle engine—the longest lead time component" would begin in FY1972, and that "airframe design and development will proceed on an orderly step-by-step basis leading to detailed design or initiation of development in FY1972." The OMB space staff objected to this language as reflecting a commitment to the shuttle that had not been made, and suggested that "the Administration preserve flexibility" by "making no commitment to proceeding with the development of the entire shuttle system" and "making no commitment to an FY1972 decision on initiation of development of the airframe." The OMB Evaluation Division, headed by Assistant Director William Niskanen, was even stronger in its objections, telling Rice "it is important that the commitment to finance an advanced space engine not imply a commitment to the space shuttle." Niskanen suggested that the language "in all sections of the budget document" should describe "this engine as an advanced lower-cost space engine rather than as a shuttle engine."

This difference in views persisted into January as the budget documents were being sent to the printer. The OMB staff noted that while "NASA is firmly convinced that the lower-cost earth to orbit launch vehicle will be at least partially reusable and hence a 'shuttle,'" it would be "desirable from our position" to use "a term with broader meaning than 'space shuttle,' which could cover low cost expendable rockets." The staff noted that "the key issue is not really the term 'shuttle,' but rather achieving an understanding on Dr. Low's part that the Administration is not now committed to a reusable space shuttle." The staff predicted that NASA would "strongly resist" a change in budget language.[36]

This prediction was accurate. Low considered the staff suggestion as a reversal "of the words Don Rice and I agreed to concerning the space shuttle" and was upset to discover that at one point the "words space shuttle had been completely deleted from the President's budget and, in their place, the words future launch vehicle had been inserted." Low met with Rice on January 9. He told Rice he "fully understood the extent of the commitment (or lack thereof) by this Administration to the space shuttle, but that I also understood that such a commitment would be forthcoming if our studies so indicated during the spring and summer." Low suggested that "Rice apparently agreed with me, but mentioned that he had internal problems within OMB and that the evaluation group in OMB had insisted that far more restrictive language be included." Low and Rice "argued about this for some time"; Rice finally agreed that the language Low wanted "would be reinstated in the budget book." When NASA received its official budget allowance declaration from OMB on February 19, included was the statement, echoing Low's preferred language, that "shuttle airframe development should proceed on an orderly step-by-step basis which may lead to continued detailed design or initiation of development of a specific design, depending on the progress in studies now underway."[37]

The final NASA FY1972 budget request that President Nixon sent to the Congress was for $3.271 billion in budget authority (compared to the FY1971 budget of $3.298 billion) and $3.152 billion in outlays (compared to the FY1971 budget of $3.368 billion). Although there would be $200 million less to spend during FY1972 than a year earlier, the overall FY1972 budget authority for 1972 and projected for future years would be basically the same as for FY1971, thus arresting the half-decade long cuts in NASA funding. As he met with Tom Paine in January 1970, Richard Nixon had indicated that he might be willing to approve a NASA budget of as much as $3.9 billion for FY1972, but continuing economic and fiscal problems had made such an increased allocation for NASA politically and fiscally impossible. Reflecting on the final budget, George Low suggested that "although I am personally disappointed that we did not do better, the general feeling around NASA appears to be that we did considerably better than people had expected us to do."[38]

While NASA may have "done better than people expected," a decision crucial to the space agency's future remained unmade. That was whether NASA would get presidential approval to proceed with the space shuttle as its major program during the 1970s. NASA's hope, embodied in the budget language that Low had fought to preserve, was that such approval would come at the end of the ongoing shuttle studies in June 1971. Then NASA would quickly invite bids on developing the shuttle airframe and select the winning contractor by the end of 1971. However, not including funds for airframe development in the FY1972 budget request almost certainly meant that this plan was not viable. While White House approval of funds for developing the new rocket engine intended for shuttle use was a significant step to shuttle approval, there remained major obstacles, budgetary and technical as well as political, to a final go ahead. NASA's uncertainty about its future continued, and 1971 became a make-or-break year for what was left of the space agency's post-Apollo aspirations.

Chapter 9

National Security Requirements Drive Shuttle Design

When NASA in its September 30, 1970, budget proposal to the Office of Management and Budget OMB) characterized the space shuttle as "cost-effective," it was responding to pressure from the budget office to demonstrate that the combination of the costs of developing and operating the reusable shuttle would, over the period of shuttle use, produce a cost savings over the use of existing or new expendable launch vehicles to launch the same missions. This requirement was unprecedented; in the 12 years since NASA had begun operations, it had never been required to show that one of its programs could be justified in economic terms. The NASA leadership, once it had decided to defer the space station and to justify the shuttle as a general-purpose launch system, concluded that it had no alternative but to accede to the cost-effectiveness requirement. NASA quickly recognized that meeting this requirement would require the shuttle being used to launch essentially all U.S. payloads. In particular, military and intelligence satellites launched by the national security community comprised almost half of the U.S. demand for space launches, and there was no way that the shuttle could be cost effective unless that community abandoned its own launch vehicles and committed to use the shuttle once its feasibility had been demonstrated.

This put the national security community in a strong bargaining position. Knowing that NASA needed its commitment to use the shuttle, the community could both set out a demanding set of performance requirements for the shuttle to meet and refuse to share in the cost of shuttle development, claiming it already had perfectly adequate launch capability. This was the path that was followed from early 1969 to the final approval of the shuttle. While NASA if it had not had to respond to national security requirements might well have chosen another shuttle design, its leaders decided that they had no choice but to meet those requirements. Throughout the shuttle study process, and particularly in the critical year of 1971, it was the ability of the shuttle to launch all or almost all national security as well as NASA payloads that defined the shuttle design NASA would advocate.

National security requirements defined three shuttle performance characteristics:

1. *Payload bay dimensions*: The shuttle would carry its cargo in a "payload bay." The width and length of the payload bay would determine the size of the cargo that could be carried.
2. *Payload weight*: The lifting power of the shuttle was usually expressed in how many pounds of payload it could launch to various orbits. The weight of payloads that the shuttle could take to various orbits was in turn linked to how many future missions could be launched by the shuttle. The heaviest payloads anticipated for the shuttle were national security missions.
3. *Cross range*: This was the ability of the shuttle to maneuver sideways from a "straight ahead" path as it returned to Earth. There were a variety of speculative national security missions for the shuttle that required cross range of over 1,100 nautical miles (nm).

This chapter gives only minimal attention to the detailed technical issues involved in defining a space shuttle design that would meet these national security requirements; those issues have been treated in several other studies.[1]

Shuttle Studies Begin

The concept of a reusable space plane to carry people and equipment into orbit has a long history, and both NASA and the Department of Defense in the 1960s devoted significant attention to whether such a vehicle was technologically feasible.[2] But the first high-level designation of such a concept as a "space shuttle" came from NASA's Associate Administrator for Manned Space Flight George Mueller as he addressed the British Interplanetary Society in August 1968. Mueller projected that "the next major thrust in space will be the development of an economical launch vehicle for shuttling between Earth and the installations, such as the orbiting space station, which will soon be operating in space." Mueller was of course aware of the various studies of reusable space vehicles, and realized that the space station program he saw as a major next step in space development would not be economically feasible unless there was a low-cost transport to "shuttle" crew and supplies to and from such an outpost. Mueller's concept for such a system was a fully reusable vehicle capable of "airline type" operations.[3]

Mueller decided to fund several of what NASA designated Phase A feasibility studies to carry out an initial examination of the technical feasibility of what was at that point called the integral launch and reentry vehicle (ILRV). NASA set out an initial set of performance requirements to guide these contractor studies. They included the capability to carry up to 25,000 pounds of cargo or ten passengers to the 270 nm, 55 degree orbit then being planned for a space station. The payload bay was to provide a volume of at least 3,000 cubic feet. The ILRV was to be able to launch within 24 hours of the decision to do so, and to be capable of returning from orbit to a designated runway within a day after a deorbit decision. To achieve such a return, a cross-range capability of 450 nm was specified. NASA initially told its contractors to assume a flight rate of 8 to 12 missions to a space station per year; the use of the system to launch other NASA missions or national security missions was at this point not part of the space agency's thinking.[4]

NASA's ability to design a space shuttle solely to meet its own requirements was short-lived. One of the first decisions of the Space Task Group (STG) as it began its review of the U.S. space program in March 1969 was to direct NASA and the Department of Defense to jointly investigate whether a single, lower-cost vehicle could meet the needs of both organizations. A charter for the joint study was signed in early April by NASA Administrator Tom Paine and Secretary of the Air Force Robert Seamans. NASA's George Mueller and Air Force Assistant Secretary for Research and Development Grant Hansen were named as the study's co-chairs.

There were a number of formal and informal meetings during the April–June period between Mueller and Hansen to discuss top-level shuttle requirements. At one of these meetings, Hansen's top assistant Michael Yarymovych told Mueller that, if NASA wanted national security community support for the shuttle, the vehicle would have to carry payloads up to 60 feet long and would have to be able to operate from the Vandenberg Air Force Base on the California coast. After a California launch, the shuttle would have to be able to carry out a one-orbit mission without overflying the Soviet Union, so that it would not be exposed to potential Soviet interference, and then be able to return to land at Vandenberg. During the shuttle's 90-minute or so orbit, the Earth would have rotated eastward some 1,100 nm, and thus the shuttle would have to have at least that amount of cross-range maneuvering capability to be able to land back at Vandenberg. Yarymovych told Mueller "we'd support the shuttle, but only if he gave us the big payload bay and the cross-range capability." Mueller knew that this would mean changing the shuttle design that he and his NASA engineers preferred, "but he had no choice."[5]

Following his meetings with the Air Force, Mueller called together the ILRV study contractors to inform them that the requirements originally specified for their studies had to be changed in light of national security preferences. He told the group that the vehicle should now be able to launch 50,000 pounds of payload to the space station orbit, rather than 25,000 pounds, and should have a payload bay providing 10,000 rather than 3,000 cubic feet in volume, which was translated into a bay 15 feet wide and 60 feet long.[6]

DOD/NASA Study Bullish on Shuttle

NASA completed its initial report for the STG on future space transportation requirements in mid-May; the report concluded that "fully reusable or near fully reusable systems offer the maximum potential for an economic and versatile space shuttle system that could readily satisfy a vast majority of future space transportation requirements." Also, a "reusable space shuttle would provide a broad range of capability in space operations—a capability that is the keystone to the success and growth of future space flight developments for exploration and exploitation of near and far space."[7] The separate Air Force study effort was finished in the same mid-May time frame. The next step in the process was integrating the two studies into a single report, to be submitted to the STG by June 15. Lead responsibility for assembling the final study report was assigned to a national security community support contractor, The Aerospace Corporation; Aerospace staff member Don Dooley led the report-writing effort.

The Air Force was not the only national security organization participating in the shuttle study. Also deeply involved was the National Reconnaissance Office (NRO), the organization created in 1961 to develop and operate the highly classified intelligence satellites that provided crucial national security information to the nation's leadership. In 1969, the very existence of NRO was classified, and thus NRO participation in the shuttle study could not be publicly acknowledged. The director of the NRO was a civilian, usually holding a high-level Air Force position, such as undersecretary or assistant secretary, but because the NRO was classified this responsibility was not acknowledged. The Aerospace Corporation supported not only the space elements of the Air Force but also the activities of the NRO, and thus was well positioned to reflect the interests of both organizations.[8]

A Very Optimistic Assessment of Potential Shuttle Missions

The "Joint DOD/NASA Study of Space Transportation Systems" was submitted to the STG on June 16, 1969. The three-volume report was (and still is) classified "Secret." A separate "Summary Report" shared the same classification for 30 years, but was declassified in 1999; the following information is extracted from that declassified document.[9]

The study team provided an extremely positive assessment of the potentials of the space shuttle and reusable upper stages to carry payloads from the shuttle to higher orbits; the combination was called the Space Transportation System (STS). Its report concluded that "the development of an STS is needed to provide a major reduction in operating costs and an increased capability for national space missions."

Space Shuttle Missions

The report identified four "basic mission areas":

1. *Satellite placement, servicing, and recovery.* In this mission area, a shuttle would deliver large satellites to low Earth orbit. Such satellites could be checked out in orbit before being deployed, and a future shuttle mission could rendezvous with a satellite "to replace non-operating or outdated" equipment or to return the satellite to Earth for refurbishment.

2. *Launch of propulsive stages, propellants and payloads for high energy missions.* In this mission area, a shuttle would launch payloads destined for transfer from low Earth orbit to synchronous orbit or other destinations requiring additional propulsion. The shuttle would carry another new system, known as an "orbit-to-orbit shuttle" or "space tug," to carry out such transfers.
3. *Space station/space base logistical support.* In this mission area, tied to NASA's ambitious post-Apollo plans, the space shuttle would serve as a logistics system "capable of routinely transporting numbers of personnel and significant amount of discretionary cargo to and from low earth orbit." For example, "to sustain operation of a 50-man space base would require on the order of 70,000 pounds of cargo and passengers every three months." The shuttle could also return to Earth "significant amounts of return cargo such as tapes, film, and processed material."
4. *Short-duration orbital missions.* This was the most operationally challenging type of shuttle mission. In purposely opaque language the report noted that the space shuttle could make possible "special purpose orbital missions of a unique nature," lasting from just one orbit up to seven days, to support "programs of space systems operations, earth sensing or sky viewing." A shuttle could also place in orbit "self-contained mission modules which possessed their own crews to operate specific mission equipment." Such modules could either operate from within the shuttle's payload bay or be left in orbit to be recovered and returned to Earth on a subsequent shuttle flight.

The report noted that "in times of crisis our national leadership requires accurate information for decisions. This information could be crucial to the survival of the United States. The possible locations of crises are worldwide: Southeast Asia, Korea, the Middle East, and Czechoslovakia are but current examples." In 1969, the only way that national decision makers could get rapid photographic evidence of a situation in a far away crisis area was through an overflight by the U-2 or supersonic SR-71 spy planes, an action that was a violation of national sovereignty and subject to possible interception. The NRO was in 1969 operating a photo-intelligence surveillance satellite called Corona and another, higher resolution satellite called Gambit, but those two systems recorded images on photographic film. That film was returned to Earth in a capsule dropped from orbit and recovered by a waiting aircraft, and it could take from several days to weeks for the final film product to reach the desks of decision makers.[10] The DOD/NASA report suggested that a "mission-equipped" shuttle "could return accurate information on a crisis located anywhere in the world or an assessment of an attack to national leaders within the shortest time from launch." To carry out such a mission, the report discussed "a single-pass [one orbit] request surveillance mission with return to Washington, DC." That mission would require a cross-range capability of 1,400 nm. Such a space flight would not be a violation of sovereignty according to the practice recognized by the United States and the

Soviet Union since the early 1960s and formalized in the 1967 Outer Space Treaty—that outer space was not subject to national sovereignty. This practice had been interpreted to mean that flying over a particular nation while in outer space was not a violation of its sovereignty.

Another short duration mission possibility mentioned in the report was "the interception and inspection of objects in space." The report noted that "future unknown satellites could operate for days or weeks, posing a threat ranging from intelligence gathering to delivery of a nuclear weapon," and suggested that "a national ability to intercept, inspect, and determine the purpose of (as well as destroy, if necessary) unknown satellites is vital."[11]

The DOD/NASA report projected a shuttle flight rate between 1975 and 1985 of 30 to 70 flights per year, based on "only those flights required for existing, approved, or high priority planned missions." Expanding the "mission model" to include flights related to post-Apollo lunar exploration by NASA and other prospective DOD missions could increase the flight rate to 140 missions per year. At such flight rates, the cost of launching a payload to low Earth orbit, the report suggested, could be reduced from approximately $800 per pound to $50–$100 per pound; a similar reduction from $10,000 per pound to less than $500 per pound for payloads going to synchronous orbit was forecast. The report predicted additional cost savings from "major improvements in payload environment, methods of operations, and through return of payload from orbit," and noted that "the full potential" of a space shuttle "can only be realized if it is indeed a means of *low* cost transportation."[12]

The report concluded that shuttle development "does not require a breakthrough in technology." Costs of developing the shuttle designs then being considered were estimated to be between $4 and $6 billion. All designs examined had a 15 × 60 foot payload bay and would be able to carry 50,000 pounds to a 100 nm polar orbit (an orbit that would go from south to north, crossing over or near the Earth's poles) after being launched from California. The vehicle would also be able to return a heavy payload from orbit, allowing satellite refurbishment and re-launch. The 15-foot width of the payload bay was required for "space station logistics support, propulsive stages, and satellites such as... surveillance systems." The 60-foot length of the payload bay was required for "ocean surveillance spacecraft, stage-plus-payloads for synchronous missions, or two medium altitude surveillance satellites." A cross-range capability of 1,500 nm was "the selected design value."[13]

The report concluded by noting that "a fully reusable system has inherent advantages compared to a partially reusable system." It added that "unless the stage and one-half partially reusable system [an option that at that time was being considered during the NASA Phase A studies and would in 1971 be adopted as the final shuttle design] is found to have substantial advantage in cost, schedule, or reduction in technical risk, a fully reusable system should be selected."[14]

The extremely optimistic—indeed, unrealistic—tone of the DOD/NASA report, with its projection of a high space flight rate and the ability to launch on demand and its conclusion that there were no technological barriers to designing a space shuttle that would launch anticipated missions at a major reduction in cost while at the same time offering unique capabilities for new missions, set the baseline for the policy-level discussions of the space shuttle over the next several years. In a period of a few months in early 1969, the shuttle concept had expanded from being only a supply vehicle for a space station, to be launched 8 to 12 times a year, to a system that could launch up to 140 times a year, carrying out all government space missions. This very high launch rate (almost three launches per week!) was well beyond the bounds of realism, but suggests the aspirations of some of those involved in the DOD/NASA study. The projected low cost of shuttle operations remained a major selling point, and the validity of the report's call for a large payload bay and substantial cross-range were key issues in the debate over shuttle approval. Thus the June 1969 DOD/NASA report marked a key milestone in the space shuttle decision process.

Which Payloads Drove Shuttle Design Requirements?

There is little controversy with respect to the influences that originally led to setting the desired width of the shuttle payload bay at 15 feet. They were both NASA's space station crew and cargo payloads and a potential new upper rocket stage—a space tug—for moving national security and other payloads from the shuttle's low Earth orbit to higher altitudes, particularly geosynchronous orbit.*

It is also now clear which payload defined the need for a 60-foot long payload bay. In an 1997 interview, Hans Mark, who as both Under Secretary and Secretary of the Air Force during the administration of President Jimmy Carter (1977–1981) had concurrently served from August 1977 to October 1979 as the director of NRO, commented that "the shuttle was in fact sized to launch HEXAGON." This photo-intelligence satellite, also known as KH(Keyhole)-9 and nicknamed "the Big Bird," was under development in 1969 as the successor to the Corona satellites, which had been operating since 1960 to provide broad area photographic surveillance of various regions of the world. Hexagon was a very large object, only ten feet in diameter but almost 60 feet long. The satellite would weigh over 30,000 pounds when fully loaded with film for its four entry capsules that would return exposed film to Earth.[15]

The existence of Hexagon was in 1969 classified at a very high level, above "Top Secret"; thus it could not be mentioned in the DOD/NASA report, which bore only a lower-level "Secret" classification. As Mark suggested, Hexagon was used to "size" the payload bay; originally there were no plans to actually launch it on the shuttle, since the Hexagon program

---

*After NASA decided in 1970 that a future space station would be assembled from shuttle-launched modules and other components, the 15-foot width also became a requirement related to the size of those space station elements.

would be reaching the end of its likely service life as the shuttle began operational flights in the late 1970s or early 1980s.[16] Air Force and NRO planners judged that whatever system would be the follow-on to Hexagon would likely be equally as large, and Hexagon thus could serve as a surrogate for that future system in determining an appropriate payload bay length.

Less clear is which potential Air Force or NRO missions drove the requirement for a shuttle to have a high cross-range capability. No prior actual national security space system had been required to maneuver to return to Earth, since all were expended after completing their mission. However, the Air Force had pursued from the late 1950s until it was canceled in 1963 a research program called Dyna-Soar, which involved developing a small glider-like winged vehicle that would be launched into orbit on an expendable booster and would have cross-range capability upon its return to Earth. The idea of a piloted space system that could be brought back to a secure base after a one-orbit or short-duration mission remained attractive to national security planners, but that idea had not gone through the typical rigorous review to establish it as a firm national security requirement. Air Force Secretary Robert Seamans suggested that the cross-range requirement was advocated by "operational types," not the top Department of Defense, Air Force, or NRO leadership.[17]

The DOD/NASA report had mentioned a "single pass" mission with an unspecified launch location and requiring 1,400 nm of cross-range to return to a location near Washington, DC, presumably so that the intelligence products obtained during the mission could be rushed to top-level decision makers. None of the subsequent discussions of national security shuttle flights discussed such a mission profile; it seemingly reflected the aspirations of those who prepared the 1969 report. Missions taking off and landing at Vandenberg Air Force Base on the California coast (or some other Western launch site[18]) were much more prominent in later discussions. If the space shuttle were to carry out a one-orbit mission launched from Vandenberg, the shuttle would have to have at least 1,100 nm of cross-range to return to a secure runway at that Air Force base.

A clue to the character of missions that required high cross-range can be found in studies performed by NASA in 1973, after the shuttle entered its development phase. By then, NASA had already done considerable work in designing "reference missions" for two uses of the shuttle—placing a satellite in geosynchronous orbit and resupplying a spacecraft in low Earth orbit. In 1973 NASA developed two new reference mission scenarios for single-orbit shuttle flights from Vandenberg Air Force Base. These reference missions were "representative of Air Force requirements on the shuttle." One of the two missions, designated 3A, would deploy a satellite into a 104 degree, 100 nm polar orbit; the shuttle would return to Vandenberg after one orbit. The satellite to be deployed would weigh 32,000 pounds and was ten feet in diameter and 60 feet long; it would almost certainly be the follow-on photo-intelligence satellite to Hexagon. It would be deployed less than 24 minutes after launch. NASA noted that "the mission of the payload is beyond the scope" of the reference mission description, likely referring to its intelligence objectives. The second mission, designated 3B, after carrying out a rendezvous within 25 minutes of launch, would retrieve a similar satellite and return it to Vandenberg after a single orbit.[19]

So were satellite deployment or retrieval the missions that defined the needed shuttle cross-range capability? Or was it also, or even primarily, the hope of national security planners to be able to fly an on-demand mission in polar orbit to get crisis-related information on what was happening at a flashpoint anywhere in the world, such as the mission landing in Washington, DC, mentioned in the DOD/NASA report? This latter speculation is supported by a letter drafted in late 1971 for then NASA Administrator James Fletcher to send to Deputy Secretary of Defense David Packard as NASA sought DOD support for the shuttle program. The draft letter suggested that "the shuttle could be maintained on ready alert, making possible rapid responses to foreseeable and unexpected situations"; such a mission could examine "unidentified and suspicious orbiting objects"; enable "capture, disablement, or destruction of unfriendly spacecraft"; and make possible "rapid examination of crucial situations developing on earth or in space."[20]

The DOD/NASA report also mentioned launches of "self-contained mission modules which possessed their own crews to operate specific mission equipment." Might these "mission modules" have carried the human-operated KH-10 very high-resolution camera system, code named Dorian, developed during the 1960s for the Manned Orbiting Laboratory (MOL) program? That program was canceled on June 10, 1969, just as the DOD/NASA shuttle report was being prepared. The MOL combined a capsule based on NASA's Gemini spacecraft, to be used during launch and reentry, and a two-segment module containing the Dorian camera system and crew quarters. The 1971 NASA draft letter said, "the shuttle could be equipped to perform the MOL mission for seven days on station...Alternatively, the shuttle could transport MOL-like equipment in a self-supporting module to the desired orbit for operation over a longer period of time." Such missions would most likely have been launched into polar orbit so they would overfly all areas of the world, and would return to Vandenberg at their completion, thus requiring cross-range capability.[21]

The need for high cross-range was throughout the shuttle debate a point of contention between NASA and the national security community. In reality, requirements for national security missions requiring high cross-range were never formalized and more or less evaporated during the 1970s. Well before that time, however, NASA had decided that a shuttle having significant maneuvering capability as it returned from orbit was needed to survive the heat of entry into the atmosphere. So while the national security cross-range requirement initially drove NASA to a particular shuttle orbiter design, one with delta-shaped wings and the thermal protection needed to resist high temperatures during a maneuvering entry, NASA likely would have adopted a similar design even if that requirement had not been levied in 1969. Whether NASA would have gone forward with a shuttle having a

15 × 60 foot payload bay and powerful enough to launch the most heavy national security payloads is not as clear; in the final days of the shuttle debate in December 1971, NASA put forward a somewhat smaller and less powerful shuttle as its proposed design.

Mueller Tries to Go His Own Way

In his new instructions to NASA's Phase A study contractors on May 5, George Mueller had changed his original guidance to include the capability to launch 50,000 pounds rather than 25,000 pounds to the space station orbit and to provide 10,000 rather than 3,000 cubic feet of volume in the shuttle payload bay. But he did not direct the contractors to focus their study effort on vehicles capable of providing the cross-range desired by the national security community. Mueller was very aware that NASA's "chief designer," Maxime (Max) Faget, director of engineering and development at the Manned Spacecraft Center (MSC) in Houston, preferred a shuttle concept with straight wings and limited cross-range. Faget had designed the Mercury spacecraft and helped design the Gemini and Apollo spacecraft, and was a powerful force within NASA's engineering community. As the DOD/NASA report was being approved for submission to the Space Task Group, Mueller insisted that the preface to the report include the following statement: "If it is later determined that a specific performance characteristic imposes severe penalties on technical risk, cost or schedule, the necessity for fully achieving that characteristic will be assessed." It is likely that the cross-range requirement was in Mueller's mind as he inserted this reservation into the report.[22]

Mueller on August 6 mandated that the "space shuttle will be developed utilizing fully reusable systems only." This directive came as NASA was pushing Mueller's integrated plan, with its emphasis on low cost based on reusability, as the basis for the recommendations in the STG report. This was an influential order. NASA and industry studies for the following two years focused only a two-stage shuttle with a fully reusable "booster" stage lifting a fully reusable spacecraft, designated an "orbiter," off the launch pad and accelerating it to a high velocity; then the orbiter's engines would fire to accelerate it the rest of the way into orbit. Mueller's ambitious objective of full reusability ruled out of the Phase A studies several promising concepts that were not fully reusable; those concepts reemerged only after NASA abandoned its hope for a fully reusable shuttle in mid-1971, discovering that it was both too expensive to develop within projected budgets and likely too technically risky.[23]

The Air Force Is Concerned

That Mueller was not fully committed to a space shuttle design responsive to the performance requirements proposed by the NASA/DOD report soon became evident to the Air Force. In a September 15, 1969, memorandum to Secretary of the Air Force Seamans, Air Force Chief of Staff General John Ryan suggested that Mueller had "redirected the activities of the NASA and responsive contractors to a Space Transportation System/Space Shuttle which is knowingly inadequate for the Air Force." This harsh judgment was based on Mueller's August directive to those studying shuttle designs and Mueller's comments at a September 10–11 meeting attended by shuttle study contractors and Air Force representatives. At that meeting, Mueller had indicated that designs with a payload of 20,000 pounds to the space station orbit, not the 50,000 pounds minimum, which was the national security requirement, should be studied. He also identified cross-range "as desirable but not required." Mueller was reported as saying that the Air Force position regarding cross-range and payload weight was "soft."[24]

Seamans was in a difficult position. On one hand, in his role as STG member he had taken a "go slow" stance with respect to shuttle development; in his comments at the August 4 STG meeting and the letter he had given Vice President Agnew at that meeting, Seamans had recommended that "we embark on a program to study by experimental means including orbital tests the possibility of a Space Transportation System that would permit the cost per pound in orbit to be reduced by a substantial factor." Seamans added "it is not yet clear that we have the technology to make such a major improvement." On the other hand, Seamans recognized that NASA was not taking his advice and instead was pushing for rapid development of an operational shuttle. Given the possibility that a shuttle not meeting national security requirements might be approved, Seamans proposed an action to make sure that those requirements were accommodated in whichever shuttle design was eventually approved. In November 1969, Seamans wrote NASA Administrator Paine, suggesting "a senior-level management policy board" to guide the shuttle program; such a board would "insure that the interests and objectives of both the DOD and NASA are fully represented and maximum cooperation between the agencies is achieved." The board, said Seamans, "would be essentially the Board of Directors for the STS development and would be concerned with requirements, technology, funding, and management." Given what was happening under George Mueller's direction at NASA, Seamans added "I am convinced that such a policy board is necessary."[25]

In his letter Seamans referred to the Gemini Program Planning Board as a desirable model for the board he had in mind. That board had been set up in 1963, after Secretary of Defense Robert McNamara had attempted to seize control of NASA's Gemini program. Seamans, in 1963 NASA's associate administrator, had been on the other side of the table negotiating with McNamara to create an arrangement that retained NASA's lead role in Gemini while still providing a channel for making sure that the program also served DOD interests. As a senior DOD official six years later, he wanted to make sure that whatever shuttle NASA might propose also served national security interests.[26]

The Threat of Withdrawn Support

Paine accepted Seamans's suggestion, which came close to being a demand, given the perceived importance of national security community support if the shuttle program were to move forward. A charter for a NASA/Air Force STS Committee was signed on February 17, 1970. The committee, "in order that the STS be designed and developed to fulfill the objectives of both the NASA and the DOD in a manner that best serves the national interest," was to conduct a "continuing review" of the STS program and make recommendations "on the establishment and assessment of program objectives, operational applications, and development plans." The agreement noted that the shuttle program "may involve international participation and use" and would be "generally unclassified." The agreement stated definitively that shuttle development "will be managed by NASA."[27]

The committee met six times during 1970, four times in 1971, and once in early 1972. The NASA/Air Force STS Committee turned out to be primarily a forum for the national security community to keep pressure on NASA to propose a shuttle design that met the community's requirements. There was throughout those two years the not-so-veiled threat to withdraw DOD support for the shuttle if NASA did not do so. NASA reflected that pressure in the requirements it established for its continuing shuttle design studies. As the shuttle entered the decisive 1971 year, NASA was proposing a shuttle that would meet all national security requirements, and continued until the final days of 1971 to insist that only such a "full capability" shuttle was worth developing. This position eventually prevailed. The national security requirements established in 1969 thus had a pervasive impact on the final design of the space shuttle.

Chapter 10

A Time of Transitions

Acting Administrator George Low in January 1971 characterized the NASA Fiscal Year (FY) 1972 budget request sent to Congress as one of transition from the program of the 1960s to the programs of the 1970s. This was indeed the case, as the budget request formalized canceling two Apollo missions and deferring space station development, and suggested that at least in principle the Nixon administration intended to move forward with a space shuttle program as the central U.S. space effort in the 1970s.

This was only one of the transitions taking place in the first months of 1971. The White House finally selected Tom Paine's successor as NASA administrator. He was Dr. James Fletcher, the president of the University of Utah. Fletcher's nomination was submitted to the Senate in February, he was confirmed in March, sworn in by the president in April, and took over NASA in the first days of May. Fletcher and George Low, who stayed on as deputy administrator, became a very effective team in leading the space agency through the tortuous process over the second half of the year, ultimately resulting in presidential approval of the shuttle that NASA wanted to develop.

In another potential transition, a White House initiative created something of an identity crisis for NASA. President Nixon and his advisers were interested in developing technology-based solutions to major societal problems, and seriously considered transforming NASA into a general applied science agency—a "new NASA"—to take on that responsibility. Fletcher and Low assessed the desirability of NASA's assuming such a role while still also maintaining its aeronautics and space responsibilities, and decided to respond positively if asked by the president to take on added missions.

Finally, there was a major transition in NASA's thinking about the character of the space shuttle program it would put forward for presidential approval. At the start of 1971, NASA Associate Administrator for Manned Space Flight Dale Myers had decided to press forward with a two-stage fully reusable shuttle design meeting all national security requirements. But by mid-year the combination of Fletcher and Low recognizing that NASA was very unlikely to get White House approval of the funding required for such a development and growing concern among NASA's engineering staff regarding the technical challenges associated with simultaneously developing both the shuttle booster stage and the shuttle orbiter led NASA to abandon the fully reusable design. There followed a rather frenzied search for an alternative that presented the best combination of development and operating costs to make the shuttle cost-effective while still preserving all shuttle capabilities that NASA and the national security community sought.

NASA Gets a New Administrator

As George Low had led NASA through the process of developing the agency's FY1972 budget request, at the White House Peter Flanigan continued his search for a person to take on the administrator's job on a permanent basis. By late 1970 two promising candidates had been identified—Frank Jameson, president of Teledyne-Ryan Aeronautical Corporation in San Diego, California and James Fletcher, president of the University of Utah in Salt Lake City, Utah. Neither had been on the White House radar screen a few months earlier. The White House ran background checks on the two. Director of personnel Fred Malek reported the results to Flanigan on January 6, 1971, noting that there had been "no attempt to contact the candidates" and "no attempt to determine their political philosophy." Of Jameson, Malek reported that he was known as "an accomplished and marketing-oriented executive" and "an extroverted, hale, hearty, and well-met type of individual," but "not generally well regarded for his administrative skill." This led Malek to suggest "if we are seeking a tough minded, control-oriented, inside executive, to really manage the agency, Frank Jameson would not seem to be a top choice." With respect to Fletcher, Malek reported that he had "a unique combination of management and technical skills," was "intelligent, articulate, and a proven leader of technical people," and was "reported to have an uncanny ability to embrace a large spectrum of diverse business and technical activities simultaneously."[1]

The suggestion of Jameson for the NASA position had come from House Minority Leader Gerald Ford (R-MI). Supporting Fletcher was Senator Wallace Bennett (R-UT). In addition to their Utah and Mormon connections, Fletcher and Bennett were related by marriage; Bennett's daughter was married to Fletcher's brother. In early February, Bob Haldeman asked Flanigan "what's the status of NASA? Gerry Ford is pushing Jameson. Have we got a candidate yet or is that still hanging fire?" Flanigan responded a few days later that "Gerry Ford has been informed... that Jameson is not getting the position. Subject to clearance Jim Fletcher will."[2]

On February 17, Flanigan formally recommended to President Nixon that he nominate Fletcher as NASA administrator. He told Nixon that of "a large number of candidates proposed for the post," Fletcher "appears to be by far the strongest." Flanigan noted that in his role as president of the University of Utah Fletcher "has had unusual success in running the university while placating both radical and conservative students." He also noted that Fletcher, a physicist and engineer with a doctorate from the California Institute of Technology, had served for many years as a member of the President's Science Advisory Committee (PSAC). He alerted the president that Fletcher had just been offered the position of chancellor at the new University of California campus in San Diego, and thus it was important "to assure Fletcher now that he is our first choice." He closed his memorandum by noting "Fletcher's high business, management and technical qualifications would seem to be an ideal blend for a NASA Administrator." It is not clear whether Richard Nixon saw the memorandum and told Haldeman he approved the choice of Fletcher, or whether Haldeman made the choice himself without bothering the president, something that happened on occasion with respect to issues of secondary interest to the president. At any rate, the initial in the "Approve" box on the Flanigan memo was Haldeman's, not Nixon's.[3]

The White House sent Fletcher's nomination as NASA administrator to the Senate on February 27. Although easy Senate confirmation seemed likely, the nomination soon ran into trouble with the president. On March 9, veteran CBS correspondent Daniel Schorr on the evening's nightly news program reported that Fletcher had advised President Nixon to take more time before endorsing a proposed antiballistic missile system called Safeguard. News anchor Walter Cronkite said that Schorr had gotten his information from overhearing a Fletcher conversation. Nixon was enraged by this report; his reaction was caught in his newly installed taping system. Meeting with Flanigan on the morning of March 10, which was the day of Fletcher's Senate confirmation hearing, Nixon said "I am going to withdraw his nomination today unless that [the Schorr report] is denied." Regarding Fletcher, Nixon said "I have never met the son of a bitch. I shook his hand once in my life... I am not going to have the new director of NASA, that good job, not meeting this flatly... We want him to say that he is in support of the ABM program. He has got to say that or I will withdraw his nomination this afternoon. I mean it, we are going to get tough around this place." Nixon's anger soon passed, and Fletcher's nomination was not withdrawn.[4]

The Nixon Tapes

Beginning on February 16, 1971, most White House conversations involving President Richard Nixon were recorded. Nixon through his chief of staff Bob Haldeman ordered the Secret Service to set up a taping system. Seven microphones were installed in the Oval Office—five on the president's desk and one on each side of the office fireplace. Two microphones were located in the Cabinet Room. In April 1971 Nixon's hideaway office in the Old Executive Office Building next to the White House was also wired for recording. Telephones in the Oval Office, the hideaway office, and the Lincoln Sitting Room in the

White House family residence, where Nixon often spent his evenings working, were also tapped. Other than the Cabinet Room, where a switch had to be thrown to turn the tape recorders on and off, the recording systems were sound activated and linked to a locator device worn by the president, and thus only operated when the president was in the room. Only the president, Haldeman, a few of Nixon's close personal assistants (but not Ehrlichman, Kissinger, and Shultz of his inner circle), and the Secret Service knew about the recording system.

Nixon had been advised by former president Lyndon Johnson to reinstall a recording system; Johnson had used one during much of his presidency. There were two main reasons for having a taped record of presidential conversations. One was protecting against others misrepresenting what was said in meetings with the president; the other was as a source for the eventual memoir that Nixon, like all former presidents, would write after leaving office.

The Nixon tape recordings are difficult to understand. The tapes used and the slow speed at which they were recorded were not appropriate for voice recording. Various sounds, ranging from ringing telephones to the sound of coffee cups being set down, obscure parts of many conversations. Those conversations and meetings were often rambling, filled with incomplete sentences, and unstructured, making them difficult to follow. Participants interrupted each other and finished each other's sentences. (More information regarding the Nixon tapes can be found at www.millercenter.org/presidentialtapes.)[5]

The author, assisted by George Washington University student Luis Suter, made a "best effort" attempt to transcribe conversations relevant to the space program on the Nixon tapes. Any errors of transcription are the author's responsibility.

In contrast to Frank Jameson, who in his White House interviews apparently argued for bold new initiatives in space, Fletcher was known from his years on PSAC to be somewhat skeptical of the value of human space flight. Fletcher was aware of the proposal to develop a space shuttle; as a member of the PSAC, he had been exposed to the thinking of the committee's panel on space science and technology about the need for low-cost space transportation and was aware of the content of the Space Task Group report. During PSAC deliberations, Fletcher had asked "why is it necessary to have a manned system to get [a payload] to and from space?" Before Fletcher was nominated, several members of Congress supportive of human space flight sent a telegram to the White House opposing Fletcher's selection as NASA head, saying he was "a negative person on [manned] space," but their objections did not prevail.[6]

After his nomination was announced, Fletcher was quickly invited to attend, two days later on March 1, a large White House dinner in honor of the *Apollo 14* astronaut crew—Alan Shepard, Edgar Mitchell, and Stuart Roosa—whose mission had taken place from January 31 to February 9. NASA's Willis Shapley remembered that as he went through the receiving line before the dinner, he noticed a person "sitting by himself with his little pad of pink papers." Someone told Shapley that the person was Fletcher, his new boss. Shapley went to say hello, and later commented "that was Fletcher to a tee. He was always trying to find out who the key people were, keep his notes as to who was worthy of note... He was using his chance at this White House thing just to get a look at the people that were going to be significant to him in the future."[7]

The next day, Fletcher and Low spent several hours together discussing the state of NASA. Low's "first impression" was that Fletcher was "excellent" and "that he will be very good for NASA." In several subsequent meetings during March and April, Low and Fletcher discussed the major issues facing NASA and plotted the approach they would take to managing the agency. They "spent considerable time discussing the space shuttle." Fletcher indicated that he understood "that the space shuttle decisions this summer will be some of the most important decisions that he will make early in his career at NASA." According to Low, Fletcher came into NASA "negative on manned space flight. He was selected... by the people close to Nixon for being the kind of person who would support an unmanned space program... He came in probably not to support" the space shuttle, but he "very quickly turned himself around." Fletcher admitted that he "changed from the time I was on PSAC to the time I came to NASA," recognizing that "neither the president nor I wanted to go ahead with a program that didn't really have a manned element to it." Before he decided to support the shuttle, Fletcher pushed hard on Low, Dale Myers, and Myers's deputy Charles Donlan, who was overseeing shuttle studies, to convince himself that a shuttle, rather than some other human space flight system like a recoverable capsule, was the best option for NASA to advocate.[8]

Low early on also shared with Fletcher his growing concern that NASA should not take on the task of developing a two-stage fully reusable shuttle "without having clear-cut support for a space agency budget in excess of $4 billion." The two discussed "the possibilities of other shuttle concepts and of phasing the orbiter and booster separately." They also discussed whether there really was a need for more Apollo flights; they agreed both to examine this question after the *Apollo 15* flight, scheduled for July. Given the sensitivity within NASA of possibly not going forward with the *Apollo 16* and *Apollo 17* missions, they agreed to not to raise the question with anyone else in NASA.[9]

The Senate confirmed Fletcher as NASA administrator on March 11, but Fletcher did not plan to arrive at NASA until around May 1. Flanigan recommended that President Nixon personally swear in Fletcher, noting that there had been a "long hiatus" since Tom Paine had left the agency

James Fletcher is sworn in as NASA administrator by Judge James Belson as President Nixon and Fletcher's wife Fay stand by. Science adviser Ed David and NASA Deputy Administrator George Low are visible behind the president. (NASA photograph 71-H-791)

and that "interest in NASA and Congressional support has been declining." This suggestion was accepted, and Fletcher came to the Oval Office on April 27, 1971, to be sworn in as NASA's fourth administrator as his family looked on.[10]

Fletcher reported for duty at NASA on May 3. He made his first speech as NASA administrator on May 20; his venue was the annual meeting of the Aerospace Industries Association. The speech was attended by senior people from all parts of the space industry, anxious to hear what the new head of NASA had in mind. Fletcher noted that it was "not the time for me to attempt to make definitive policy statements." He observed that NASA was "in a period of uncertainty" since its major programs for the 1970s were "still in the study stage." Fletcher claimed that he had "been backing the shuttle concept for a number of years" and that he was prepared "to advocate it vigorously." He added "if we are going to put most of our eggs in the shuttle basket, it had better be the best basket the American... aerospace industry can devise." Fletcher said "we will take as much time as we need right now to be sure we make the right decisions" regarding which shuttle design to develop, on what schedule; he added "let's not go off half-cocked on the shuttle." In a statement that likely was troubling to the companies working on shuttle study contracts, Fletcher indicated that "we are not committed at this time to a two-stage fully reusable concept for the shuttle." Rather, NASA would "continue to consider the various possibilities as cold-blooded engineers." He argued that "we are not trying to justify the shuttle as a money-making project, but as a new capability of great promise."[11]

Fletcher's speech was a mixture of the preexisting NASA arguments for the shuttle and his own ideas. In particular, the emphasis on the shuttle as offering important new capabilities, not primarily as a means of reducing the cost of space activities, was his. He had not been previously involved in the shuttle program, and thus was free to indicate that he was not committed to the shuttle design that had been central to NASA's thinking in preceding months. Within the first few weeks of his time at NASA, Fletcher had become convinced that the shuttle was NASA's most important program for the 1970s. Now it would be up to Fletcher and Low to convince the White House that the space shuttle deserved presidential support.

Richard Nixon, "Exploring the Unknown," and Ending Apollo

Richard Nixon liked grand concepts. Such was the case with respect to space. Nixon frequently mused about the importance to U.S. interests and national vitality of "exploring the unknown"; he connected the space program with that impulse. A particularly full example of Nixon's thinking about space exploration came in a March 9, 1971, Oval Office meeting with a group of current and past NASA astronauts who had been touring college campuses to gauge reactions to administration policy. He told them

> I know what people say, we are being jingoistic. America stays number one and so forth. In the history of great nations, once a nation gives up in the competition to explore the unknown, or once it accepts a position of inferiority, it ceases to be a great nation. It happened to Spain. It happened in the 20th century to the French and then to the British. And it could happen to the United States. That is what it's all about, and so when we look at...the space program, whether it's Mars or whether it's the shuttle or who knows what it is. I don't care what it is, but the main thing is we have to go, we have to go, we've got to find out.
>
> The majority of the people in all of the polls show that they are against the SST [supersonic transport], they are against the space program. They just want to sort of settle down...If the United States just didn't...have the problems of going to space, then what a wonderful country this would be. And the answer is it wouldn't be at all. It would be a terrible country. It would be a country big, fat, rich, but with no sense of spirit...If an individual does not want to do something bigger than himself, he is selfish. That's what space is about.[12]

Nixon's line of thinking was somewhat different when he was talking to a person not strongly involved with the space program. For example, on the morning of March 24, 1971, he met with several senators in a last minute (and unsuccessful) effort to avoid the Senate voting that afternoon against the supersonic transport program. Reflecting on his meetings, Nixon told his Congressional liaison Clark MacGregor that "the United States should not drop out of any competition in a breakthrough in knowledge—exploring the unknown. That's one of the reasons I support the space program." Without pausing, he added "I don't give a damn about space. I am not one of those space cadets."[13]

Congressional refusal to continue funding for the supersonic transport was deeply disappointing to Richard Nixon, and may have reinforced his belief in the importance of the space program as a means of symbolizing America's commitment to leadership in "exploring the unknown." John Ehrlichman observed that "Nixon died very hard on the SST; he had a commitment to that which had to do with chauvinism." To Nixon, the United States "had to be at the leading edge of this kind of applied technological development. And if we weren't, then a great deal of national virtue was lost, and our standing in the world."[14]

However, remaining first in space in Nixon's mind did not include repeated trips to the Moon; in fact, he was much more interested in eventual human trips to Mars and at least once mused about exploring the moons of Jupiter. He had been talked out of canceling *Apollo 17* at the end of 1970, but in May 1971 returned to that idea, this time including also canceling *Apollo 16*. Meeting with Ehrlichman on May 13, Nixon said "I personally think [we should] stop at probably five Apollos, no more...The reason for the space program, the best reason, is not going to the moon but is the fact that we are exploring the unknown. I don't know what the hell is up there. We've got to continue to explore just for the sake of it." Later the same day, he told Ehrlichman "the one [part of the NASA program] that seems to me to have the least appeal are more Apollo shots. Why in the hell would they have to go up there and take a look around the damn thing again?" On May 18, he asked Ehrlichman "did you get those moon shots knocked off?" Ehrlichman replied "we're working on it." Nixon suggested "do your best." Finally, on May 26, Nixon told Ehrlichman "we have got to get a way to get off those damn moon-shots...There can't be any after July [the date for the *Apollo 15* mission]. And we all agree, none after July." Referring to the *Apollo 13* mission, he said "I don't want risk any more."[15]

In response to Nixon's interest in canceling the last two Apollo missions, Ehrlichman told Office of Management and Budget (OMB) Director George Shultz "the President would like us to review and analyze the NASA budget and future program with an eye to cutting the number of Apollo shots." OMB's Don Rice responded, providing estimates of the budget savings and job losses associated with canceling both missions and with canceling only *Apollo 17*. That latter action would save $101 million in FY72–74 and result in the loss of 9,000 jobs; canceling both missions, $192 million and 15,000 jobs. Rice commented that "California, Long Island, and Cape Kennedy would be hardest hit" by the job losses. Ehrlichman used Rice's information in a memorandum to the president, noting that job reductions resulting from canceling the two Apollo missions would be "centered principally in the South and Southern California."[16]

Aerospace unemployment was by this point becoming an important political issue for the White House in advance of the 1972 presidential campaign. Meeting with science adviser Ed David on February 22, Nixon indicated that he wanted David and his external advisors to direct particular attention "to the unemployed from the space and defense industries." The president met with OMB Deputy Director Cap Weinberger and Flanigan on May 5 to discuss "what could be done about high unemployment areas with specific emphasis on California." Nixon "indicated a very great concern about the California area and the high level of unemployment among technically-trained individuals." He directed his associates to review federal programs to identify those "which could be moved either in time or in place...to areas of high unemployment." Weinberger and Flanigan agreed to meet with a number of government agencies, including NASA, to pursue this directive. During the rest of 1971, aerospace unemployment, particularly in California, would be an influential factor in shaping White House space decisions.[17]

By mid-1971 Richard Nixon's interest in trips to the Moon had definitely waned. When *Apollo 15* was launched on July 26 at 9:34 a.m. EDT, the White House put out a statement that Richard Nixon had watched the launch with great interest; in fact, he was still asleep.[18]

A "New NASA"?

In February 1971, retired Air Force General Bernard Schriever had told George Low that NASA might be "the only agency that can see to it that the country continues to develop the very advanced technology that is needed for our security and our survival." Schriever was planning "to go to the President with a proposal that would maintain this capability within NASA, the Defense Department, and industry, by devoting some effort to advanced civilian technological problems." Schriever in 1969 had been asked by president-elect Nixon to become NASA administrator but had demurred; however, he still maintained good access to the top levels of the White House.[19]

It is not clear whether Schriever followed through on his initiative, but the idea of broadening NASA's mission was in President Nixon's mind as he was briefed on a possible major initiative to desalinate (remove the salt from) the ocean or other salty water so that it could be used for purposes such as agricultural irrigation or even human consumption. Meeting with Ehrlichman and Shultz on May 6 to discuss a possible desalination program, Nixon suggested: "Terrific. Put it in NASA...What if we change the name of NASA to the Experimental Space Agency. They have very bright guys...Don't leave it over there with that Department of Interior with those damn geophysicists. Geologists, I mean." The desalination briefing was repeated during a May 11 cabinet meeting. Haldeman reported that the briefing "really got him [Nixon] all excited, and he's charging away now with that as his great new program. He wants to put a real crash effort behind it, put it under NASA or someplace where we can really get something going...He's been interested in this to some degree before, but the presentation at the Cabinet meeting obviously cranked him up."[20]

Meeting with Ehrlichman, Haldeman, and Shultz following the cabinet session, Nixon was still enthusiastic, saying "build the biggest [desalination] prototype that we possibly could in Southern California... Take the appropriation, what is it, 27 million for this year?... Let's move it up to 100 million dollars and... put the scientific effort to have it done in places like, maybe NASA." Ehrlichman added, "put it in NASA and take it out of their budget... Cancel the rest of the moon program, and save a lot on the Spacelab [Skylab] and Mars. We're not going to do any more lunar landings. We're going to take all that money, you know 500 million dollars, and we're going to put it on desalting possibilities." Nixon chimed in: "I think the landing on the moon thing, see what we can do in terms of bugging that out. They'll squeal but I need to put up the money [for desalination]. I can deal with the astronauts." The conversations continued throughout the afternoon.

> *Ehrlichman*: Supposing we would say to the new head of NASA, that he has been concerned about presiding over a finite operation, [but] here is an open door now to certain permanent new [missions].
> *Nixon:* Can we name it something other than National Aeronautics and Space?
> *Ehrlichman:* We're working on that.
> *Nixon:* If we put some research projects in a few places, wonderful. Put a lot of them in California.
> *Shultz:* Why not take full advantage of everything about this? In broadening NASA's horizons we can finally do that. They like the idea of a well defined mission in space and aeronautics, but they are gradually being brought to think a little bit more broadly.
> *Nixon:* We can put it in terms of taking them to a mountaintop. We bring them in, we say, look, you have shown how it can be done, in other words we give you a project and we say go off and do it. Now we're going to give you this one [desalination], and you go out and do it. And that's the best way to get the teams [working], and you know how they get, they go "Ra-Ra-Ra" and they wear the blue shirts with... letters and things.[21]

The idea of changing NASA's name to reflect a new purpose for the agency got White House attention soon after these conversations. Ehrlichman wrote Shultz on May 17, reminding him that "the President would like serious consideration given to changing the name of NASA to something designating a more domestic orientation." The same day, speechwriter Bill Safire wrote Haldeman, saying "the idea of redefining the mission of NASA to include desalting water and other breakthroughs is great; the idea of calling it the National Applied Science Agency is horrible." He observed that "we seem to feel bound to the acronym NASA, as if it were a trade name with high consumer acceptance too valuable to change. Baloney." Safire added "if we are to widen the mission, let's do it in a way that identifies the agency as our own, reflecting our own exciting view of the future." Among Safire's suggestions for a new name: "The Discovery Agency," "Center for Exploration of the Unknown (CENEX)," and "National Scientific Breakthrough Agency." But, he suggested, "let's get the NASA people, who are an imaginative bunch, to focus on a name for their new agency." He added a caveat to that thought: "no 'technology' or 'applied science' or other words that turn technicians on and turn people off."[22]

NASA was informed of these discussions at a May 17 meeting between Fletcher, Ehrlichman, and Flanigan and then in a letter from Shultz asking NASA to discuss how it would diversify into other high-technology areas. Fletcher met with Shultz on May 25 for a broad ranging discussion of NASA's future. Fletcher reported to Low that Shultz "was wondering whether we could do anything in NASA to solve some of the other problems which you [Low] and I have discussed at some length." Fletcher and Shultz had discussed "the value of technology in developing productivity in the country and also in the possible effect it might have in influencing the balance of trade." Fletcher found Shultz "very lucid" and "not entirely inflexible...neither sold that NASA should do a great deal more nor sold that they shouldn't be, and at this point has an open mind." Low in advance of Fletcher's meeting with Shultz had prepared a memo providing his ideas on why "it might make sense to assign to NASA the government-wide responsibility for the application of technology to national needs," because "NASA has demonstrated a capability to solve difficult technological problems and to apply systems management and know-how in the solutions of these problems." Low saw two alternatives: (1) "NASA could provide its services to other agencies"; or (2) "NASA could do these things in its own right as part of an expanded NASA mission." Low thought that, despite problems associated with the transfer of missions and programs from other agencies to NASA, which would cause bureaucratic conflicts, the second alternative "would be much more likely to succeed." Low's suggested name for a redefined NASA was the "Aeronautics, Space, and Applied Technology Administration."[23]

On June 9, Low directed Edgar Cortright, the director of NASA's Langley Research Center in Hampton, VA, to "undertake a study...to determine whether NASA has the capabilities to undertake the solution of non-aerospace technological problems; what types of problems NASA should consider; how NASA would work on those problems; and what implementing action would be required." Cortright was to report back in "approximately one month."[24]

Rethinking the Space Shuttle

One study of space shuttle development comments that during 1971 "pressures of financial stringency penetrated every aspect of the Shuttle program. Few high-technology development programs, if any, have been subjected to the kind of fiscal pressures and controls which the Shuttle Program endured, and it was during this period that they had the greatest impact on the design process." Indeed, "the fiscal and political environment influenced the detailed engineering design decisions on a month to month, and at times, a day to day basis."[25]

This pressure was already in the background as NASA's Associate Administrator for Manned Space Flight Dale Myers and his top associates decided in January 1971 to direct NASA's contractors to restrict their studies to a shuttle design that could meet all national security requirements. Myers convened a January 19–20, 1971, meeting in Williamsburg, Virginia, attended by all those involved in shuttle studies. At the meeting, Myers announced the requirements that would guide the remaining months of the ongoing shuttle studies. Performance requirements included:

- The ability to launch 65,000 pounds into a due east 100 nautical mile (nm) orbit, which equated with the ability to launch 40,000 pounds into a polar orbit, a national security requirement;
- Nominal cross-range of 1,100 nm, the least amount acceptable to the national security community; up to that point, NASA's contractors had been studying both a delta-wing orbiter design capable of 1,500 nm cross-range as well as one with straight wings and only 200 nm cross-range;
- Engines capable of generating 550,000 pounds of sea-level thrust. NASA had allowed its Phase A and Phase B contractors also to examine the use of an engine with 415,000 pounds of thrust, and most industry studies had preferred that option. Myers's directive removed that choice. The more powerful engine would be required to launch the heaviest NASA and national security payloads;
- The ability to return payloads weighing up to 40,000 pounds, also a national security requirement.[26]

Although the cross-range requirement had originated with the Department of Defense (DOD) and in the early stages of shuttle studies had been resisted by NASA, by this time many of those within NASA and industry involved in shuttle design efforts acknowledged the limitations of the straight-wing orbiter design, which was the preference of NASA's Max Faget, and recognized that a high cross-range vehicle had a number of operational advantages in terms of dissipating energy during return from orbit and of getting the shuttle orbiter to an appropriate landing site from various orbits. Myers's January 1971 directive eliminated the straight-wing design from further consideration; whatever shuttle design NASA would choose would have delta-shaped wings.

Phase B Study Results

The Phase B preliminary design studies of a two-stage, fully reusable shuttle being carried out by North American Rockwell and McDonnell Douglas continued until mid-1971. There were a wide variety of orbiter and booster designs considered and cost estimates also varied widely as industry engineers struggled to meet the requirements set out by NASA. There was one

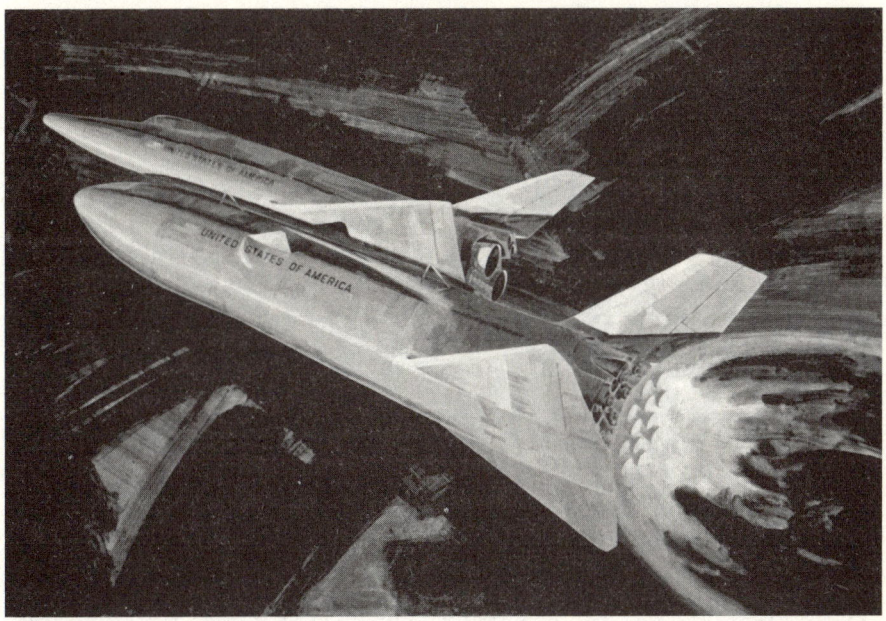

A 1971-vintage artist's concept of a two-stage fully reusable space shuttle. (Illustration courtesy of Dennis Jenkins)

constant: the shuttle designs being considered involved developing two large and expensive vehicles. For example, one version of the North American Rockwell orbiter was 206 feet long and had a wing span of 107 feet, about the size of the four-engine Boeing 707 commercial airliner then in use. The booster was 269 feet long and had a wing span of 143.5 feet, about the size of the Boeing 747 jumbo airliner then just entering commercial service; it had 12 rocket engines to provide the initial power to lift itself and the orbiter off the ground to what was called a "staging velocity." The vehicles would then separate and the booster's two-person crew would fly it, using a dozen air-breathing jet engines, back to a runway landing. The orbiter, also operated by a two-person crew, upon separation would fire its two engines of the same design as those used on the booster and accelerate to orbital velocity. One contractor's estimate of the cost of a fleet of four boosters and five orbiters flying 445 missions through FY1989 was $9.6 billion.[27]

Low Has Reservations

George Low, at that time still NASA acting administrator, reluctantly approved Myers's January decision to orient the Phase B studies to a full capability fully reusable shuttle, although he "had hoped that Myers would be able to come up with a phased program, where we would first develop the orbiter to be launched on a [expendable] Saturn IC stage"; a reusable booster stage would be developed several years later. Such an approach would mean giving up, at least for the first few years of shuttle operation, the goal of full reusability and the accompanying very low operating costs that had been at the core of the shuttle's attractiveness for potential users.

Low's thoughts about phasing the shuttle development program dated back several months. As he had argued with OMB in the fall of 1970 for full funding for the shuttle in the FY1972 budget, Low recognized that if future NASA budgets remained at the same low level as what was being proposed for FY1972, there was no way to fund the development of a two-stage, fully reusable space shuttle without taking up an unacceptable share of the overall NASA budget. On the day after Thanksgiving 1970, Low had called to his home Willis Shapley, Dale Myers, and Charles Donlan. Low noted that "we held the meeting because of our collective concern that the shuttle program, as now constituted (two-stage fully reusable vehicle), would cost more than we could afford on an annual basis in the middle of the 70's." He added

> A phased program, wherein we would first procure only the orbiter and launch it on a modified [Saturn] S-1C stage and only subsequently build a booster, would make more sense from the point of view of annual funding. It might also make more sense technically because we would face only one major problem at a time. At the same time, we could also adopt a Block I/Block II approach, wherein many of the "nice to have" features would be reserved for Block II and would not be incorporated into Block I. In other words, the Block I vehicle would have the potential for cross-range, but only Block II would fly with cross-range.[28]

These ideas did not get translated into NASA policy for some months; in the interim, studies of the shuttle went forward based on Dale Myers's January 1971 requirements.

Fletcher Makes the White House Rounds

In his first month in office, Jim Fletcher made the rounds of White House people concerned with space issues, and found them skeptical about the prospects of approving the shuttle as NASA was then planning it. On May 4, the day after he arrived at NASA, Fletcher had lunch with Nixon assistant Peter Flanigan and science adviser Ed David. Fletcher reported to Low that with respect to the shuttle "Ed David took a rather negative view" and was "beginning to get cold feet about deciding to go ahead this fall." David's reservations included that he was "not yet convinced of the economic value" of the shuttle. His primary concern, however, was "political"; he feared that if "we hit Congress with something this large at this particular time, it might become another SST." The Senate had canceled federal funding for that program in March 1971. David was also concerned, as were Fletcher and Low, about the large fraction of the NASA budget required for the shuttle program, "not leaving much room for [other] new programs along the way." By contrast, "Peter Flanigan was not negative on the shuttle at all and was willing to be convinced." Both David and Flanigan indicated that a NASA budget of $5 billion "was too large a peak for political salability." David had heard rumors of the president's interest in canceling both *Apollo 16* and *Apollo 17*, and reminded Fletcher, referring to his intervention with the president the preceding December to avoid canceling *Apollo 17*, "that his name was signed in blood on this one" and that NASA "had better fight very hard for it since he had stuck his neck out so far." David also suggested that "NASA ought to think seriously about alternatives to the shuttle."[29]

David's uncertainty about shuttle approval was also reflected in a brief memorandum regarding the shuttle he prepared at John Ehrlichman's request in early May. Ehrlichman's Domestic Council staff had identified the decision on whether to proceed with the shuttle as a major policy issue for 1971; Ehrlichman asked David to let him know "what has transpired and what the planning and time factor might be with regard to where we stand or plan for a decision on the shuttle." After summarizing the current state of the program, David told Ehrlichman that "there is no commitment to development of the space shuttle system by the Administration, but it is clear that a decision will be required this fall if the shuttle is to proceed." He added "personally, I am not yet in a position to support or to oppose the shuttle program. A great deal of important information remains to be assembled." David also noted that "with the encouragement of OMB," he was convening a "special panel" of the President's Science Advisory Committee "to conduct a detailed review of the space shuttle program" and that "Dr. Fletcher, the new NASA Administrator, shares my desire to take a hard look at the program."[30]

Fletcher also met with Bill Anders, executive secretary of the Space Council, who had been a confidential conduit to George Low of sensitive information regarding White House thinking on NASA issues. Anders was concerned with respect to filling the several-year gap in human space flights between the end of the Apollo and Skylab programs and initial space shuttle flights. Anders suggested to Fletcher that NASA could fill the gap by launching four left-over Apollo spacecraft on Earth-orbiting missions.[31]

An OMB "Bombshell"

On May 17, Fletcher received a letter from Don Rice, the OMB assistant director with space responsibilities, after the two had met on May 7. To Rice, the letter was the one action that should have made clear to NASA that it had to adjust its plans to the reality of continued budget constraints. At the meeting, Fletcher and Rice had agreed that NASA and OMB would "work together to develop a realistic NASA plan for the future." Rice agreed to provide NASA with a five-year projection of the budget that NASA was likely to receive, "allowing NASA and OMB management to consider the relative priorities of alternative programs." Rice realized "that the most difficult aspect of this approach to five-year planning would be to secure agreement between NASA and OMB on the range of overall agency totals which could be considered 'realistic' for the five-year period." He suggested that the then-current NASA budget of $3.2 billion per year might be an appropriate expectation. Rice later reflected that his letter "hit [NASA] like a bombshell," since a $3.2 billion annual budget for the next half-decade would make it impossible for NASA to develop the shuttle it was then planning. Dale Myers described Rice's letter as "the single stroke of the pen that knocked out the first stage booster, because you just couldn't get the two [stages] into the budget."[32]

At this point, NASA had been hoping that its annual budget might increase to $4 billion—perhaps even $5 billion—by the mid-1970s. Rice's suggestion that NASA plan for an annual budget at a significantly lower level was a sobering reminder that the agency had to constrain its future ambitions. Among other things, it implied that NASA could not budget more than $1 billion per year for shuttle development and still maintain the balanced program that was presidential policy.

By the time he finished his White House visits, Fletcher had come to share George Low's concern that NASA was not on a sustainable path with respect to the shuttle program. Fletcher, reported Low, "in fact, has asked for the development of a shuttle program that will fit within a $4 billion overall NASA budget." To Low, an important question was "is there a phasing of the shuttle or, alternatively, a cheaper shuttle that will not reach the very high expenditures in the middle of the decade?" Low worried that despite "pushing this point for about six months now, we have not yet been able to come up with an answer. Perhaps there is no viable answer." In a thought that would reoccur several times in the following months, Low suggested that perhaps there was no alternative but "the choice of foregoing the shuttle altogether for the 1970's and starting it in the 1980's."[33]

Shuttle Economics

NASA from 1969 on had stressed that the overriding objective of the shuttle program was to lower the cost of space launch and operations. By emphasizing the lower cost aspect of shuttle use rather than the new capabilities it would provide and its role in maintaining a human presence in space, NASA left itself open to having the shuttle evaluated on economic grounds. Indeed, the Bureau of the Budget (BOB) in March 1970 had asked: "Is full scale development of a new launch system to reduce the cost of payload in orbit economically justifiable?" That question set in motion a process of economic analysis that would parallel shuttle technical studies throughout 1970 and 1971. The impact of the economic analyses would come to be seen, in George Low's words, as "important and unfortunate."[34]

In his cover letter transmitting this question, BOB Director Robert Mayo told NASA to use a 10 percent "discount rate" in comparing future benefits of the shuttle program with the current investment required to obtain them and with other uses of that amount of funds.[35] One way of looking at the discount rate is as representing the equivalent for a government program of the interest that a private investment would have to earn in order to be justified. The discount rate used determined how much in future benefits was required to justify a current investment. A 10 percent discount rate was associated with a particularly risky government program, one with uncertain future benefits; this was the category in which BOB placed the space shuttle program.

Critical to showing a high level of future benefits from the space shuttle was a high flight rate, since each shuttle flight would save money compared to the use of an expendable launch vehicle to launch the same mission. Additional savings would also come from the reduced costs of payloads for many missions. Thus it soon became evident that to justify the shuttle in economic terms, there had to be very active U.S. civilian and national security space programs in the 1980s, and the shuttle would have to launch essentially all flights in those programs. The shuttle would also have to be complemented by a space tug so that the combination could carry out the many missions to geosynchronous and other high orbits. Thus developing a plausible "mission model" for future space activities was a key starting point for an economic analysis; the benefits from flying the missions in that model, when compared with a forecast of the one-time costs required to develop a shuttle and the anticipated lower costs of operating it, would allow a judgment of whether the shuttle was a justifiable investment at the 10-percent discount rate.

Initial Economic Studies

NASA from the time it received the BOB letter asking for a study of shuttle economics had planned to have that study carried out by an outside contractor with impeccable economic credentials. Tom Paine decided that "no one would believe NASA's results." There was, however, an interim in-house NASA study managed by Robert Lindley of the Office of Manned Space Flight. Lindley had been one of the first people within NASA to suggest that "payload effects"—the cost savings from reusing or repairing satellites and initially designing them for the less demanding characteristics of a shuttle launch—might be as important a benefit from shuttle development as lower launch costs. In terms of overall space program costs, payload development accounted for 80 percent of total costs; launch, only 20 percent, and thus lowering payload costs could have a greater impact than lowering launch costs. Lindley's study produced positive results, but Paine was correct. It had little credibility when it was submitted to the new Office of Management and Budget (OMB) in August 1970.

NASA selected Mathematica, Inc. of Princeton, NJ to lead an independent study of shuttle economics. Mathematica had been founded by prestigious economist Oskar Morgenstern of the Institute for Advanced Studies; there he had worked with mathematician John von Neumann to develop game theory, an approach to analyzing situations in which actors with conflicting interests pursue independent courses of action. Morgenstern had founded Mathematica to pursue practical applications of this approach. At

Mathematica, a young Austrian-born economist named Klaus Heiss was put in charge of the space shuttle study. Mathematica was supported in its analytic efforts by the Aerospace Corporation, which developed various mission and cost models, and Lockheed, which performed technical analyses of payload effects. The first meeting among Mathematica, NASA, and OMB took place on July 9, 1970; the firm's contract was for an 11-month study to be completed at the same time as the shuttle Phase B studies in June 1971.[36]

Results of Mathematica Study

OMB Director Shultz wrote NASA's Low on February 27, 1971, to reiterate the need for the economic analysis as an input into the shuttle decision process. Mathematica's final report arrived at OMB only on July 23, even though the report was dated May 31. A major reason for the delay in submitting the report to OMB were Mathematica–NASA interactions on what the report would say. Heiss commented that "there was tremendous pressure by NASA on us to come out and say okay, the two stage shuttle is a good thing." Drafts of the executive summary were read by NASA "12, 15 times" and Mathematica was asked "why do you need this paragraph in and this is gratuitous and this is sort of not warranted." Mathematica "stuck with it." The report's language, noted Heiss, was "carefully balanced." In particular, Heiss said that the report's wording was intended to imply that "it's not clear that the two-stage system is really the one that could accomplish this [fly the projected 1980s missions] at the least cost."[37]

The Mathematica report analyzed 23 different "scenarios"—forecasts of future launch demand. One of those scenarios was based on the mission model provided by NASA and DOD, which called for 736 missions over the period 1978–1990, an average of 56 missions per year. The costs of developing a shuttle and the facilities required to operate it was estimated by Mathematica to be $12.8 billion and the incremental cost of a shuttle launch was set at $4.6 million (in 1970 dollars). For this scenario, Mathematica estimated that there would be almost an $8 billion savings in launch costs and an $18 billion savings in payload costs compared to the use of existing expendable boosters. The report concluded that at a 10 percent discount rate, the "allowable" development cost (the maximum amount that would produce economic benefits) of a fully reusable space shuttle, including the space tug needed to move payloads to orbits the shuttle could not reach, was

- $15.4 billion for a flight rate of 56 launches per year, the NASA/DOD mission model;
- $14.6 billion "for the historic flight level of the unmanned U.S. space program of the 1960's," corresponding to 51 launches per year; and
- $12.9 billion at a flight rate of 46 launches per year.

Since NASA study contractors were estimating that shuttle development (not counting the space tug) would range between $9 and $12 billion, on a purely economic basis the case for investing in the shuttle clearly required a high flight rate for the 1980s, including not only NASA but national security and commercial missions; it would take over 40 launches per year for the investment in the shuttle to break even financially.[38]

While there was a relatively firm basis, based on past experience with developing large aircraft and other space systems, for estimating the costs of developing and testing the shuttle, there was no comparable experience in estimating the cost of operating a reusable space system or the payload savings that could result from its use. Yet those costs were an important element in determining the system's economic payoff. This was a significant flaw in Mathematica's analysis, since the cost per launch and payload savings inputs to its analysis were based on what one senior NASA engineer would describe only as an "optimistic guess."[39]

Mathematica went on to note that the fully reusable shuttle was potentially "not the only system" that could achieve these significant payload cost savings, that "other technically acceptable systems should be studied," and that "the task of identifying the *best* reusable Space Transportation System among all viable alternatives" had not begun. These points resonated with Don Rice and his staff at OMB, who had been pushing NASA, without much success, to look at alternatives to the fully reusable shuttle. The report commented that "the economic justification of a reusable Space Transportation System is independent of the question of *manned versus unmanned* space flight," suggesting that the shuttle should be seen as a means of achieving space program objectives, not an end in itself. The report said that a shuttle should be developed "only...if can be shown, conclusively, what it is to be used for and that the intended uses are meaningful to those who have to appropriate the funds, and to those from whom the funds are raised."[40]

The Mathematica report was hardly a ringing endorsement of the case for developing the two-stage, fully reusable shuttle. Its appearance in mid-1971 was an important added input to the moves already underway within NASA to make major changes in the agency's approach to developing the space shuttle.

Beginning to Explore Alternative Shuttle Designs

NASA in mid-1970 had issued, along with the two Phase B preliminary design contracts to North American Rockwell and McDonnell Douglas, three smaller study contracts to examine alternative shuttle designs. While the Lockheed and Chrysler studies were managed by the Marshall Space Flight Center in Alabama, a Grumman/Boeing study was managed by the Manned Spacecraft Center (MSC) in Houston. Houston used this study contract as a means of getting industry analysis of various ideas with respect to shuttle design emerging both from within NASA and from the various study contractors. In particular, Grumman began in late 1970 to examine a shuttle orbiter design in which the fuel tanks holding the very low temperature liquid hydrogen needed as fuel for the orbiter's advanced shuttle engines were moved from inside the orbiter fuselage to the vehicle's exterior and discarded when the fuel was expended. The idea of expendable fuel tanks was not new; several of the 1969 Phase A contractors had initially examined their use, but George Mueller in August 1969 had mandated that the studies from that point on would only consider fully reusable designs. Because liquid hydrogen is light in weight but accounted for three-fourths of the volume of shuttle propellant, the hydrogen tanks had to be large, and removing them from the vehicle's internal structure made possible shrinking the size and weight of the orbiter by some 30 percent. Having expendable fuel tanks, Grumman suggested, would lower orbiter development costs by more than $1 billion while not adding significantly to per flight costs; in the budgetary context of 1970–1971, this was a very attractive prospect. NASA on April 1, 1971, added an additional task to the two more in-depth Phase B studies, asking North American Rockwell and McDonnell Douglas to examine an orbiter with two external hydrogen fuel tanks.

As industry studies continued in mid-1971 and NASA's in-house engineering design team at MSC also focused on a smaller, lower cost, less complex orbiter, the idea of using a single large external and expendable propellant tank containing both hydrogen fuel and oxygen oxidizer gained increasing acceptance, and became a part of the consensus orbiter design that was emerging from Houston's efforts. The cost of discarding the external tank on each flight was seen as acceptable in terms of the overall costs of both developing and operating the shuttle, given the savings in development costs resulting from designing a smaller orbiter and the relatively low increase in the cost of each flight associated with using an expendable propellant tank.[41]

A Smaller Payload Bay?

One challenge in designing a smaller orbiter using an expendable propellant tank or tanks was maintaining the 15 × 60 foot payload bay required to launch the largest national security payloads. As NASA began to explore what it called the "drop tank" design, Dale Myers on May 25 wrote Grant Hansen, asking him to "determine if Air Force requirements [which included National Reconnaissance Office payloads] could be accommodated" in a 12 × 40 foot payload bay. He added that "if this is not possible, I would appreciate some thoughts as to what missions *must* exceed these dimensions and what alternate launch capabilities could be used."[42]

Hansen's reply was negative in tone, saying that a shuttle with the smaller payload bay would "preclude our full use of the potential capability and operational flexibility offered by the shuttle" and would "degrade the payload cost savings" that were an important part of the national security interest in the shuttle. Maintaining the Air Force Titan III expendable boosters to launch the largest national security payloads would mean that "the potential economic attractiveness and the utility of the shuttle to the DOD" would be "severely diminished." Hansen estimated with the shorter payload bay "71 of the 149 payloads forecasted for the 1981 to 1990 time period in option C and 129 of the 232 payloads forecasted in Option B of the mission model will require launch vehicles of the Titan III family." Hansen also noted the negative consequences of a narrower payload bay, especially in terms of the use of a large "transfer stage" to carry national security payloads to geosynchronous orbit.[43] This response reflected the continuing national security community pressure on NASA to maintain a shuttle design with a large payload bay, even as NASA was seeking an approach to minimize shuttle development costs.

A New Approach to Developing the Space Shuttle

In early June, George Low noted that "during discussions with Dale Myers, we had repeatedly decided to look for a phased program approach, but had been unable to establish the technical feasibility of such an approach." However, "during the past two or three weeks, because of the smaller orbiter made possible by moving the hydrogen tanks outside of the orbiter airframe," a phased approach was "beginning to look like a technical possibility. Dale Myers and his centers are moving out to establish technical details for this approach." He added, "in the meantime, von Braun's group is putting together NASA long range plans, incorporating the phased shuttle development, so that the peak funding during the 1970s need not exceed $4 billion." Fletcher and Low met with von Braun and his planning staff on May 26. At that meeting, von Braun had reported that "a reasonable shuttle alternative from both developmental and cost savings standpoints" appeared to be the orbiter with an expendable propellant tank, initially launched on an expendable booster, with "subsequent development of a fully reusable booster for use with that orbiter."[44]

This advice reinforced the sense that Fletcher had gathered from his White House meetings and exposure to NASA's thinking on the shuttle in his first month at NASA—that simultaneous development of both elements of a fully reusable two stage shuttle was not a viable approach in either budgetary or technical terms. He had told industry representatives on May 20 that he was not committed to the two-stage reusable approach. George Low had been thinking along the same lines since at least the preceding November. Fletcher, Low, and Myers decided in mid-June to investigate a phased approach; in doing so, they were in essence making a major decision—to give up hopes of developing simultaneously both elements of a shuttle system. Commenting on the influences that led to this decision, Fletcher suggested that three-fourths of the pressure for change came from financial constraints such as the $3.2 billion annual budget proposed in Don Rice's May 17 letter to Fletcher, and one-fourth from "our own technical concerns" regarding the fully reusable design. With respect to the latter concern, Charles Donlan, who had been designated shuttle program director at

NASA headquarters, later commented that "It was not until the phase B's came along and we had a hard look at the reality of what we mean by fully reusable that we shook our heads saying 'No way you're going to build this thing in this century.'...Thank God for all the pressures that were brought to bear not to go that route." Shuttle program manager Robert Thompson at the Manned Spacecraft Center in Houston agreed, saying that the fully reusable shuttle was "a bridge too far."[45]

On June 16, NASA announced that it would be "examining the advantages of a 'phased approach' to the development of a reusable space shuttle system in which the orbiter vehicle would be developed first and initially tested with an interim expendable booster." In addition, the NASA press release said, quoting Fletcher, "we have been studying...the idea of sequencing the development, test, and verification of critical new technology features of the system" such as its rocket engines, thermal protection, and electronic systems. Fletcher added "we now believe that a 'phased approach' is feasible and may offer significant advantages."[46] To give its contractors additional time to explore the implications of such an approach, NASA extended its study contracts, which were due to expire on June 30, by four months. Recognizing that with the adoption of the phased approach Mathematica's analysis of the economics of a fully reusable shuttle had been overtaken by events, NASA also gave Mathematica a contract extension to examine the economics of alternative shuttle systems.

Even as he announced this shift in plans, Fletcher was pessimistic about the future of the shuttle program. Writing to leading space scientist James van Allen, who was scheduled to testify before Congress in opposition to the shuttle, Fletcher suggested that "the political cards are so heavily stacked against this program...that no opposition from the scientific community is necessary. I think you are shooting at a dead horse...My feeling is that those who oppose the shuttle program—and there are good reasons for opposing certain portions of it—would be wise not to say anything now and let nature take its course."[47]

NASA's shift in direction did not please all potential users of the shuttle. In particular, in response to the June 16 announcement, Secretary of the Air Force Robert Seamans suggested that "because of the extensive effort that has gone into the evolution of the current shuttle baseline, I believe it is a system that can perform our needs." He suggested that phased development "would reduce the potential utility of the shuttle for DOD for an indefinite period." Seamans urged NASA to make every effort to stay with "a reusable booster and orbiter with the 15 × 60 foot payload bay." The continuing national security pressure on NASA to develop a shuttle meeting that community's needs was a factor that could not be ignored.[48]

Also responding to NASA's June 16 announcement that it was examining a phased approach to shuttle development, OMB's Don Rice on July 20 noted that "in light of continuing fiscal constraints," such a move was "very appropriate." But Rice wanted more than just deferring booster development. He urged NASA as it rethought its strategy for the shuttle to place emphasis "on defining approaches which will substantially reduce the overall investment cost of the future space transportation system." Rice wanted NASA to examine "alternative, lower cost systems" such as "expendable systems, partially expendable systems, the stage and one half concept." Rice noted that "while the economic analyses conducted to date have been very useful, they have not covered the full range of alternatives," a point that Mathematica was also making in its report. Rice wanted additional economic analysis with respect to alternative systems in terms of "estimated payload savings, realistic mission models, and alternative payload characteristics." He would later reflect on "the difficulty of getting any attention paid to alternative [shuttle] designs," noting "how hard it was to get an examination of alternative specifications of what you wanted to accomplish and the systems design that reasonably derived from that."[49]

With its de facto decision to abandon concurrent development of a shuttle orbiter and booster, NASA had once again adjusted its plans to the reality of what kind of post-Apollo space program the Nixon administration might be willing to approve. Already the ambitious plan set out in the Space Task Group had been stillborn, and NASA had abandoned hope of developing simultaneously a large space station and the space shuttle. With the adoption of an expendable propellant tank design and particularly a phased approach to shuttle development, NASA was making a third major adjustment, giving up for at least some years, if not forever, on its plans to develop a fully reusable two-stage shuttle. The June 16 announcement opened the door to an intense and broad-ranging effort in the next several months to identify a shuttle system design that represented the best compromise among several conflicting objectives. They included:

- keeping the annual shuttle development budget at or less than $1 billion per year, the budget level that would fit within an overall NASA allocation of $3.2 billion per year that Don Rice had suggested was an appropriate target;
- minimizing the cost of shuttle operations so that the cost per flight was as low as possible;
- maximizing the number of future missions that the shuttle would fly, in order to spread the cost of shuttle development and operation across a robust mission model and thus make the investment in shuttle development economically sound; and
- retaining the capabilities that would convince the national security community to commit to using the shuttle and would allow NASA to plan for a future shuttle-launched space station.

Between June and December 1971, there was "a frantic search for the most cost-effective and technically sensible" shuttle design; in that search there were dozens of alternate configurations and development approaches considered. In the words of one close observer, during those months "everyone became a shuttle designer."[50]

Chapter 11

A Confused Path Forward

NASA's move toward phased development of the space shuttle was a clear indication that the shuttle studies to that point had failed to identify a shuttle design that would both fit within the anticipated budget during the 1970s and that NASA's engineers were confident could be successfully developed. This realization put the space agency in a rather difficult position. A year had been spent studying shuttle designs that turned out to be neither politically nor technically acceptable. Yet Jim Fletcher and George Low were convinced that a decision to go ahead with the shuttle had to be made by the end of 1971 if NASA were to hold together the engineering design and development teams, both within the agency and in its contractors, required to undertake the shuttle program. They found themselves, six months before that deadline, without a specific shuttle design to put forward for approval. Fletcher and Low at several points in summer 1971 gave serious consideration to pulling the plug on seeking approval for shuttle development, instead putting forward some alternative, less ambitious human space flight effort during the 1970s. Ultimately they rejected this fallback position and decided to press forward with the attempt to find a shuttle program that both made sense in terms of NASA's future ambitions and was acceptable to the White House. Meanwhile, there were several related developments that would influence the eventual outcome of the shuttle decision process.

International Participation in the Shuttle

Once NASA in 1970 made the decision to defer the space station and focus its hopes on the space shuttle, potential European contributions to shuttle development became central to its planning for international cooperation in the post-Apollo program. Preliminary discussions between NASA and European space officials suggested that Europe might contribute up to 10 percent of the costs of developing the shuttle. Three possibilities for that contribution emerged. One was Europe building a portion of the shuttle airframe, such as the vehicle's vertical tail. Another was Europe contributing a "research and applications module," also called a "sortie module" or "sortie can," which would fit into the shuttle's payload bay and serve as a facility for scientist astronauts to carry out on-orbit research. The third, which became Europe's preferred option, was its taking on responsibility for the space tug needed to move payloads from the shuttle to higher orbits. This last possibility was troubling to the national security community, which was leery of depending on a foreign-made system to position its sensitive and often highly classified satellites. How to reconcile national security uses of the shuttle and international participation in the effort was a continuing issue.

The prospect of significant European participation in shuttle development had been troubling to Tom Whitehead for some time; as the March 1970 presidential statement on space had been drafted, he had been skeptical of any specific commitment to space cooperation. Whitehead, by 1971 the director of the new White House Office of Telecommunications Policy, was no longer working for Peter Flanigan on NASA issues, but occasionally became involved. In a February 1971 memorandum, Whitehead took a very skeptical position with respect to NASA's attempts to engage Europe in the U.S. post-Apollo program. He noted "NASA is aggressively pursuing European funding for their post-Apollo program. It superficially sounds like the 'cooperation' the President wants," but asked "is this what the President would *really* want if we thought it through?" Whitehead was concerned that "if NASA successfully gets a European commitment of $1 billion [to the shuttle program], the President and the Congress will have been locked into NASA's grand plans because the political cost of reneging would be too high." He suggested that "the kind of cooperation now being talked up will have the effect of giving away our space launch, space operations, and related know-how at 10 cents on the dollar."[1]

Issues of international space cooperation were discussed in a February 22 Oval Office meeting attended by the president, science adviser Ed David, and Nixon assistants Flanigan and John Ehrlichman. Excerpts from the conversation at the meeting include:

> EHRLICHMAN: "Well, Mr. President, you have urged that we get international involvement in the space program...[You have said] let's get an actor up there from a foreign government. But that's been interpreted to a large extent by NASA, as bringing foreign countries into the development of the space shuttle...To the extent that we have developed a very significant technology here which is all ours, it would seem to some of us that we risk giving that away for a pretty small amount of money."
>
> FLANIGAN: "I am all for getting their astronauts up there and letting them walk around...We get a lot of visibility. But I wonder if for a little bit of money we aren't selling our heritage."
>
> NIXON: "Well then, don't do it...What I want is symbolism. Nothing more. Give us a little cosmetics...What you are doing for cosmetics, do for cosmetics. Let's appear to be very liberal."[2]

There were continuing talks with Europe regarding participation in the shuttle program through 1971 and 1972, but the potential for international cooperation was not a major factor in the 1971 debate over whether to approve shuttle development. In June 1972 the United States would give Europe a "take-it-or-leave-it" choice of contributing a research and applications module. Europe decided to take that offer; the result was the program that came to be known as Spacelab.[3]

Incidentally, Nixon's "what I want is symbolism" comment certainly applied to another space cooperation initiative under discussion during 1971. This was the idea of a space rendezvous between a leftover Apollo spacecraft and a Soviet spacecraft. George Low had traveled to Moscow in January 1971 for a round of discussions with his Soviet counterparts regarding the feasibility of such an undertaking, which had little substantive justification but would help the Nixon administration symbolize a changed U.S.-Soviet relationship. Approval for what became the *Apollo–Soyuz* Test Project would have to come soon if the cooperative initiative were to move forward; funds for that effort would have to be allocated at the same time as a commitment to shuttle development was made.[4]

Bill Anders and the Space Council

Since the beginning of the Nixon administration in 1969, the National Aeronautics and Space Council at the principal's level had met infrequently, and its staff had not become closely involved in policy decisions related to human space flight. There had been proposals to eliminate the council during 1970, and in mid-1971, its future remained very much in doubt, although by that point the council staff members had developed good working relationships with their peers in the White House and the Office of Management and Budget and had become involved in policy choices related to NASA's robotic space science and application programs and aeronautics program and to other government aeronautics and space activities.

Although the council's executive secretary, Bill Anders, had carved out a personal role as adviser on space issues to the Office of Management and Budget's Deputy Director Cap Weinberger, he was somewhat frustrated by the marginal role being played by the Space Council and its staff in the decisions regarding future human space efforts. He shared his frustration with Arizona Senator Barry Goldwater; Goldwater relayed that concern to Richard Nixon during a June 17, 1971, Oval Office meeting:

> GOLDWATER: "I hate to burden you with a problem, but this young Bill Anders, who I think is one of the smartest boys around, I spoke to him today and I think he's thinking of quitting... He's in charge of the Space Council."
>
> NIXON: "I've got to use him someplace else. He's bright as a tack... Let's put him in that new deal [the 'new NASA'] where we're trying to develop the new, the water and all that sort of thing, the NASA management approach and so forth. Anders has got to be held."[5]

Anders told George Low in early August that he had "about decided that a staff function without an active council had reached its point of diminishing returns" and that "he might propose to the White House that the National Aeronautics and Space Council should be abolished." (He would make such a proposal in late 1972.) By the end of August, Anders had also become "extremely pessimistic" regarding White House staff attitudes, especially within the Office of Management and Budget (OMB) and the Office of Science and Technology (OST), with respect to the human space flight program. Meeting with NASA Administrator Fletcher, he suggested that NASA "drop the shuttle completely and focus on evolving a space station out of Skylab." Anders thought that "a vastly trimmed down manned space program, presented simultaneously with the closing of one of our centers, might make NASA more credible (and incidentally, more popular) with the 'White House.'"[6]

As it became clear that the FY1973 budget process would be contentious, Tom Whitehead suggested that Vice President Agnew loan Anders to Peter Flanigan's office to help Flanigan work with David, Don Rice, Al Haig (Henry Kissinger's deputy at the National Security Council), and Whitehead "to square away coordination between the various elements of the Executive Office and the White House in the space area." Whitehead thought that Anders's help in "getting the various Executive Office agencies working along the same track" and "tiding us over a bit of confusion among all the players" was "almost essential."[7] Although Whitehead's suggestion that Anders temporarily become part of Peter Flanigan's staff was not pursued, Anders was one of those over the next several months working to bridge the gap between the views of OMB and OST on one hand and NASA on the other, hoping to arrive at a sensible presidential decision on the space shuttle.

NASA and Applying Technology to Societal Problems

The White House idea of turning NASA into a general-purpose applied technology Agency persisted through summer 1971. A first draft of Edgar Cortright's internal study of broadening NASA's role into other areas of technology was ready by late June. The study concluded that there were indeed many areas where a high-technology approach was needed and that "NASA, and only NASA, could really bring many of these problems to an early solution." Problems addressed included "environmental monitoring, health care services, transportation needs, and urban needs," among others.[8]

Also in July, the White House Domestic Council established a subcommittee chaired by science adviser David to take a government-wide look at the issue of applying technological solutions to national needs; NASA participated in that effort. George Low's understanding of the Domestic Council plan was "to first worry about the problems, and to define the organization to solve the problems later on." NASA supported the subcommittee's efforts in the areas of short-haul air transportation systems, a global environmental system, a wide-band communication system, and, to a lesser degree, ground-based transportation and health services. Low found working in the interagency framework "very frustrating in that other agencies are, by and large, impossible to work with. Everybody wants to play in their own little sandbox and, particularly, wants to keep NASA out of that sandbox." Low was becoming convinced that "if anything is to be done" with respect to applying technology to national problems, "it will have to be done by NASA under a Presidential mandate." Fletcher agreed with Low, indicating his "pessimism about the possible success of the current interagency exercise."[9]

William Magruder, who had been in charge of the canceled supersonic transport program at the Department of Transportation, moved to the White House as a "special consultant to the president" to take charge of what was becoming known as the "New Technology Opportunities" program. Magruder broadened the scope of the effort beyond looking at the technical issues that had been the focus of the David subcommittee, examining issues such as balance-of-trade, antitrust, and other nontechnical aspects involved in the kind of effort being contemplated. Magruder's goal was to define a number of major initiatives to be included in President Nixon's January 1972 State of the Union Address. He told NASA Administrator Fletcher that he "had the distinct impression that the President would like to give the whole job to NASA." Responding to that possibility, Fletcher drafted a letter to Magruder in late September, suggesting that "it might be wise to place the 'soluble' [solvable?] problems in NASA, but begin to develop new capabilities in other agencies, particularly those in which NASA is not particularly qualified. NASA might be given the responsibility for outlining a government-wide program through its systems analysis capability."

Fletcher and Low by this time had decided that it would be a good move for NASA to try to take the lead in this new area. Discussing tactics on how to achieve that outcome, Fletcher thought that NASA should "not enlarge our contacts much beyond response to requests...I am convinced that it has to be their [the White House's] initiative if we are to succeed in this venture; although we can respond with enthusiasm when asked, if we do too much politicing behind the scenes, word will get around somehow." But, he added "this seems like a 'sporty course' for something we really think NASA and the country ought to undertake...The risk we take is that the President will decide to go some other route because of influence from various other vested interests. At this point in time I am inclined to take that chance." Low agreed with Fletcher's ideas, suggesting that NASA "would play the role of the reluctant bride, but would be prepared to jump in if the opportunity presented itself."[10]

Uncertainty

As NASA entered the Fiscal Year 1973 "budget season" by submitting its proposed budget on September 30, 1971, its prospects remained very uncertain. On one hand, NASA might both receive presidential approval for a major new space effort, the space shuttle, and at the same time be asked by the White House to take on a much broadened role as the nation's applied technology agency. On the other hand, NASA might not gain approval for the shuttle program, and some other approach might be chosen for the new technology opportunities effort. The short-term stakes for the organization's future were very high.

Enter a New Actor: the "Flax Committee"

Both OST and those dealing with space issues in OMB had recognized at the start of 1971 that a decision on whether or not to proceed with NASA's space shuttle program would almost certainly be made during the consideration of NASA FY1973 budget request that fall. The two organizations decided that it would be useful to have an external group assess the technical aspects of the NASA shuttle concept and the shuttle's relationship to likely future space program activities. The Mathematica study was already underway to assess shuttle economics; there was a felt need for a parallel technical assessment by an expert group outside the government. To take on this task, science adviser David decided to constitute a "special panel" of the President's Science Advisory Committee (PSAC).

David discussed a potential chair for the panel with George Low in April. David's first inclination was to have Gene Fubini be the chairman. Fubini was a well-known "gadfly" in the Washington technical community, respected for his technical acumen and insight, but notorious for his arrogance and quick temper. Low suggested that "Fubini would not be the right person because (a) he is too flighty and jumps too much from one thought to another, and (b) he really does not have any aircraft background to contribute to this." Low suggested several alternatives among the respected leaders of the aerospace community. Low soon learned that David had selected one of his suggested people, Alexander Flax, as chair; Flax was head of the national security think tank Institute for Defense Analysis and from 1965 to 1969 had been Assistant Secretary of the Air Force for research and development and simultaneously director of the National Reconnaissance Office. Low observed that "Ed David really wants to be helpful in establishing this panel."[11]

Fubini was made a member of what quickly came to be known as the "Flax committee." Other members of the group included both individuals with the aerospace engineering background needed to assess shuttle design and potential users of the shuttle's capabilities. The Flax committee scheduled its first meeting for August 13–15 at the National Academy of Sciences summer conference center in Woods Hole, Massachusetts. In advance of the meeting, OMB's Don Rice sent David a list of questions that he hoped the Flax committee would address. In turn, David forwarded the questions to NASA so that NASA was prepared to respond as they interacted with the Flax committee. Among the ten questions were:

- "What are the major high risk technology areas?"
- "What trade-offs are implicit in the manned vs. unmanned operation of the space shuttle?"

- "How sensitive are the cost benefit relationships to changes in the rate of activity and assumptions about the lifetimes of unmanned satellites?"[12]

Also in advance of the Woods Hole meeting, Fubini met with Low. He told Low that the committee would work "under the premise that the Science Adviser cannot recommend to the President the cancellation of the United States manned space flight program"; the group should thus ask "what is the manned space flight program that (a) is acceptable from a budget point of view; (b) is clearly a step beyond what has already been achieved; and (c) is not dead-ended?"[13]

After the first Flax committee meeting, Dale Myers reported that in synthesizing his committee's initial reactions Flax had suggested that the committee "felt that...the broad implications of the shuttle had not yet been addressed." The committee thought that "the peak funding and perhaps the total funding" for the shuttle program were too expensive to be acceptable and that "the program goes on too long before there is a payoff." (At this point, NASA was still advocating in its external presentations the two-stage, fully-reusable shuttle design.) Flax had added that the committee's view was that "a smaller, lighter, shuttle vehicle seemed to be in order" and that the "60 × 15 payload compartment may be larger than we need." With respect to the reactions of other committee members, Myers observed that "Fubini felt that a program this long should have something spectacular every four years." Committee member Richard Garwin from IBM "felt that there should be more effort on big, dumb boosters, parachute recovered boosters, automatic landing, unmanned flights of the shuttle, and much greater use of the data relay satellites where you can get rid of the men in the orbiter...All actions done by men in orbit can be done by men on the ground using a data relay satellite and good data transmission." Myers said that "most of the other members of the Committee were not very outspoken." Myers paraphrased the committee's questions:

1. "Will the users really design cheaper payloads to take advantage of the volume and weight capability?"
2. "Will the launch rate stay at 40/year or greater during an era when satellite life is increasing?"
3. "Is there a firm requirement for a 15 × 60 payload compartment?"
4. "Why crossrange?"
5. "Why not build the booster first?" and
6. "Why not unmanned shuttles, with automatic landings, or parachutes?"[14]

Low met with David on August 24 to discuss the results of the committee meeting. It was David's feeling "that the Flax Committee (with Fubini leading the pack) is going to come in with some interesting options" that "include a shuttle of about $5 billion total investment running about $1 billion per year." Low told David that NASA "was thinking along similar lines but so far had not discussed them in any detail with the Flax Committee."[15] These budget targets, no more than $1 billion per year and a total investment cost for shuttle development of $5 billion, were by the end of August beginning to be recognized by NASA as defining the limits within which a proposal for presidential approval of the shuttle had the best chance of success. The question facing Fletcher, Low, and their engineering colleagues was whether a shuttle worth having, particularly one that met both NASA's institutional needs and the requirements set out by the national security community, could be developed within those budget constraints. It was clear that the reusable two-stage shuttle, with development cost estimates of $10 billion or more, could not be pursued on a $1 billion per year budget; this had led NASA internally to abandon that option during the summer.

NASA Gets a Low Budget Target

The shuttle's fate, and with it the character of the post-Apollo NASA, would be decided against the background of continuing problems in the U.S. economy and their impact on the federal budget; those problems included a high rate of inflation, an unacceptable level of unemployment, stock market declines, deficits in international trade, and threats to the U.S. dollar as the basis for international financial transactions. The economic policies pursued by the Nixon administration in its first two years in office had not been successful in reversing these negative trends.

Meeting with his budget and policy advisers on July 23, Nixon made a tentative decision to cut the budgets of "civilian agencies" by 10 percent from their FY1972 levels. Among those agencies was NASA; it was noted at the meeting that by canceling the last two Apollo missions and the NERVA program and not starting the shuttle program, some $1.32 billion could be saved in FY 1973.[16] Nixon's July decision to reduce the budgets of the civilian agencies of the government by 10 percent was reflected in the budget targets provided to NASA by Cap Weinberger in an August 2 letter. Rather than the $3.2 billion per year budget that Don Rice in his May 17 letter had indicated was a reasonable expectation, Weinberger told NASA that its budget authority for FY1973 would be $2.835 billion, with a limit on outlays during the year of $2.975 billion. Weinberger told NASA that "the President emphasized at his 1973 budget planning meetings that it is essential that we do not exceed the overall budget totals he has decided upon" and thus that NASA was required to "submit your budget at or below these figures."[17]

Fletcher, upset at these low budget targets, quickly met with Weinberger to get a fuller understanding of the thinking behind them. Weinberger first told him that "things were tough all over" and that NASA should come in with a budget at the targeted level. But when Weinberger was told that the $2.8 billion budget target recommended by his staff meant the end of human space flight, he told Fletcher that it might be possible to bring the budget up to the $3.2 billion level of FY1972.[18]

Another Rethinking of the Space Shuttle

Fletcher and Low found themselves in a quandary with respect to how to proceed in developing NASA's budget request for FY1973, given the low budget target and the feedback from the Woods Hole meeting. NASA's internal planning, led by Wernher von Braun, had been assuming a FY1973 budget at approximately the $3.7–$3.8 billion level, fully $1 billion above the OMB target, with the budget gradually increasing to $4 billion per year; von Braun was warning that it would be very difficult to carry out an ambitious shuttle program with that budget outlook. This was a message that Fletcher and Low did not want to hear. They briefed David and Flanigan at the White House on the von Braun plan and got a noncommittal reaction. Flanigan "stated that a transportation system alone, without clearly understood objectives (a transportation system to where?) would not get the support" that NASA needed. The "only bright spot" in the meeting was David's comment "that it would be inconceivable for any President at any time in this age to stop manned space flight."[19]

Low's assessment of NASA's situation as of August 1971 was sober.

> In the course of planning for Fiscal Year 1971 and 1972, we assumed each year that the current year was a bad year but that things would get better on the next year. In effect, we pushed a funding bow wave ahead of us. My view today is that we should no longer build a future program on promises, but that we should, instead, assume that the NASA budget will be confined to the $3 billion level (say up to $3.3 billion) for the next several years...We should drop the shuttle right now and come up with a different manned space flight program.
>
> This program should be based on an evolutionary space station development, leading from Skylab through a series of research and applications modules to a distant goal of a permanent space station. We should also set for ourselves a distant goal of a lunar base. The transportation system for this manned space flight program would consist of Apollo hardware for Skylab; a glider launched on an expendable booster for the research and application modules; and finally, the shuttle but delayed 5 to 10 years beyond our present thinking.[20]

Alternative Space Transportation Approaches

As he thought through the path that NASA should follow, Low in August had outlined for his senior colleagues his ideas on "the desired Space Transportation System for the 1980's." He rejected both developing a full-sized, two-stage reusable shuttle and pursuing an approach using a reusable "ballistic" spacecraft, a capsule without wings, launched on an expendable booster and parachuted back to Earth. This approach was based on modifying the two-person Gemini spacecraft used in the mid-1960s to carry six or more people, and was becoming known as "Big-G." Low focused on a "mini-shuttle approach wherein a smaller shuttle vehicle is first developed and launched on an expendable booster. The recoverable booster and the desired full-scale shuttle are phased in at a later date." The mini-shuttle would have a 15 × 40 foot payload bay (so that it could carry research and application modules and eventually space station modules), upgraded Saturn J-2 engines, and a disposable hydrogen/oxygen propellant tank. It could carry 40,000 pounds (rather than 65,000 pounds) to a due-East orbit. The initial version of this mini-shuttle would make use of existing technology in its on-board electronic systems. It would be propelled to staging velocity by an expendable booster, then fire its engines to accelerate to orbital velocity. In successive stages of development, an advanced shuttle rocket engine could replace the J-2 engines and a recoverable booster, not necessarily piloted, could be used.

Low also considered a "glider approach." This vehicle, Low suggested, would be winged but smaller, with a 12 × 40 foot payload bay, carrying 30,000 pounds to orbit. It would have small engines for maneuvering in orbit and to initiate return to Earth, but no large rocket engines. It would be propelled to orbit by an expendable booster. Low did not have "enough information in hand to lead to a firm recommendation between the glider approach and the mini-shuttle approach." He suggested that NASA "take a further look at both the glider and the mini-shuttle before we decide to limit our work to one or the other." Low noted that Dale Myers preferred the mini-shuttle approach, suggesting that a glider would only send astronauts "whirling about the Earth" to no evident purpose, while he, Fletcher, and von Braun favored the glider.[21]

Phase B Extended and a New Approach to the Shuttle

While Low and others at NASA headquarters in Washington were considering a glider or smaller shuttle, NASA's engineers, particularly at the Manned Spacecraft Center (MSC) in Houston, and the shuttle study teams at NASA's contractors were examining alternative ways of moving forward with an affordable program while still retaining the operational capabilities of the full-size shuttle in terms of payload capture and cross-range. They also were resisting the phased development approach advocated by NASA headquarters, which involved postponing development of a reusable booster. The engineering team at MSC had during the summer converged on an orbiter design that seemed to meet all requirements. This design, designated MSC-040, had triangular-shaped delta wings, a 15 × 60 foot payload bay, and a single expendable propellant tank containing both hydrogen fuel and oxygen oxidizer mounted under the airframe belly. That design would turn out to be the basis for the shuttle orbiter that eventually would be approved for development.

On September 14, the NASA human space flight leadership called its contractors together at MSC to discuss various changes in study direction. One shift of lasting significance was that all contractors were told to use the MSC-040 orbiter design as the baseline for further studies. NASA also directed the contractors to study a "phased technology" approach as a way of reducing short-term and peak funding requirements for shuttle development. In this approach, a "Mark I" orbiter using the MSC airframe design would be developed first; it would use existing technology, mainly derived from the Apollo program, as much as possible in areas such as thermal protection, on-board electronic systems, and rocket engines. After a few years, a "Mark II" orbiter would be developed, incorporating advanced technology in terms of the thermal protection needed for demanding cross-range missions, a new high-pressure space shuttle main engine, and state-of-the art electronics. Only the Mark II orbiter would be able to meet all NASA and national security requirements. This approach spread out over a longer time the total cost of orbiter development, thereby lowering the annual budget required but resulting in a higher overall program cost.

There was a major political problem with the Mark I/Mark II approach to shuttle development. NASA in July had announced that it had selected the Rocketdyne Corporation to develop the new rocket engine for use in the shuttle. But the Mark I orbiter would use modified Apollo-vintage J-2 engines, and thus the new engines would not be needed for several years; this put NASA in a potentially embarrassing position vis-à-vis Rocketdyne, a California-based company, at a time when the White House was eager to see all possible high technology government contracts go to that state. (Rocketdyne's main competitor for the engine contract, Pratt & Whitney, with its rocket engine facility in Florida, lodged a formal protest regarding the contract award, implying that political considerations had played a role in Rocketdyne's selection.) There was some merit to that argument. When Richard Nixon learned of the protest and the possibility that the engine contract might be taken away from a California company, his request was "if the contract does not go to the California firm, the White House should review the matter and possibly cancel the contract."[22]

What Shuttle to Propose?

NASA Administrator Fletcher also met with science adviser David on August 24, separately from the David–Low meeting on the same day, to discuss the best approach to getting a shuttle program approved by the White House. While he had developed a sense of trust with David, he was not sure "how much we could trust OMB" if NASA came in with a budget proposal at the $3.2 billion level through the 1970s, as Don Rice had suggested in May, given the low budget target OMB had provided in early August. Fletcher, as had Low, told David of NASA's internal discussions of a shuttle program that could be carried out for less than $1 billion per year and a total investment of $5 billion. David advised him to keep that thinking confined to a few people within NASA, and instead to let David propose the low-cost shuttle to OMB, with NASA resisting that proposal. This gambit, thought David, would put NASA in a "better bargaining position." As Fletcher saw it, the issue was that "OMB can't entirely be trusted to commit to any kind of program and that if we [NASA] agreed too easily to the low-cost shuttle, they might try to work us down to a smaller budget yet." David felt that "there was not anyone in OMB who could be completely trusted—not that they were dishonest, but that their sole function was to put a ratchet on the budget and [they] couldn't make a commitment on anything."[23] Clearly, OMB and OST were not working harmoniously with respect to space issues, and Davis was angling for increased influence within the White House decision process.

As usual, the NASA budget request for FY1973 was due at OMB on September 30. What to request with respect to the space shuttle was a major issue in formulating the budget proposal. Three basic alternatives were considered:

1. A lower-cost glider or mini-shuttle, as suggested by Low in August;
2. The Mark I/Mark II shuttle. Low noted in mid-September that "OMSF [Office of Manned Space Flight] is now focusing on a 'phased technology' Shuttle, wherein today's technology is used in a Block I version and any more expensive, more sophisticated subsystems are phased in at a later date";
3. No shuttle at all. Low had considered such an option during August, and during the final two weeks before submitting the budget recorded that "Fletcher and I debated whether we should not forego the shuttle entirely and develop instead some alternate manned space flight program."

Fletcher and Low finally decided to include "something like the Mark I/Mark II shuttle," but to delay the start of shuttle development by approximately six months to "give us more time to reach final decisions on the configuration."[24] As it had done a year earlier, NASA would request presidential approval of the space shuttle without being able to specify the shuttle design being approved.

As NASA prepared its budget request, its leaders also concluded that the space agency needed "a new justification for manned space flight." Low recognized that "during the past year we begged the issue by stating that we needed a new transportation system—a space shuttle—and it just happened to be manned." It was now time to "try to justify manned flight in its own right." This shift in emphasis likely reflected James Fletcher's influence at the top levels of NASA. Discussing the issue, the NASA leadership "agreed that the main justification for manned space flight is the 'American presence in space' and not the fact that man can twirl knobs better than machines can."[25]

The NASA FY1973 budget request contained funding proposals at three levels—a "minimum recommended program" with budget authority at $3.385 billion and outlays at $3.225 billion, close to the FY1972 levels; an "alternate recommended program" at $3.54 billion in authority and $3.305 billion in outlays; and a budget at the OMB target of $2.975 billion in outlays. To reach that lowest figure, NASA would cancel *Apollo 16* and *17* and not start the space shuttle, actions that would cause "irreparable damage." NASA argued that "this nation must continue to fly men in space. Man will fly in space, and for many reasons the United States should not forego its responsibility—to itself and to the free world—to take part in this great venture." This was a theme that would reappear throughout the next few months and would be important to convincing Richard Nixon to approve the shuttle.

The budget letter contained "detailed figures on the effects of the various program options on unemployment," an issue that NASA knew was of increasing interest to the White House. The letter pointed out that "the NASA program is labor intensive: small changes in program funding now have immediate, and nearly one for one, effects on employment." With respect to the space shuttle, NASA told OMB that "the single most important consideration" was how to "achieve it with lower annual funding in view of continuing severe budget constraints." NASA's plan was "to select the optimum shuttle configuration, considering both technical and budgetary factors, next spring, to select the contractor next summer, and to proceed then with development leading to the first manned orbital flight in 1978 or 1979."

NASA described "a promising configuration that would substantially reduce the funds required prior to the first manned flight." This was the Mark I/Mark II approach, although those designations were not mentioned in the letter. NASA noted that "the reduction in shuttle development funding will come at the expense of somewhat higher operational costs initially and of delaying by several years the full realization of the planned operational capabilities." The NASA letter argued that delaying a decision on the shuttle configuration past spring 1972 "would be expensive and unsettling to the aerospace industry, which is forced to maintain a capability to respond to the government until our decisions are reached."

The budget letter noted that NASA was proposing "run-out costs, in future years, at or below the FY1973 level" of $3.385 billion in new budget authority that NASA was requesting. This "constant budget" was an invention of NASA's Willis Shapley, and allowed NASA to propose a six-year program "at a level between $500 million and one billion [per year] *below* the financial plans presented to OMB and the Congress at the time the FY1972 budget was approved." NASA's hope was that OMB would not only approve a new start on the shuttle, but would agree in advance to a constant budget level for the next several years that would provide the space agency with some stability after the several prior years of budget uncertainty.[26]

In its budget submission, NASA was asking OMB, and ultimately President Nixon, to approve development of a still rather fuzzily defined space shuttle. Fletcher, Low, and their associates fully realized that it would not be easy to get OMB and then the president to approve that request.

What NASA Did Not Know

As NASA submitted its budget request, it did not know that President Nixon had already made a tentative decision that NASA's budget for FY1973 would be in the $3.3–$3.4 billion range, with a strong bias toward approving space shuttle development. That decision originated with OMB Deputy Director Cap Weinberger and had been approved by the president. But that information had not been communicated to the White House technical and budget staffs, much less to NASA, and thus had little impact on NASA's interactions with OMB and OST over the next four months.

Weinberger had discovered as he met with Fletcher on August 5 that the budget target for NASA that had been recommended by his staff would mean the eventual end of the U.S. human space flight program. This was not an acceptable option to Weinberger, and on August 12 he had sent a thoughtful memorandum to President Nixon. That memorandum is worth quoting at some length.

> Present tentative plans call for major reductions or change in NASA, by eliminating the last two Apollo flights (16 and 17), and eliminating or sharply reducing the balance of the Manned Space Program (Skylab and Space Shuttle) and many remaining NASA programs.
>
> I believe this would be a mistake.
>
> 1. The real reason for sharp reductions in the NASA budget is that NASA is entirely in the 28% of the budget that is controllable. In short we cut it because it is cuttable, not because it is doing a bad job or an unnecessary one.
> 2. We are being driven, by the uncontrollable items, to spend more and more on programs that offer no hope for the future: Model Cities, OEO [Office of Employment Opportunity], Welfare, interest on the National Debt, unemployment compensation, Medicare, etc. Of course, some of these have to be continued, in one form or another, but essentially they are programs, not of our choice, designed to repair mistakes of the past, not of our making.
> 3. We do need to reduce the budget, in my opinion, but we should not make all our reduction decisions on the basis of what is reducible, rather than on the merits of individual programs.
> 4. There is real merit to the future of NASA, and its proposed programs. The Space Shuttle and NERVA particularly offer the opportunity, among other things, to secure substantial scientific fall-out for the civilian economy at the same time that large numbers of valuable (and hard-to-employ-elsewhere) scientists and technicians are kept at work... It is very difficult to re-assemble the NASA teams should it be decided later, after major stoppages, to re-start some of the long-range programs.
> 5. Recent Apollo flights have been very successful from all points of view. Most important is the fact that they give the American people a much needed lift in spirit (and the people of the world an equally needed look at American superiority). Announcement now, or very shortly, that we were cancelling Apollo 16 and 17... would have a very bad effect, coming so soon after Apollo 15's triumph. It would be confirming, in some respects, a belief that I fear is gaining credence at home and abroad: that our best years are behind us, that we are turning inward, reducing our defense

commitments, and voluntarily starting to give up our super-power status, and our desire to maintain our world superiority. America should be able to afford something besides increased welfare, programs to repair our cities, or Appalachian relief and the like.
6. I do not propose that we necessarily fund all NASA seeks—only that if we decide to eliminate Apollo 16 and 17, that we couple any announcement to that effect with announcements that we *are* going to fund space shuttles, NERVA, or other major, future NASA activities.
7. I believe I can find enough reductions in other programs to pay for continuing NASA at generally the $3.3–$3.4 billion level.[27]

Richard Nixon read Weinberger's memorandum and wrote on it a cryptic message, "I agree with Cap." He also wrote "OK" next to point 7. What exactly he meant by these notations was not clear. A month later, one of Haldeman's staff provided some clarification, telling OMB Director Shultz that the "the President read with interest and agreed with Mr. Weinberger's memorandum of August 12, 1971, on the subject of the future of NASA. Further, the President approved Mr. Weinberger's plan to find enough reductions in other programs to pay for NASA at generally the 3.3–3.4 billion dollar level."[28]

If the NASA leadership had known of Weinberger's memorandum and Nixon's response, they likely would have been much less nervous about the outcome of NASA's negotiations with OMB over the FY1973 budget. The Weinberger memorandum represented one of several points in 1971 when it could be said that a decision to approve space shuttle development had been made. But if there was such a decision made on the basis of the memo, it was to approve the *idea* of a space shuttle, not a specific shuttle design. NASA in its budget submission left itself vulnerable to continued debate over what shuttle design merited presidential approval by its admission that it would take another six months to make the configuration choice. That debate was not long in coming.

Chapter 12

Debating a Shuttle Decision

With the September 30, 1971, submission of NASA's Fiscal Year (FY) 1973 budget request to the Office of Management and Budget (OMB), the process that would most likely result in an up-or-down decision on approving space shuttle development or pursuing some less ambitious post-Apollo human space flight program entered its final stage. Even though Cap Weinberger had suggested in his August 12 memo that the NASA budget should be set at a level that would allow the beginning of space shuttle development and President Nixon had indicated "I agree with Cap," that news had not been communicated to lower levels in the White House or to NASA. The result was a fragmented and contentious debate over shuttle approval.

Over the next three months, as NASA's Jim Fletcher and George Low sought support in the national security community and the aerospace industry for NASA's position that a "full capability" shuttle orbiter able to launch all U.S. payloads should be approved for development, OMB's Rice and his staff were joined by science adviser Ed David, his Office of Science and Technology (OST) staff, and David's advisory Flax committee in opposition to an ambitious space shuttle program. Others in the Executive Office of the President, such as Tom Whitehead, now at the Office of Telecommunications Policy, Bill Anders at the Space Council, and Peter Flanigan and his assistant Jonathan Rose in the White House, tried to mediate the conflict between NASA and OMB/OST and to move the process toward a productive outcome.

The debate over what should be the next step in human space flight, although conducted in the context of decisions with respect to the president's FY1973 budget proposal, was not intimately tied to NASA's budget level for that year, since NASA had requested only $228 million for the space shuttle in its budget submission. Rather, it was fundamentally about what kind of space program the United States would carry out in the coming decade and beyond. Approving a new start on the full capability shuttle would imply that once the shuttle was flying the United States would use it as the basis for an active national space effort, even if it were far less ambitious than what the Space Task Group had proposed in 1969. Choosing a more modest shuttle option or an alternative to the shuttle such as an unpowered glider would signify the Nixon administration's intent to reduce even more NASA's post-Apollo ambitions with respect to the future of human space flight.

Battle Lines Are Drawn

The first step in the budget decision process was a set of "hearings" at which NASA met with those in OMB and other parts of the Executive Office of the President working on space issues to present the thinking behind NASA's budget request. The OMB hearings took place on October 6, 7, and 8, with the human space flight program being the focus on October 7. Low led the NASA delegation to the hearings and reported that they were generally positive and friendly in tone. Low told Rice and the OMB staff that at the budget level of approximately $3.2 billion that NASA had requested, the agency's priorities with respect to human space flight were, in order: flying Skylab, starting shuttle development, flying the *Apollo 16* and *17* missions, carrying out a docking mission with a Soviet spacecraft, and adding additional "gap-filler" missions using left-over Apollo hardware. But if NASA were held to a budget less that $3 billion as proposed in the OMB August target, said Low, NASA would give priority to flying the remaining two Apollo missions and would "re-examine what to do about future manned space flight."[1]

There was one exception to the collegial tone of the October 7 hearing. William Niskanen, head of the OMB Evaluation Division, made two provocative suggestions. The first was to finance the NASA program through revenues raised by selling the material returned from the Apollo missions to the Moon. The second was to have the private sector, using its own financial resources, develop the next generation space transportation system, and then sell transportation services to NASA and the Department of Defense (DOD) to recoup that investment. The staff of the Evaluation Division was in general even more skeptical of the value of NASA's human space flight program than were OMB's mainline budget examiners, and Niskanen a year earlier had been an opponent of any hint of a commitment to the space shuttle. Niskanen was a student of conservative economist Milton Freidman at the University of Chicago and libertarian in political and economic philosophy, advocating a very limited role for government; this perspective made him an opponent of major new government programs justified on economic grounds.

Reacting to the first of Niskanen's ideas, the Space Council's Anders commented that "unless the President himself ordered us to consider the selling of lunar material for profit, we should not even discuss the subject because it would be embarrassing to the Administration." With respect to a commercial shuttle, Low told Niskanen that "the reason for not doing it is that it simply won't work: if the idea is to cancel the space program, this might be a way to do it." Whereupon, Low reported, Niskanen and his staff left the room, "but not without making a fairly strong threat about NASA's budget."

OMB's Rice was personally sympathetic to the idea of not going ahead with the shuttle, noting that Niskanen and his group "wanted to kill it, just kill it off," but that he had "adopted the view fairly early on that while that may well be the desirable thing to do from a broader public interest point of view, I didn't believe that the President was in fact going to take the country out of manned space flight." This skeptical perspective would lead Rice in the following weeks to seek the least costly shuttle program possible, putting him in direct opposition to NASA's insistence that only a large space shuttle made sense. Rice's background was in the type of systems analysis that had been developed at the Rand Corporation (which he would later head) and applied during the 1960s under Robert McNamara at the DOD; both Rice and Niskanen had worked at DOD then. Rice had pushed NASA to take a "whole systems" approach to evaluating the shuttle and possible alternatives for space transportation in terms of their cost-effectiveness. This approach, with its emphasis on quantitative measures, gave primary importance to economic factors. NASA's Fletcher, believing that the primary reason for going ahead with the shuttle was the new capabilities it would offer and the intangible values associated with human space flight, was skeptical of a systems analysis approach to evaluating the shuttle, believing that "you can make systems analysis prove anything you want...it was just a lot of hocus-pocus," since it could not assign a quantitative value to those new capabilities or to the value of the shuttle in terms of national prestige and international cooperation. NASA's Willis Shapley described Rice as "a strong believer in the religion of systems analysis" who took the shuttle issue on "to prove...that you could really get a better decision by really giving this the full systems analysis treatment." Shapley, himself a long-time Bureau of the Budget staff person before joining NASA, observed that "analysis becomes a weapon in a controversy rather than a search after some abstract kind of truth."[2]

A Skeptical Perspective

In preparation for the director's review, Dan Taft, head of the OMB space unit, prepared a lengthy paper on "The U.S. Civilian Space Program: a Look at the Options" that at its core reflected the budget office's long-held skeptical view of the value of human space flight. The options paper recognized that the "key issue" with respect to FY1973 budget decisions was "the future role of man in space." It noted that "historically, [the] primary reason for man in space has been the international technological image of the U.S.," and asked "are our historical reasons for keeping man in space still sufficient to justify keeping man in space? If so, how much extra should the U.S. be willing to pay for manned flight relative to an unmanned program which could produce comparable scientific and practical benefits?" The paper observed that

> The contrast between President Nixon's [March 7, 1970, space] statement and former President Kennedy's 1961 address on space provides an interesting illustration of the change in attitude of the national leadership towards

the space program. In contrast to President Nixon's call for a balanced and orderly space program, former President Kennedy's address conveys a sense of urgency, international competition with the Soviets, and the battle "between freedom and tyranny."

With the passage of time and the achievement of successful programs, the importance of international competition and world opinion seems to have diminished...And yet, the significance of international competition in space is not over...With the Soviets steadily continuing their manned space program, would the U.S. be willing to terminate manned space flight?[3]

The paper declared "the objective of the future space transportation system is to reduce the total investment and operating costs (launch vehicles plus payloads) of space operations." New capabilities provided by the shuttle, a point that NASA was advocating, did not enter into OMB's evaluation. The paper concluded that "at the 10% discount rate, all of the shuttle options save less systems cost" than a new expendable launch vehicle. To Taft, "only the need to resupply a Space Station begins to justify investing in a reusable shuttle capability." Recognizing the reasoning behind NASA's 1970 decision to give priority to shuttle development, the paper presciently commented: "In a sense, a commitment to a shuttle is an implicit commitment to a subsequent space station program." Given that station development had been deferred to an undefined future date, this perspective led to the conclusion that there was no justifiable reason for approving shuttle development in the FY1973 budget.[4]

Taft's paper set out "an illustrative future space program." That program would complete the remaining scheduled Apollo and Skylab manned space flights, but would "postpone the space shuttle indefinitely." It acknowledged "the possibility that the shuttle might become more economically attractive and be initiated in the 1980's," but until then a slow-paced human space flight program would use expendable launch vehicles. With the deferral of shuttle development, NASA's Marshall Space Flight Center in Huntsville, Alabama, could be closed; reducing NASA's institutional base by closing Marshall was a particular OMB objective. Taft's proposed program would reduce NASA's budget to $2.6 billion per year by the mid-1970s.[5]

Weinberger Disagrees

As Caspar Weinberger prepared for the director's review of OMB staff recommendations with respect to the NASA budget, he was uncertain about what precisely Richard Nixon had meant when he wrote "I agree with Cap" on Weinberger's August 12 memorandum. On October 19, Weinberger asked Nixon's chief of staff Bob Haldeman to have the president clarify his intent. Haldeman discussed the issue with the president the same day.

Haldeman: "So you do want to cancel [Apollo]16 and 17?"
Nixon: "Yes, I do want to cancel them, and do other things."

Haldeman: "Do we want to follow his point, coupling [the cancellation with] the announcement that we're going to fund the space shuttle?"

Nixon: "That's right, and let the other two [Apollo missions] go...the other two shots....I just don't think we should take the risk of a possible goof off in the damn thing."

Haldeman: "The other thing you could do is postpone them."

Nixon: "Postpone and then cancel them, if you could get away with it...That's right, no shots in '72."[6]

Haldeman reported to Weinberger that "the President did agree with your feeling that a public announcement now of the cancellation of Apollo XVI and XVII would have a bad effect," but nevertheless Nixon "does want to eliminate" the missions, "at least in calendar year 1972," and had directed that "steps should be taken immediately to implement that decision." Nixon also agreed with Weinberger's point that "if we decide to eliminate Apollo XVI and XVII that we couple any announcement to that effect with announcements that we are going to fund space shuttles, NERVA, or other major future NASA activities." Weinberger in reply told Haldeman that *Apollo 16* "was scheduled for mid-March 1972 to secure data on some of the oldest events on the moon" and that *Apollo 17* was scheduled for December 1972 (after the presidential election, as agreed in January 1971; it seems as if neither Nixon nor Haldeman remembered that agreement) and would provide "the first opportunity for a geologist astronaut to visit the moon." He noted the modest cost savings if the missions were canceled, and said that if both missions were eliminated "we would lose about 3,800 jobs by June 1972 and about 6,200 by December 1972." Weinberger concluded, repeating an idea from his August 12 memo, that "if it is decided to cancel either one or both Apollo missions, it could be announced that we were doing so in order to concentrate our resources on other NASA-planned high-priority space objectives, because the prior Apollo moon explorations were so successful."[7]

The combination of Weinberger's thinking in his August 12 memo, Nixon's reaction to the memo, and Nixon's October guidance to Weinberger boded well for eventual shuttle approval. But the battle that would lead to that approval was just beginning. In a 1977 interview, Cap Weinberger recalled that "the OMB staff was against the shuttle, and I was for it, and that produced a very substantial amount of discussion and debate...In previous years it apparently was not necessary to get to a decision point, but in that particular year [1971]...it was an active part of the budget, and after the various arguments and presentations, I supported it...After the so-called director's review, I indicated to the staff that I disagreed with them, that I would recommend that particular item, and they protested and we had many more arguments." Weinberger added that "I had personal feelings that this was something we should be doing...If I had not taken as strong a position as I did in favor of it, that ultimately just the force of inertia would have prevailed." A major influence on Weinberger's positive views on the shuttle and the space program in general were his interactions with Bill Anders, who "stoked his [Weinberger's] enthusiasm for space at any opportunity."[8]

Weinberger's support for the shuttle at the director's review did not translate into approval of the Mark I/Mark II shuttle program that NASA was proposing. At the director's review the alternatives being examined by the Flax committee, including a smaller shuttle and a glider, were also discussed. While Weinberger indicated that he would recommend to the president going ahead with some form of a shuttle program, he was told by his staff "if we wanted to do it, it could be done less expensively, I was delighted to hear that, and...they went back and worked with NASA to work out a different configuration. In other words, did we have to have a vehicle that could carry that much on each trip, or couldn't we have a smaller one that could make more trips. Why did we need one this big?" He added "I could have cut it off at the director's review, and insist that we are going to do it the way NASA wants it. But the opportunity to do it at a lower cost on additional analysis appealed to me." Even so, Weinberger added, "I never had any doubt in my own mind but, one way or the other, I wanted to do it." Weinberger's willingness to let the OMB staff consider shuttle designs different from the full capability shuttle that NASA was proposing, itself still not very well defined, opened the door to a very confusing debate, as multiple versions of a shuttle were examined. According to Weinberger, the OMB professional staff had "a certain degree of pride. The staff doesn't like to be overridden, and they were firmly convinced that they were right, and that this was not a thing the government should be doing."[9]

NASA, as was the practice, was not formally notified of the results of the director's review, and thus did not realize that there had been a decision that some sort of shuttle program would be recommended to the president by Weinberger. On a very confidential basis, Anders told Low only that the director's review was characterized by "general discussion only, and no decisions were made." With respect to the shuttle, Anders reported that "there appears to be a general acceptance that the United States must stay in the manned space flight business," but "the assembled group still felt that the Mark I/Mark II Shuttle was too much" and that "the Flax Committee has recommended something less than the Mark I/Mark II."[10] The results of the October 22 OMB director's review thus only muddied the waters with respect to eventual shuttle approval. Weinberger had indicated that he would recommend developing a space shuttle to President Nixon, but his willingness to allow his staff to define an alternative, less expensive shuttle design than what NASA was proposing meant that the character of the shuttle he would recommend was still very much up for grabs.

Flax Committee and a Space Glider

One of the other key actors over the November–December period was science adviser Ed David. When Fletcher first discussed the shuttle program with David in May 1971, he had found David rather negative with respect to the wisdom of moving ahead with that program, at least as NASA was then defining it. By late September, David was still "negative about the wisdom of a shuttle in view of the political pressures, both from the public and the Congress," was "receptive to the idea that we needed some kind of a new booster for the '80's," but was "not sure that the shuttle is the way to develop that booster." David's main concern was "assuming that we do need a manned space program, is the shuttle the best program we can come up with?"[11]

As discussed in the previous chapter, David had created an ad hoc panel of the President's Science Advisory Committee (PSAC), chaired by Alexander Flax of the Institute of Defense Analysis, to advise him on the shuttle program. Flax made an interim report on the committee's deliberations to David on October 19. Flax noted that the committee "was far from achieving any degree of unanimity regarding the attractiveness, utility, desirability, or necessity of the shuttle system or, for that matter, on the virtues of alternatives to it." He added that "most of the members of the Panel doubt that a viable program can be undertaken without a degree of national commitment over the long term analogous to that which sustained the Apollo program. Such a degree of political and public support may be attainable, but it is certainly not now apparent." He observed that "planning a program as large and as risky (with respect to both technology and cost) as the shuttle, with the long-term prospect of fixed ceiling budgets for the program and NASA as a whole, does not bode well for the future of the program." Given this reality, "most Panel members feel that serious consideration must be given to less costly programs which, while they provide considerably less advancement in space capability than the shuttle, still continue to maintain options for continuing manned spaceflight activity, enlarge space operational capabilities, and allow for further progress in space technology."

The 23-page summary of the committee's views made a number of sage observations regarding the shuttle program and possible alternatives:

- The space shuttle program...represents a technical synthesis which, to a remarkable degree, integrates into a single vehicle system and proposed mode of operation the means for potentially achieving improvements and advances relevant to virtually all foreseeable space program objectives...If an enthusiastic, optimistic, and expansionary view is taken of the probable growth of the nation's military and civilian space programs over the next twenty years...the development of the space shuttle as proposed by NASA is undoubtedly the most important and valuable major new space program which could be undertaken at this time.
- The Mk I/Mk II approach [is] a very dubious course of action.
- The Panel has been impressed by the large amount of effort which has been put into the cost analysis of the shuttle program and into the study of the economic cost–benefit justification for the program. Nevertheless, we are unconvinced that such analyses have sufficient credibility to serve as

a primary basis for deciding to undertake such an expensive and high-risk program... We believe that a decision to proceed with a program such as the space shuttle should be based on an assessment of new capabilities it would provide and whether they serve the national purpose to a degree sufficient to justify the costs.
- Prudent extrapolation of prior experience would indicate that estimated development costs may be 30 to 50 percent on the low side. In consideration of the technical and operational risks and uncertainties and the sensitivity of potential savings from the space shuttle system to the resulting uncertainties in development, production, and operational costs, it is clear that there is little incentive to embark on the program if the aim is primarily to achieve the possible economic benefits. Rather, if the program is to be undertaken, it must be primarily for the purpose of acquiring new capabilities, aggressively pursuing new opportunities in space, and assuring continuing national leadership in space technology and space activity.[12]

It is not clear how widely these observations were known at the upper echelons in the White House or much influence they had on the shuttle decision process, although they certainly were incorporated into OMB and OST attitudes. The Flax committee had considered several alternatives to the full capability shuttle that "met to some degree the requirements for a continuing manned program and for further progress in space and space vehicle technology." But NASA in its interactions with the committee took the position that none of the alternatives merited approval. The NASA position "effectively left only two alternatives for the next ten years: either (1) proceed with the shuttle now or soon, or (2) drop manned spaceflight activity after Skylab A and the possible Salyut visit... Most of the Panel rejected these 'all or nothing' views."[13]

The committee gave particular attention to three alternatives, although several others were briefly mentioned in Flax's report. The three were:

1. *To defer decision on the shuttle*: "This alternative contemplates the possibility that with further studies, analyses, and technology advancement, uncertainties and risks in the shuttle technical and cost areas can be reduced to a point of greater acceptability and that the national climate for generating the requisite of commitment to the program may be improved over the next year or two."
2. *To develop a ballistic recovery system*: – This approach would forego "technological innovation in launch and recovery" by developing a spacecraft that would be launched, as had Mercury, Gemini, and Apollo, on an expendable launch vehicle and would return from orbit using parachutes to slow it for a water or land landing, rather than flying back to a runway landing. The leading candidate was the "Big Gemini," which was "billed as a growth version of the Gemini recovery capsule, but, which to all intents and purposes, is a new spacecraft design based on Gemini technology." This new spacecraft could carry nine people to orbit and

back, rather than the two-person crew during the mid-1960s Gemini program. Such an approach, thought the committee, "would be justified only if a slow-paced manned spaceflight program were contemplated (2 to 4 manned flights per year)."

3. *To develop an unpowered but winged orbital vehicle, a "space glider"*: Such a vehicle would have a much smaller cargo bay (10 × 20 feet rather than 15 × 60 feet) and less payload capacity (10,000 pounds versus 65,000 pounds) than the NASA-proposed shuttle. The space glider would be launched on an expendable booster, probably a version of the Titan III, and be able to return from orbit to a runway landing. The committee was positive in its view of the glider because such a vehicle "could provide a more convenient and lower cost means of recovering men from space missions; it would insure greater safety in unscheduled aborts from orbit; it would entail making progress in reentry vehicle technology... It would allow the acquisition of experience in payload recovery,... maintenance, refurbishment and replenishment; and finally, it would lead to the accumulation of a body of data on the techniques and operational characteristics and costs of reusable orbital recovery vehicles."[14]

As the Flax committee was carrying out its deliberations, NASA's Fletcher and Low had met with Flax and, separately, Fubini, to get as much perspective as possible on the committee's thinking, on the grounds that both the committee's views and Fubini's individual perspective "will have a lot to do with the kind of shuttle we will be able to sell to OMB." In Low's judgment, Flax was "in complete agreement with NASA's position, but has a great deal of difficulty with the scientists on his committee," while Fubini "is pushing strongly for a glider as opposed to an orbiter." These meetings led to an October directive to manned space flight head Dale Myers that "he must study all of the alternatives in great detail so that those that are discarded will be discarded not through arm-waving, but through facts." Myers and his space flight teams at the Manned Spacecraft Center and Marshall Space Flight Center were convinced that some form of large shuttle was the only reasonable path to pursue. Even after Low's directive, they spent little time studying concepts such as the space glider or the "Big G," which they did not believe were productive ways to proceed. Myers would later comment "we probably were the guys that were dragging our feet."[15]

Mathematica and the TAOS Concept

In late October, there was an unexpected external intervention in the shuttle decision process. Mathematica, the Princeton-based company that NASA had selected to carry out an independent analysis of the cost-effectiveness of a space shuttle, had submitted its final report with respect to the two-stage fully reusable shuttle concept in summer 1971. But this submittal came after

NASA had already decided to abandon the fully reusable approach and to examine alternative shuttle designs. The Mathematica report had made the point that while the two-stage fully reusable design was marginally cost-effective, it was not necessarily the optimum shuttle design from an economic perspective. NASA had decided to extend Mathematica's work to examine the economic dimensions of the alternate shuttle concepts during the extended study period.

The person in day-to-day charge of the Mathematica effort was economist Klaus Heiss. During September, Heiss visited with two of the study contractors, McDonnell Douglas and Grumman, to get information on the alternatives being examined by the two companies. Each firm had been allowed by NASA to allocate 10 percent of its study effort to a shuttle concept in which an orbiter with an external propellant tank was carried to orbit by the power of its own engines combined for the initial few minutes of the flight with the much higher thrust of one or two conventional rockets attached to the orbiter or its propellant tank, all engines firing from the launch pad on. McDonnell Douglas had labeled its concept rocket-assisted takeoff (RATO); Grumman, thrust-assisted hydrogen-oxygen (TAHO) takeoff. Heiss got cost and other data on those configurations and other designs under study from the two companies and also from a third study contractor, Lockheed. He used that information as input to the complex computer-based model that Mathematica had developed for its shuttle-related work. (Heiss did not interact with the fourth shuttle study contractor, North American Rockwell, because he "was convinced from the beginning that they would win the competition." Apparently, he was aware of the bias toward awarding the shuttle contract to a California firm.) Heiss discovered that "whatever space program [mission model] you used and even if you changed interest rates from five percent to ten percent to fifteen percent, again and again and again the same configuration came out" as economically preferred—the RATO/TAHO approach. He labeled this concept TAOS (thrust-assisted orbiter shuttle).[16]

Heiss faced a dilemma with respect to what to do with that finding. The second Mathematica report was not due until the end of January 1972, and by that time a decision on the space shuttle design might have been reached. He was aware of the conflicts between OMB and NASA over shuttle approval, and thought that his findings could help resolve the debate. Heiss told Bob Lindley that "I'm going to do something that maybe I'm not supposed to, but since it's so clear...I'm going to write up my conclusions in fifteen or twenty pages and send that to [NASA Administrator] Fletcher." Heiss chose not to route his analysis through Dale Myers, believing that Myers and his team were still trying to find a way to get approval for some version of a two-stage shuttle in order to have enough work to occupy both Houston and Huntsville.[17]

The Heiss memorandum, dated October 28, 1971, was titled "Factors for a Decision on a New Reusable Space Transportation System." It was co-signed by Oskar Morgenstern, Mathematica's head. The memo led off with three conclusions, all emphatically stated in capital letters:

1. *A REUSABLE SPACE TRANSPORTATION SYSTEM IS ECONOMICALLY FEASIBLE*, ASSUMING THAT THE LEVEL OF UNMANNED U.S. SPACE ACTIVITY WILL NOT BE LESS THAN IT HAS BEEN ON THE AVERAGE OVER THE LAST EIGHT YEARS.
2. *AMONG THE MANY SPACE SHUTTLE CONFIGURATIONS SO FAR INVESTIGATED, AND WHICH ARE DEEMED TO BE TECHNOLOGICALLY FEASIBLE*, A THRUST ASSISTED ORBITER SHUTTLE (TAOS) *WITH EXTERNAL HYDROGEN/OXYGEN TANKS EMERGES AT PRESENT AS THE* ECONOMICALLY PREFERRED CHOICE.
3. THE DEMAND FOR SPACE TRANSPORTATION *IN THE 1980'S BY THE NATIONAL AERONAUTICS AND SPACE ADMINISTRATION, THE DEPARTMENT OF DEFENSE, BUT PARTICULARLY BY COMMERCIAL AND OTHER USERS IS THE BASIS FOR THE ECONOMIC JUSTIFICATION FOR THE TAOS PROGRAM.*

The memorandum noted that "in part the choice of the current Mark I-Mark II approach was forced by a peak funding requirement for space shuttle development of, say, $1 billion per year. In this approach, however, several important parts of the system would be postponed in some configurations *while other configurations with the same total funding requirement assure an early IOC [initial operating capability] date not only of the space shuttle alone, but also of the space tug.*" It suggested that "the non-recurring costs of TAOS are estimated by industry to be $6 billion or less" and noted that the TAOS configuration would promise "the *same capabilities* as the original two-stage shuttle." Heiss added that "the most economic TAOS would use the *advanced orbiter engines* immediately" and that "the cost per launch of TAOS can be as low as $6 million or less." The memo thus concluded that "TAOS practically assures NASA of a reusable space transportation system with *major objectives achieved.*"[18]

It is difficult to judge the impact of the Heiss memorandum on the ultimate decision regarding the shuttle program. A version of the TAOS concept was indeed the shuttle configuration selected for development. Heiss suggests that "as soon as Fletcher read" his memo, he concluded "that's the solution to this problem" and "ran all over town with it," going first to OMB and saying "this group of outside people finds that this makes sense, so why do you fool around with this negative attitude?" Fletcher himself suggested that the Mathematica work reflected in the memo "did influence the decision in the sense that if it had come out negative, we'd have been in trouble." But, he added, "the Mathematica stuff all along was really supportive of our decision, not determinative."[19] The memorandum did not make its way to those managing the shuttle studies at the Manned Spacecraft Center, who were interacting directly with their study contractors in evaluating the final shuttle configuration. The TAOS concept they ultimately adopted likely reached them through those interactions, not as a result of the Heiss intervention. As the shuttle debate continued in the last two months of 1971, there were few, if any, references in the interactions between NASA, OMB, OST, and the White House to this memorandum or to the economic analyses it reported. It seems as if the Mathematica memo was one, but only one, of the influences that converged on the concept of a "thrust assisted" shuttle orbiter as the best technical choice for a new space transportation capability.

Rallying Support for the Shuttle

Soon after submitting its budget proposal to OMB, the NASA leadership set about seeking support for the space shuttle from the aerospace industry, members of the Congress, and the DOD. Fletcher and Low in October held a meeting with the top leadership of the companies involved in the shuttle studies to explain to them NASA's current plans and the reasoning behind them. The executives welcomed this information, and told NASA that "it was imperative to move out with the shuttle as soon as possible." Low noted that "the meeting was frank and open, and perhaps the first of a kind in NASA history."

With respect to the Congress, Low thought that "support will be a little more difficult to obtain because there really is no center of power within either the Senate or the House." As NASA leaders began to visit individual members of Congress, they discovered that since they were "now deeply involved in so many other things, that most members would just as soon not hear about NASA until after the first of the year."[20]

Seeking DOD Support

Fletcher lunched with Deputy Secretary of Defense David Packard on October 19. It was Packard, with a background in high-technology industry, who was the most senior DOD official dealing with space issues, rather than Secretary of Defense Melvin Laird. Fletcher found that Packard had "two general points to make" with respect to the shuttle. The first was that Packard personally felt "very uneasy" about the three requirements laid down by those at lower levels within DOD that were driving the shuttle design—"the cross-range requirement, and payload [weight] requirement, and the size requirement." Packard "felt that the cross-range requirement might have been an artificial one" and "that if it were causing difficulties, it could easily be modified." Fletcher assured Packard that the payload bay width "came primarily from NASA and not the Air Force, but that the length probably came from the Air Force." Packard "knew quite well which program caused the length difficulty" (the successor to the then highly classified Hexagon photo-intelligence satellite program) and suggested "that something could be done about it." Fletcher and Packard also agreed that "the payload [weight] requirement was somewhat arbitrary at this point."

The fact that Packard suggested that there was flexibility in the national security requirements had levied on the shuttle was likely surprising to Fletcher, since both the DOD representatives on the DOD/NASA Space Transportation Systems Committee and Air Force Secretary Bob Seamans and Assistant Secretary for Research and Development Grant Hansen had been adamant in their pressure on NASA to meet those requirements. DOD support was seen by NASA as a key to White House approval of the shuttle, and this had been a major driver of NASA's determination to pursue a shuttle design that met all the DOD requirements. So Packard's flexibility was not exactly an asset in the final stages of the shuttle debate; rather, it suggested that the top leadership in the Office of the Secretary of Defense, including Packard and Director of Defense Research and Engineering Johnny Foster, were not yet fully committed to supporting NASA's preferred shuttle on national security grounds.

By October 1971 NASA's engineers had come to recognize that "whereas the initial request for a 1500 n.m. [nautical mile] cross range capability originated as an Air Force requirement, it became evident with increased depth of study that a substantial degree of aerodynamic maneuvering capability at hypersonic and supersonic speeds is fundamental to the operation of the orbiter." So even if DOD were to relax its cross-range requirement, NASA would still want a delta-winged orbiter that was capable of such maneuvers.[21] In contrast, Packard's suggestions that "something could be done" about the DOD-imposed payload bay length requirement of 60 feet and his view that the payload weight was "arbitrary" would influence NASA's thinking during the final stages of negotiations over shuttle design.

Packard's second point was that NASA's approach to selling the shuttle "was all wrong." Packard suggested that the real reason for the shuttle "has to do with national security and an intangible thing which might be called 'men's presence in space.'" Packard suggested that he and Fletcher put together a team "to develop a rationale for the shuttle." He thought "it is probably desirable to write a letter to the President indicating recent progress on the shuttle development, incorporating perhaps the rationale ... and asking for a chance to explain it to him in person." In reporting this conversation to Low, Fletcher indicated that it was important for NASA that any rationale developed on the basis of NASA–DOD effort "includes all of the essential points that NASA wants to make" and "doesn't become unduly military in its flavor."[22]

Following his conversation with Fletcher, Packard quickly convened a meeting to begin the process of developing a revised shuttle rationale. Attending it were Fletcher and Low from NASA, Packard, Foster, Seamans, and Under Secretary of the Air Force John McLucas, who was also the director of the National Reconnaissance Office. As a result of the meeting Foster, thought to be a recent convert to supporting the shuttle, was charged with preparing a paper to be used within the executive branch and the White House to support the shuttle. Low suggested that "this single event is probably the most important in NASA's ability to move out with this [shuttle] program. Without DOD support, we would not have been able to do it. If Fletcher and Laird together can go to the President to seek Shuttle support, we just might get approval."[23]

Working with the White House

In the aftermath of the OMB director's review on October 22, George Low focused his attention on making NASA's case to OMB, the Flax committee, and the science adviser's office, while Fletcher was working to gain the support of those at the policy and political levels at the White House. Having observed Low's actions from outside NASA, Klaus Heiss would later comment that "George Low was the key creative figure...when crucial decisions came, they were George Low's decisions...He had enough engineering and other judgment that people respected him...He was crucial to NASA at that time."[24]

Fletcher lunched with Whitehead and Anders on November 5 "to discuss how NASA could better relate to OMB and the White House staff." Whitehead felt that "there are only two ways to bring together the diverging views of the White House staff." One was "to let Peter [Flanigan] act as our [NASA's] advocate in White House circles and, in particular, with the President. To do this we would have to keep Peter better informed." A second option was "to call the essential constituents together and thrash through what we felt is a program responsive to the President's desires (which, incidentally, coincide with the national interest)." If this were done, "when the time came for a battle with OMB or to confront the President with alternatives, there might be a reasonable degree of support from the White House staff." Whitehead thought that "at the present time Henry [Kissinger] is very much an advocate of space, but more particularly the Manned Space Program; that Peter and Ed [David] were neutral; and that OMB, as possibly represented by George Shultz, is in favor of a continued reduction, year by year, in NASA's total budget." Whitehead believed that "these views need to be reconciled in favor of an agreed upon national program which makes sense."[25]

In mid-November, Anders gave Low a rundown of the positions on the space shuttle of key White House players:

Weinberger: is a real space buff. The only one in OMB really positive toward the NASA program. Causes Rice to over-balance in the opposite direction. Everybody lower in OMB is negative.
Rice: the most knowledgeable opposition comes from Rice. Feels that NASA is out of control; however, he will probably support a glider on a TITAN III.
Ed David:...noticeably quiet, measuring his words, and repeatedly saying he represented science and that other factors are involved...Not really plugged into the President.
Flax: Fubini is really running the Flax Committee. Flax apparently states that no program as large as the Shuttle will gain continuing support. We need a less costly program...Flax is driving David toward the glider and not vice

versa...David will support the Orbiter with the parallel staged pressure fed booster [the TAOS concept] if Flax so recommends.

Whitehead: Whitehead could be helpful in making Flanigan a meaningful communication link to the President...Whitehead's main motivation now is to improve the Fletcher/Flanigan communications link. Whitehead can be extremely helpful in selling the NASA desired Shuttle approach...Believes in a $3.5 billion NASA.

Rose: [Jonathan Rose was Whitehead's replacement as Peter Flanigan's assistant tracking space issues] is the California unemployment buff in the White House. Tries to be helpful and sees Flanigan all the time. He defers to Whitehead when Whitehead is present.

Flanigan: states that the Shuttle story is improving; however, he is by no means convinced that there should be a Shuttle. Is strongly influenced by Whitehead, Rose, and David.

Peterson: [Peter Peterson was White House international economic counselor] is the most negative of all about NASA. Perhaps the most dangerous opposition we have within the White House. Believes that the space program is the place to take money to stimulate technology. Asked why not take $1 billion out of space and who needs manned space flight.

Ehrlichman: asked the question, "Given the public attitude on space, why not put money in aeronautics?" However, he is very much concerned about the aerospace industry and will probably go along with whatever OMB/OST/Flanigan recommend.[26]

Continuing Discussions of the Path Forward

The Space Shuttle extended study contracts expired on October 31 and NASA for a second time extended the contracts for another four months. The focus of the continuing effort was still the Mark I/Mark II orbiter sequence with various means of boosting it into orbit. Low reported in early November that "the shuttle configuration is beginning to be focused on a considerably smaller orbiter with external hydrogen and oxygen tanks (but with the same payload size and weight), and with a pressure-fed recoverable booster that might be parallel staged... It may be possible to buy a shuttle for an investment cost (including the high pressure [space shuttle main] engine of less than $5 billion with cost per flight of the order of $10 million... Solid rocket motors also look promising." On the basis of these study results, Low suggested that "if NASA were left to its own devices, I think we are now in a position to make a decision to move out with contractor selection and to proceed with the work. I believe it is important to get a decision on this soon and within the FY1973 budget process, unless the decision is the wrong decision." The wrong decision, in Low's view, "would be a glider on a Titan III." But NASA had "not yet done adequate analysis of the glider," primarily due to the resistance from Dale Myers and his space flight team, and thus NASA "should not absolutely discard it. The next several weeks will tell the story." Also, observed Low, "NASA is not left to its own devices, and it appears that everyone wants to have their fingers in the pot." Low also noted that "the only organized effort to either support or not support the shuttle is the so-called Flax Committee." It seemed to Low to be important for NASA to influence the thinking of Flax and his associates.[27]

Final Flax Committee Meeting

To that end, Low and Dale Myers met with Flax on November 12, in advance of the November 17–18 meeting of Flax's committee. It was still Flax's view that "the next manned space flight program should involve some technological advance and that operational costs are not all that important," since whatever system was chosen would not be flown frequently. Throughout the shuttle decision process, neither NASA nor the White House and its external advisors gave careful attention to how much it would cost to operate the shuttle; this would turn out to be the program's Achilles heel. Flax also suggested "that it is Ed David's view that the shuttle is dead unless David saves it and that the only way he can save it for us is by supporting something that is much less than the previously proposed shuttle; namely, the glider."

Low by this point had developed a diagram showing a cost curve that compared the development and operating costs of various shuttle designs and the space glider; he was to use a version of that diagram and the trade-off between development and operational costs it depicted as a major selling tool in his frequent meetings during November and December. He drew the curve on Flax's blackboard and made the point that NASA "now had some very interesting developments in the range of development costs between $4.5 billion and $6 billion, with operating costs around $10 million per flight." Flax thought that "some" of his committee members might be willing to support a shuttle with those characteristics, and Low and Myers agreed that Myers would present, for the first time, "the small orbiter, together with the parallel-staged pressure-fed booster" at the November 17 Flax committee meeting. This would be the first time the TAOS configuration, the shuttle design ultimately selected, would be briefed to anyone outside of NASA. Low agreed to come to the second day of the committee's meeting to make the point that "we can buy the kind of shuttle that we are now proposing within a reasonable total NASA budget, while still at the same time having a strong science and applications program."[28]

Before the Flax committee meeting Low also interacted with committee members Fubini and Lewis Branscomb. He found Fubini "on the side of a small glider" on the grounds that "the United States should be satisfied with two or three flights per year. He sees no need for routine operations with men." That perspective, thought Low, "strongly reflects Ed David's view." By contrast, Branscomb was "very much on the other side," believing that "the United States needs routine operations, and to get these it needs a new recoverable space transportation system." Branscomb didn't care "whether or not men are on board, but... NASA has told a convincing story that men should be on board." In connection with the Flax committee session, Low also met with DOD's Johnny Foster, who had been charged with developing a statement of the rationale behind DOD as well as NASA support of the shuttle, only to discover that "Foster had not yet made up his mind on the value of the shuttle" because it was not a response to "the hiatus in United States space activity during a time that the Soviet Union was bound to have major demonstrable advances in their space flights." Low's counter-argument was that "having the shuttle well under development and on the horizon...will be a far better position to be in than not having anything to show for the future." He added "once the shuttle is available, we ought to be able to whip the pants off anybody that does not have this kind of a quick reaction, routine capability." Given his own ambivalence, Foster had made no progress in developing the shuttle rationale statement that NASA and DOD a month earlier had agreed to prepare.[29]

As Low attended the second day of the Flax committee meeting on November 18, he drew his operations costs versus development costs curve on the blackboard to make the point that "over the past six months the shuttle has become a much more reasonable vehicle in terms of development costs," since NASA was "now focusing on a shuttle that will cost between $5 and $6 billion to develop" and from $6–1/2 to $12 million to operate. Low argued that the "smaller, and much lower in development costs, glider should not be considered because it will be so terribly expensive to operate." The main questions, he suggested, were "whether the shuttle should be small or large and whether it should provide for routine operations or one or two flights per year." Low thought that most committee members understood his argument, "but if a vote had been taken right then, they would have still voted for the small glider simply because they don't believe in routine space flight operations."[30]

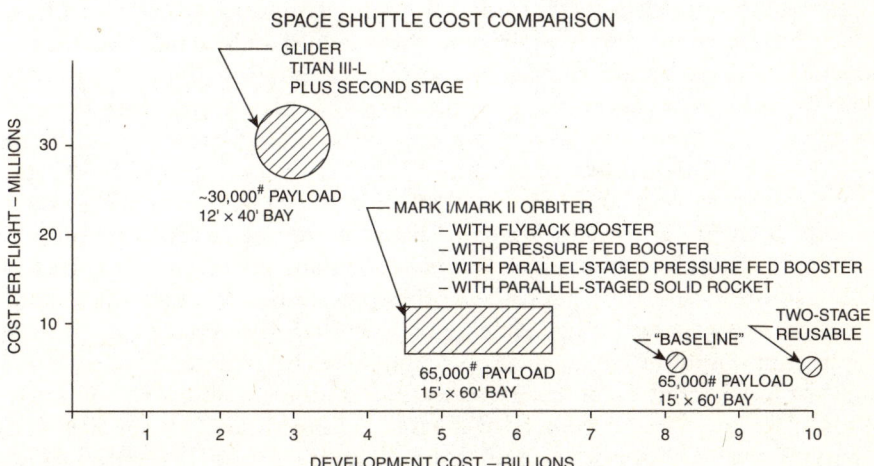

The development cost versus operating cost curve developed by George Low in fall 1971 as he attempted to gain White House support for the shuttle. What is designated as the "baseline" shuttle in this diagram is a two-stage shuttle with expendable hydrogen tanks mounted next to the shuttle orbiter's fuselage. (Diagram courtesy of Dennis Jenkins)

This was the final meeting of the Flax committee, and the group never issued a formal report. Perhaps the committee's most significant contribution was crystallizing the central issue in the shuttle debate. By this point, there was agreement that some new space transportation system was needed. The committee's deliberations focused attention on the basic issue of whether that system would include a full capability vehicle capable of launching all U.S. payloads on a routine basis or a smaller vehicle, either a powered shuttle or a glider, to be flown occasionally to test various technologies while also keeping a U.S. program of human space flight alive. As the Flax committee met for the last time, that question remained very much undecided.

Where was Wernher von Braun?

Noticeable by his absence as NASA tried to garner support for the space shuttle was Wernher von Braun, perhaps NASA's most charismatic spokesperson. Von Braun had moved to NASA headquarters early in 1970 to direct NASA's planning efforts, and thus logically he should have been one of the senior NASA officials involved in the attempt to gain White House support for the shuttle. But Fletcher and Low had discovered that "von Braun is not a supporter of the Shuttle, and in fact may be an opponent." According to Low, von Braun's skepticism was based on his conclusion that "the Shuttle will cost much more than our current estimates of Mark I/Mark II, and that NASA cannot afford to proceed with the development. To use his words, if we were given a Shuttle for a Christmas present, we would certainly use it, but, according to him, we cannot afford the cost of development."[31]

Von Braun had come to Washington with high hopes that, working together with the visionary Tom Paine, he might be able to convince the president and Congress to proceed toward a goal of eventual human missions to Mars, which had been his lifelong aspiration. President Nixon's March 1970 space statement had dampened that hope, and von Braun quickly found that in his position as head of planning for NASA, he was expected to present options for the agency's leaders to choose among, not advocate a particular course of action. When Paine announced in July 1970 that he was leaving NASA, von Braun was "just devastated." His relationship with George Low during Low's time as acting administrator was cordial but professional; "the one-on-one meetings with the administrator [Paine] ended and appointments with the acting administrator [Low] to discuss our programs became more difficult to set up as time went by." When Fletcher became NASA administrator, "it temporarily improved the climate for von Braun." Fletcher "admired" von Braun, and told him so. But given that Dale Myers and his team were leading shuttle studies, Fletcher "no more needed a 'chief architect' and planner than did George Low." Von Braun was one of those arguing in mid-1971 that NASA should give up on advocating a two-stage, fully reusable shuttle. According to von Braun's biographer, "what he could not dodge was his growing isolation at headquarters, a product of the marginalization of his planning office and his unpopular stance on space shuttle funding and design." By May 1972, von Braun decided to leave NASA for a job in industry; at his farewell party, he told a close associate, "George Low had thanked him profusely, in the name of all NASA, for fighting for a 'smaller and cheaper' shuttle." Low told von Braun: "We were not at all pleased by your warning words, but finally accepted your advice.... If you had not raised the red flag at that time, I'm certain the entire shuttle would be dead by now." Von Braun described that conversation as his "happiest moment during my time at headquarters."[32] But in the heated debate over shuttle approval in the fall of 1971, Wernher von Braun was nowhere to be seen.

Canceling *Apollo 16 and 17*

By the end of October, NASA had learned of the possibility that *Apollo 16* and *17* might be canceled, though it is not clear that the agency knew that the cancelation directive came directly from President Nixon. Fletcher wrote a long letter to Cap Weinberger on November 3, putting forth the case for not canceling the missions. He told Weinberger that "if broader considerations, nevertheless, lead to a decision to cancel Apollo 16 and 17, the consequences would be much more serious than the loss of a major scientific opportunity. Unless compensatory actions are taken at the same time to offset and minimize the impact, this decision could be a blow from which the space program might not easily recover." Fletcher proposed as a rationale for canceling the missions "that, in these times of pressing domestic needs, the manned space program should be earth-oriented instead of exploration and science-oriented." Not surprisingly, he suggested as an offsetting action "an early go-ahead for the space shuttle." Science adviser David chimed in at the end of November, urging the president to retain the mission in the NASA program, telling Nixon that "the cost of completing these missions is $118 million in FY 73, less than one-half of one per cent of the total cost of the Apollo investment... These missions will provide over *fifty per cent of the total productive time on the lunar surface*" and that "further cancellation at this time would be seized upon not only by skeptics in the science and engineering communities but also by many staunch supporters of the Administration as unwarranted and unwise." Apparently David had told his associates that he would resign if the two missions were canceled.[33]

New Technology Opportunities

As NASA and OMB debated shuttle approval, the possibility of NASA taking on a broadened role in applying technology to national problems was still alive. At the White House, Bill Magruder continued to examine a wide range of possible initiatives. In late October, Low reported that "there is still the question as to whether or not NASA should undertake the management of all of the efforts no matter what the subject." Low got a report from a NASA staff person sent to work with Magruder that "the White House is all geared up to do this and that the President himself is interested in NASA taking on the job."[34]

Low was understandably worried about where the funds to undertake new technological initiatives might come from. Magruder in a mid-November telephone conversation with Low reported "that many people are saying that the money should come from the space program." Magruder suggested that he and science adviser David were NASA's "best friends," arguing that taking the money from NASA "would defeat the original purpose of putting to work the unemployed aerospace engineers." Also, "a cut in the space program would have an instantaneous [negative] effect on unemployment, while the new technology initiatives would only have a slow buildup in employment." Magruder estimated that the cost for his effort in its first year "would be $600 to $700 million...for all except the transportation and aviation initiatives, plus another $1 billion for transportation and aviation." He thought that the funds should come from "social programs" and foreign aid. Magruder intended to set up "a small, hard-hitting interim organization" to develop various initiatives, and asked NASA to provide team leaders for most initiatives. Low noted "whether or not NASA will be asked to undertake any or all of these initiatives is still not clear"; he was worried that "if we are asked to undertake some of this work, it will be at the expense of some of our aeronautics and space work."[35]

Growing Impact of Aerospace Unemployment Concerns

As Magruder acknowledged, the New Technology Opportunities effort was in large part an attempt to find employment for those aerospace workers who had lost, or were in danger of losing, their jobs as a result of the Nixon administration's budget reductions in the defense and space sectors. This was part of a broader concern—that unemployment in states key to Richard Nixon's reelection in 1972, particularly California, would negatively impact the president's election prospects. As of late 1971, the leading candidate for the Democratic nomination was Senator Ed Muskie of Maine, and in some polls Nixon was running behind Muskie.

The unemployment issue had been a White House worry since at least early 1971. The concern was that at the time of the 1972 presidential election, the California unemployment rate might be 6.2–6.9 percent, significantly above the national average of 5–5.5 percent. By the end of August 1971, the White House had launched a "California Employment Project." President Nixon set a goal of creating 100,000 new jobs in California before the election, which would bring California unemployment down to the national average. Nixon had directed that most of those new employment opportunities would be the result of DOD actions. An individual named Fred Foy had been brought into the White House to coordinate efforts in DOD and other government agencies to target job creation in California as a high priority. John Ehrlichman remembered Foy as "a retired business executive" who would go to a community in California and "smoke out...opportunities to let contracts on an accelerated basis." Foy would report back to Flanigan, "who would pick up the phone and talk to the Defense Department and shake things loose." As a result, "they would accelerate these things and create jobs by the scores in a relatively small geographic area. The impact was dramatic." Flanigan's assistant Jonathan Rose was the White House link to the California Employment Project. Rose in an August 28 memorandum to the president indicated that a "politically loyal" employee in the DOD would provide the White House "bi-weekly reports on the status of the agency's job creation effort" and that "Governor Reagan's office has designated a team of competent economists and others" to work with the White House on this effort. That was not enough assurance for Nixon; he asked Cap Weinberger at OMB to "personally stay on the project of jobs for California and that you make sure that there is no let up in the efforts on this."[36]

It was becoming clear that the decision to approve the space shuttle would be influenced by job-creation considerations. A member of the California Legislature, Newton Russell, wrote Haldeman in early June 1971, forwarding a letter he had sent to NASA making the case of why the shuttle program "should be located in California." Among the good reasons were "unemployment, source of supply, available technical engineers" but "if you want to put it down to crass politics...California will be a key state in '72." Haldeman forwarded Russell's letter to Weinberger, who responded, saying that "because of all the factors you mentioned...I am sure that this [locating the shuttle program in California] will be the case." Weinberger noted that "there is still a problem in financing the whole project because of all of the overall budget totals," but that "I am sure that whatever is done will be largely based in California."[37]

Low noted that "on the unemployment situation, we are feeding a lot of information, first to Fred Foy, who is working in the White House on just that problem, and also to Jonathan Rose." He added that "it is clear that a small acceleration of some of the new NASA programs would have a rather dramatic influence on the unemployment situation, particularly on the West Coast." On November 3, Fletcher had written both Flanigan and Weinberger, noting the employment impacts of the space shuttle program, and especially "the substantial impacts of a possible acceleration of the Mark I/II Shuttle program." NASA in its September 30 budget submission had proposed a budget level above the "minimum acceptable program" that would accelerate the pace of the shuttle program, thereby creating jobs sooner and in larger numbers than if only the minimal program were funded. The accelerated program increased the FY1973 shuttle budget request from $228 million to $400 million, and resulted in a first shuttle flight in 1977 rather than 1978. Fletcher called particular attention to "the very sharp build up (from 5,600 to 14,300) that would occur in the last six months of calendar year 1973" with the NASA minimal budget, and, more relevant to Richard Nixon's reelection, "the very substantial increases in 1972...that are possible with the acceleration indicated."[38]

In mid-November, NASA sent Don Rice a brief report titled "California employment." The report noted that "the prime contractors for the shuttle have not been selected, but the majority of competitors are California firms and so the most favorable employment impact will be in that State." It added that "historically, NASA spends 50¢ of each dollar in California." The report noted that "the $500M contract for the rocket motor for the shuttle has already been placed with Rocketdyne (a division of North American Rockwell) in California." But if the shuttle configuration was changed or if a glider were approved, so that the shuttle engine was not needed, "Rocketdyne says it will have to go out of business." NASA provided additional employment estimates to Rose on December 1, indicating a variety of NASA-related actions, including accelerating shuttle development, that would result in increased employment.[39] NASA by this point recognized that the shuttle's employment impact could be a decisive consideration in White House thinking.

That impact did not go unnoticed at the top levels of the White House. John Ehrlichman recalled that the issue of shuttle-related employment in Southern California "was a very important consideration in Nixon's mind...I can recall conversations about that, which were highly persuasive...You must not underemphasize that element, that employment element, in Nixon's decision [on the shuttle]." Ehrlichman remembered "the quantitative breakouts of the number of jobs involved...When you look at the employment numbers, and you key them to the battleground states [Those states with electoral votes important to winning the presidency in the 1972 election], the space program has an importance out of proportion to its budget." Ehrlichman sat "in the Cabinet Room with Nixon, [Secretary of the Treasury John] Connally, and [OMB Director George] Shultz...looking at issues. We went all the way across domestic issues, the problems of veterans, the problems of the aged, space, health...and putting slides up showing where people were who were concerned about these issues. And then doing an overlay of the battleground states...It was very interesting then to see how some of these issues fell out of bed, because the people who were concerned with them were not in battleground states." In this political exercise "space was way up to the top of the list, along with one or two other issues."[40]

There was also a representative of California employment interests with direct access to the White House. Willard "Al" Rockwell, Jr., head of North American Rockwell, one of the contenders for the space shuttle prime contract and with its space operations based in California, was a long-time acquaintance of Richard Nixon and a major contributor to Nixon's election campaigns. Ehrlichman recalled that "there was kind of a direct line between Nixon and Rockwell, which was important...I knew that there was a tight relationship." Rockwell and one of his top executives, Robert Anderson, visited Flanigan, Weinberger, and Rice in late November to discuss prospects for the shuttle. Flanigan told them that "there would definitely be a shuttle program, that the government was about to make the decision but that there is still some sorting to be done." Flanigan added that "the big shuttle that NASA supported a year ago was definitely out but that NASA is still not ready to move out now...NASA is still not completely in tune with the realities of the day, but is slowly coming around." The fundamental question, Flanigan told Rockwell, was "whether the shuttle should be a research vehicle or one that is productive." This was an indication that Flanigan was aware of the NASA–OMB/OST argument about how best to proceed. When the administration decision on the shuttle was made, indicated Flanigan, there would be "a very soft pronouncement... This was so the people in the aerospace industry will clearly see Administration support for their industry, while those not in that industry will not get overly excited." Apparently Anderson in this meeting had made a case for the shuttle in terms of its being a bailout for the aerospace industry, and Flanigan had responded that this was not a rationale acceptable to the White House, at least publicly. Flanigan wrote Rockwell a few weeks later, saying that "I do hope my taking exception to what seemed to be excessive vehemence on the part of your subordinate [Anderson] regarding the shuttle did not leave the impression that we lack enthusiasm for the concept. I am quite convinced that we will come up with a viable and positive solution on the shuttle in a short period of time."[41]

NASA Makes Its Best Case

By late November, there was increasing pressure to reach some sort of decision with respect to the space shuttle. A final budget decision needed to be made in time for it to be reflected in the president's FY1973 budget request; the text of that request had to go to the printers in early January. NASA decided to make as strong as possible a case that its concept of the shuttle deserved to be approved.

The sense of urgency in getting the NASA case before White House decision makers was reinforced by reports of the initial decisions on the NASA FY1973 budget. Anders had attended a meeting at which the OMB space staff had made some tentative decisions on the NASA budget based on the discussions at the director's review; he relayed this information to Low, as usual on a very confidential basis. He told Low that the staff was recommending cancelation of *Apollo 16* and *17* "because there is no public interest." The fact of President Nixon's desire to cancel the missions was still not widely known. The OMB staff was recommending, rather than the space shuttle, a small glider, and, to make up for the employment losses from the Apollo cancelations and not starting an ambitious shuttle program, "three gap-filler missions" using surplus Apollo hardware. Marshall Space Flight Center was to be closed in 1974 and the Jet Propulsion Laboratory in 1975. Anders also had been "taking the pulse of those in the Executive Branch involved with the NASA program"; that pulse was "rapidly changing with time." He perceived "two opposing forces." One wanted "to cut NASA back to a much smaller program"; the other wanted "not to increase unemployment in the aerospace industry." He also suggested that there was "a faction in the Executive Branch that would like to cut $1 billion out of the NASA program" to start the new technology initiatives, but that "Magruder is not among those who want to cut back on space."[42] All of this added up to NASA seeing itself in a very precarious position.

Remaining Shuttle Options

George Low was finally able to meet with Don Rice in late November to bring Rice up to date on NASA's current thinking on the shuttle. Low described the meeting as "extremely good...for we communicated well." Once again, Low drew his development versus operation cost curve for Rice and used it as the basis for his presentation. He told Rice that on the basis of 18 months of contractor and NASA studies and of trading off development and operating costs, NASA had come up with "a class of [shuttle] configurations that costs much less to develop than earlier configurations, is smaller but can carry the required payload, and is still 'productive' in terms of operating costs." He suggested that "for practical purposes," the two-stage fully reusable and the baseline (a two-stage shuttle with disposable hydrogen tanks) configurations could be "discarded" because of their high development cost. He argued that "the glider, as presently proposed, also does not appear to be promising." If the glider were to carry the same payload as the full size shuttle orbiter, it would "probably not offer a significant saving in development cost, but will be expensive to operate." (This was a rigged argument, since neither the Flax committee nor OMB was suggesting a glider able to carry large payloads, and NASA had still not examined the implications of a much smaller glider.) This left, suggested Low, "the Mark I/Mark II configurations with four booster options: flyback, pressure-fed, parallel-staged pressure fed, and parallel staged solid rocket boosters." (The term "flyback" referred to the use of a modified first stage of the Saturn V Moon rocket that could be operated by a human crew and flown back to a runway after launch. The term "pressure-fed" referred to a new booster design concept, developed at NASA's Marshall Space Flight Center, in which propellant would be forced into the booster engines by gas pressure rather than fed into the engines by a large turbopump. A "parallel-staged" configuration would have both booster and orbiter engines firing at liftoff, as opposed to the usual "series-staged" approach in which only booster engines would be fired on the launch pad.) Low suggested that a space shuttle using one of these booster options could be developed for between $4.5 and $6.5 billion, with operating costs between $6 and $12 million per flight. All shuttles in this range could eventually "carry the same payload, 65,000 pounds into a due east orbit or 40,000 pounds to polar orbit, in a 15 ft by 60 ft. payload bay." Low concluded that "the most promising configuration today is the Mark I/Mark II orbiter with the parallel-staged pressure-fed booster." It is worth noting that NASA at this late point was still advocating the idea of phased technology development of the shuttle orbiter.[43]

Rice later would remark "that what sticks in my mind more than anything else was the difficulty of getting any solid attention paid to alternative [shuttle] designs...alternative in terms of mission requirements and why that mattered." He added "I still find myself a little bit incredulous to this day that there were three widely different concepts that NASA had for that system. All had the same physical capability to do work."[44] Rice was correct; NASA was strongly resistant to examining alternatives to the capabilities embodied in its preferred shuttle design. For one thing, NASA was still caught between OMB's pressure to consider a significantly smaller shuttle or a glider and NASA's perception that it had to meet national security requirements to gain the DOD support it thought essential for White House approval of the shuttle. Also, NASA's human space flight team was being stubborn, convinced that the shuttle orbiter design coming out of more than eighteen months of study was a much better choice than any of the alternatives being discussed in Washington. In a 1979 letter, Low commented that "even long after those of us in the top NASA administration had decided that a less ambitious shuttle design was 'all the traffic would bear,' it took some time to get the rest of the people in NASA who had been working on the two-stage, fully reusable shuttle to agree with this approach. Therefore, what may have appeared to some as a NASA/OMB fight, in part, was really an internal NASA debate."[45]

Making the Case to the White House

NASA's primary link to President Nixon and other senior White House decision makers was through Peter Flanigan and his assistant Jonathan Rose. George Low drafted a "best case" essay on the shuttle for Rose and Flanigan to use within the White House; it was edited by Willis Shapley and sent to Rose by James Fletcher on November 22. The essay made five points:

1. The U.S. cannot forego manned space flight.
2. The space shuttle is the only meaningful new manned space program that can be accomplished on a modest budget.
3. The space shuttle is a necessary next step for the practical use of space.
4. The cost and complexity of today's shuttle is *one-half* of what it was six months ago.
5. Starting the shuttle now will have a significant positive effect on aerospace employment. Not starting would be a serious blow to both the morale and health of the Aerospace Industry.

The paper observed that "man has learned to fly in space, and man will continue to fly in space. This is fact. And, given this fact, the United States cannot forego its responsibility—to itself and to the free world—to have a part in manned space flight... For the U.S. not to be in space, while others do have men in space, is unthinkable, and a position which America cannot accept." It suggested that the shuttle "can provide transportation to and from space each week," and that "space operations would indeed become routine." The link to an eventual space station and other long-term space activities was made explicit: "In the 1980's and beyond, the low cost to orbit [that] the shuttle gives us is essential for all the dramatic and practical future programs we can conceive. One example is a space station." The paper argued that "the shuttle helps our *international* position—both our *competitive* position with the Soviet Union and our prospects of *cooperation* with them and other nations...With the shuttle, the United States will have a clear space superiority over the rest of the world." It claimed that the shuttle could be developed for an investment of $4.5–$5 billion, with an operating cost of "around $10 million or less per flight." It noted that the shuttle orbiter "has been dramatically reduced in size—from a length of 206 feet down to 110 feet," but "the payload carrying capability has not been reduced." In terms of employment effects, "an accelerated start on the shuttle would lead to a direct employment of 8,000 by the end of 1972, and 24,000 by the end of 1973."[46]

With this paper, NASA stated its arguments for shuttle approval in what its leaders hoped would be a convincing fashion. Unlike the somewhat negative arguments in support of the space shuttle that George Low had put forward in October 1970—that shuttle development could be justified "as a versatile and economical system for placing *unmanned* civil and military satellites in orbit, entirely apart from its role in conducting or supporting manned missions" and that "with the shuttle the U.S. can have a continuing program of manned space flight...without a commitment to a major new manned mission goal"—NASA in November 1971 made a much more positive case for the shuttle as a human space flight system serving important national interests.[47] Key to this case were not only the claim that the space shuttle would make space operations routine and less expensive but also the proposition that the shuttle would advance intangible values such as U.S. space leadership and international cooperation and that it was thus essential for the United States to continue a vigorous program of human space flight based on the shuttle and its new capabilities.

As NASA put forward this case for approving the shuttle, Rice and his OMB staff in parallel were preparing a decision memorandum for Richard Nixon that took a very different tack, suggesting that the president should approve a much smaller and less frequently used shuttle than the system that NASA had in mind. That memorandum questioned the economic argument for shuttle development and assigned only limited value to potential benefits such as space leadership and new capabilities for space operations. The following few weeks would determine which point of view would prevail.

Chapter 13

Which Shuttle to Approve?

As December 1971 began, Don Rice and his Office of Management and Budget (OMB) space staff remained on a collision course with NASA. Rice had taken Cap Weinberger's guidance at the October 22 director's review as license to direct his staff not only to come up with alternative, less ambitious, and thus less expensive, shuttle requirements in terms of payload bay size and weight-lifting capability, but also to present that new shuttle concept in the context of a different program of human space flight than what NASA was proposing. Rice was convinced that the shuttle NASA preferred was "a huge overinvestment for what the country needs," and believed it was his responsibility as a steward of the federal budget to help protect the president from making that overinvestment.[1]

By mid-November, the OMB staff had drafted a decision memorandum for President Nixon on "the future direction of the U.S. civilian space program" and was circulating the draft inside the White House and Executive Office of the President for comments. The memo set forth "a description and analysis of NASA's proposed future manned space flight program and an alternative program." That alternative program "would gradually decrease NASA's annual spending from the present $3.2 billion to about $2.5 billion by 1976." Included would be a "smaller, reduced cost version of the manned reusable shuttle...NASA's larger version would not be developed now because it would probably prove too costly, uneconomical, and risky a venture."[2]

George Low on November 14 noted that "we have had no direct interaction with OMB...since the budget hearings several weeks ago...It is clear that there are opposing forces...Those who are for space for its own sake appear to be very few in number."[3] Those opposing forces would play themselves out in the following weeks as final decisions on which space shuttle to develop were made. But first President Richard Nixon twice made fundamentally the same choice—a choice that would provide the context within which those final decisions would unfold. These two presidential decisions took place in late November and early December; Nixon left the specifics of what shuttle configuration to develop for his associates to decide during the rest of December. There is no written or recorded evidence of his direct involvement in that decision, although it is probable that he was informed regarding the alternatives under consideration and informally communicated his views with respect to those options to his inner circle.

A First Presidential Decision

In his reaction to Cap Weinberger's August 12 memorandum and with the October 19 clarification of his intent, Richard Nixon had seemingly agreed that if *Apollo 16* and *17* were canceled, there needed to be compensatory actions in terms of announcing approval of new space efforts, including the space shuttle but also the NERVA nuclear rocket engine program or other NASA activities. This was not yet a specific decision to approve shuttle development, but rather an indication that the president was leaning in that direction.

President Nixon did make a significant step toward such a decision during a November 24 meeting discussing "sensitive and significant issues in the FY1973 budget." Attending the Oval Office meeting were Nixon's top assistant for domestic policy John Ehrlichman, OMB Director George Shultz, and Secretary of Treasury John Connally, a new member of Nixon's inner circle. In preparing him for the meeting, Domestic Council Deputy Director Ed Harper alerted Ehrlichman that a "complete alternative NASA program [was] being developed by OMB [and] should be ready this week. Extraordinarily important that this decision is carefully staffed out."[4]

Ehrlichman came to the meeting with a two-page list of issues for discussion. One item asked "Will the budget style be: (a) expansive? (b) austere? (c) neither?" Another question was "What economic (employment) assumptions will be displayed?" Eighteen program issues were listed; space was third, after general revenue sharing and welfare reform. As the four men got to the space issue, the following discussion occurred:

Nixon: "Space, what's the problem here?"
Ehrlichman: "Well the problem here is do we go ahead with the next two shots? [*Apollo 16* and *17*]"
Nixon: "No! If we go, no shots before the election."
Ehrlichman: "Then what would we do with all those employees?"
Nixon: "For those shots? How many, George?"
Shultz: "17,500 or something like that."
Nixon: "I don't like the feeling of space shots between now and the election."
Ehrlichman: "But thinking of this thing [the space program] in just pure job terms, it is a hell of a job creator."
Connally: "The American people are really not impressed by any more space shots."
Nixon: "NASA is saying you'll find incredible things about the Moon with these last two shots, and the American people say 'so what'?"

Shultz: "Could I try another possibility? The last shot is the one in which they have loaded a great amount of scientific stuff from the ones that have been canceled before. That shot is scheduled after the election."

Nixon: "I only see a minor waste of money. Keep the people on, but don't make the shots. I just don't feel the shots are a big deal at this time...There is also the risk you could have another Apollo 13...That would be the worst thing we could have...We are just not going to do it. There will not be any launches between now and the election. The last shot, fine. Let's go forward with the last shot."

Ehrlichman: "The southern California people have a mighty press on for the space shuttle to be located in southern California. It is a highly visible kind of thing, if we were to announce at the State of the Union or sometime that you were going ahead with the shuttle."

Nixon: "This is not a State of the Union thing. I should do it [announcing shuttle approval] out in California where you are going to put it. Jobs— right, John? Do it in terms of jobs. It ought to be in California."

Shultz: "NASA has a full thrust [shuttle] program, but there are options that are a little more modest."

Nixon: "Take the more modest option. We'll take a look later to see [if that is the right choice.] It's the symbol that we are going to go forward. We are going to be positive on space. Nobody is going to be against us if we go forward in space, and a few will be for us because we do."

Ehrlichman: "If you tell the aerospace industry that we are going ahead on the shuttle, that helps right now."

Shultz: "While the shuttle and Skylab will keep men in space to a degree, the direction of this program ought to shift away from man in space and toward doing most of these things on an unmanned basis."

Nixon: "I agree. Manned space flight becomes a stunt after a while."[5]

Ehrlichman later thought the basic decision to develop a space shuttle had been made at this meeting. His record of the discussion, prepared only on January 4, the day before Nixon was to announce his shuttle decision, said that on November 24 "the President decided to support the space shuttle providing it could be located in California." After this meeting there was little doubt that some form of space shuttle would be approved by the Nixon administration; the question was whether it would be NASA's preferred full capability shuttle orbiter design or a smaller alternative as was being suggested by OMB.[6]

There are a number of interesting elements to this November 24 discussion. The space program, including both the ongoing Apollo effort and the space shuttle, was being evaluated by Nixon and his top advisers not only in terms of its substantive value but especially in terms of its employment impacts. In particular, the space shuttle was seen as part of the ongoing White House California Employment Project, aimed at getting the most possible new jobs located in California prior to the 1972 election. Nixon continued to want to avoid the risk of another *Apollo 13* accident in the months leading up to that election, believing that such an incident could impact his reelection prospects and the 1972 summit meetings that were part of his attempt to normalize relations with China and the Soviet Union. He judged that the American public was not really interested in more trips to the Moon, which gave him a free hand to defer or cancel the remaining two Apollo missions. Ultimately, Nixon decided to go ahead with the two missions, moving *Apollo 16* so that it would not interfere with his China trip and approving *Apollo 17* once he recognized that it would take place only after the 1972 election. Perhaps most surprising, it had been an article of policy belief in the White House that Richard Nixon wanted a future NASA program including U.S. astronauts flying in space. Yet in this conversation he had said "manned space flight becomes a stunt after a while." Even so, he gave the space shuttle a qualified approval as a "symbol," saying "we are going to be positive on space." There was little consistency in the Nixon attitude toward human space flight.

The Second Presidential Decision

The OMB decision memorandum on NASA's program for President Nixon, revised on the basis of comments from various offices in the White House and Executive Office of the President, was ready on December 2.[7] The memo began with a section on why decisions were needed:

- "The lead times are gone to decide what to do after Apollo."
- "Industry wants decisions one way or the other, particularly on the Space Shuttle—on which contractors have been doing design studies for the last 18 months."
- "Adjusting space spending and turning NASA's capabilities to other domestic problems requires a 2–3 year phasing." (This was an indication that a lead NASA role in William Magruder's New Technology Opportunities effort was still a possibility.)

The eight-page memo both described NASA's human space flight program as proposed in the agency's September 30 budget request and OMB's alternative. The alternative program included "a smaller and less costly Space Shuttle," cancellation of *Apollo 16* and *17* "because we understand that is your [Nixon's] wish," and "reduction in the size of NASA's institutional base after calendar 1972." With respect to NASA's plans for the shuttle, OMB asked "since we already have the capability to put manned and unmanned payloads into earth orbit using expendable boosters, how much should we be willing to pay for a Shuttle?"

The memo noted "last year NASA was proposing a $10-$12 B [billion] Shuttle. In response to questions from OMB and OST about whether the benefits justified such a large investment, NASA has since designed a $6 B Shuttle which can do *all* the missions of the larger, more expensive one . . . (We think both costs are underestimated, perhaps by 50%.)" If NASA were given approval to develop the shuttle it was proposing, suggested OMB, "one program, the Shuttle, would dominate NASA for the coming decade, as did Apollo in the 1960's."

What OMB was proposing as a "smaller reduced cost" alternative to NASA's shuttle would involve "an investment of $4–5 billion over the next 8 years." Such a vehicle, OMB suggested, could "capture about 80% of the payloads of the redesigned larger Shuttle at about two-thirds of the investment cost." By this time OMB had accepted that there would be a space shuttle program rather than a glider or some other alternative, and was focusing on keeping the shuttle as inexpensive as possible in investment terms; there was little attention given by either OMB or NASA to an examination of shuttle operating costs, which in any event would be incurred after the Nixon administration left office. It would be necessary to "retain the reliable Titan III expendable booster to launch the few largest payloads that would not fit the smaller Shuttle. These include space telescopes and large intelligence satellites. (This may be desirable in any event since, for national security purposes, we may not want all our eggs in one basket.)" OMB added, reflecting the White House interest in California employment, that "we understand from NASA that the recently awarded engine contract with Rocketdyne division of North American Rockwell will probably be continued for the smaller Shuttle without the need for recompetition."

The OMB-proposed program also included three Earth orbital missions using launch vehicles and spacecraft left over from the Apollo program. Only one of these missions, the 1975 docking mission with a Soviet spacecraft, had been in NASA's September 30 "minimum acceptable" budget proposal. The other two would be Earth resources survey missions that had been included in NASA's September 30 "alternate recommended program," which presumed a higher budget level; OMB suggested them as a way of having one human spaceflight mission per year between 1974 and 1976, thereby avoiding a multiyear gap in U.S. human space flight activity. The smaller shuttle was anticipated to be ready for flight by 1978. With respect to *Apollo 16* and *17*, while the OMB alternative program canceled the missions on the basis that that was the president's wish, the memo actually argued for retaining the missions. Saying "if concerns about complications during 1972 [Nixon's already planned visits to China and the Soviet Union and the presidential election] can be alleviated by rescheduling Apollo 16, it would seem appropriate to retain Apollo 16 and 17 for their scientific returns and employment impacts." OMB estimated that the employment impact of adopting its proposed alternative program would be 4,000 job losses by mid-1972 and 8,000 by the end of the year, but 30,000 by mid-1975. In OMB's recommended program, the NASA budget for FY1973 would be $3.050 billion, declining to $2.975 billion by FY1976.

The "recommended next step" was for "OMB and OST to work with NASA on the reorientation of the space program." The memorandum asked President Nixon to either "Approve" or "Disapprove" four actions:

1. "Initiate reduced-cost smaller Space Shuttle program."
2. "Conduct Soviet docking mission."
3. "Conduct other manned earth-orbital missions."

4. "Apollo 16 and 17"
   - "Cancel both missions"
   - "Cancel just Apollo 16"
   - "Reschedule Apollo 16 and fly both."

Notably, OMB did not provide the president the option of approving NASA's shuttle plans.

The OMB memorandum was discussed on December 3 as Ehrlichman, Shultz, and Cap Weinberger met with President Nixon at the Southern White House in Key Biscayne, Florida. There is no recording of the meeting, since Nixon had not set up a taping system in his office at Key Biscayne, but as was his custom Ehrlichman took notes.

With respect to *Apollo 16* and *17*, Nixon suggested that it would be better to combine the two missions after the 1972 election, but that his aides should "work it out." *Apollo 16* was scheduled for March 1972, but Nixon suggested moving the launch to April to avoid any possibility of its interfering with his planned 1972 trip to China. (Nixon went to China between February 21 and 28; the *Apollo 16* mission was launched on April 16.) Nixon on November 24 had already approved going ahead with *Apollo 17*; with this discussion of rescheduling the *Apollo 16* mission, the possibility of canceling one or both of the missions, a long-held Nixon wish, disappeared.

President Nixon discusses the FY 1973 budget with his advisers. (l-r) John Ehrlichman, George Shultz, and Caspar Weinberger at his Key Biscayne, Florida, residence on December 3, 1971. It was at this meeting that Nixon made the formal decision to approve space shuttle development. (National Archives photo WHPO 7933–8)

With respect to OMB's proposal for a smaller shuttle, Ehrlichman recorded Nixon's response simply as "yes," providing that the vehicle would use the "California engine."[8] The effect of Nixon saying "yes" to the smaller shuttle was to approve the recommendation that "OMB and OST proceed to work with NASA on a reorientation of the space program." That process would take place during the rest of December.

NASA Continues to Seek DOD Support

Although NASA's Fletcher and Deputy Secretary of Defense David Packard had agreed in October that NASA and DOD would work together to develop a restated rationale for the shuttle, by the start of December little progress had been made in this effort. One problem was that Johnny Foster, DOD's director of defense research and engineering, who had been charged with preparing the rationale paper, remained ambivalent about a DOD commitment to the shuttle and to NASA's approach to selling it. Talking with George Low at a December 1 dinner party, Foster had suggested that NASA was "doing the wrong things," saying that "NASA should not let OMB impose an arbitrary cost limit on the shuttle. Dictating technical decisions through the budget process is just plain wrong." He added "it is even worse if NASA lets OMB dictate the shuttle configuration." Foster suggested that "NASA has decided to build a taxi to nowhere on faith. We should instead have a flight program that demonstrates the need" for the shuttle. Low's retort was that "the main lack was in presenting an imaginative military space program taking advantage of the new capabilities that the shuttle would represent."[9]

Foster's advice was hardly useful to NASA, faced as it was with what seemed to be an unchangeable upper limit on the budget that the White House was willing to allocate for its activities. But NASA did not give up its attempt to get DOD support; rather, it took on itself the role of suggesting the "imaginative military space program" that Low had suggested was needed. That program came in the form of a memorandum for Fletcher to send to Packard. The memo was drafted by NASA's Assistant Administrator for DOD and Interagency Affairs Jacob Smart, a retired four-star Air Force general. Smart's draft noted that "in the next few weeks the President will make decisions relating to national objectives in space" that would be of "critical importance, because the nation's military security, its political, economic and social well being in this and succeeding decades, are inextricably interwoven with what we do and what we fail to do in space." He forecast dire consequences if the United States did not maintain a position of space leadership: "the self confidence of our people would diminish, our posture in the world community will be overshadowed, and our trade in world markets will be reduced," resulting in "problems of great magnitude and complexity" which would "likely face this government, particularly DOD." As noted in chapter 9, Smart in his draft detailed a number of ways in which "the space shuttle can deliver, with few exceptions, the total traffic of presently-planned military spacecraft to useful earth orbits."[10]

Smart's suggestions for potential national security uses of the space shuttle were very similar to the ideas in the initial June 1969 DOD–NASA study of shuttle uses. They had been in the background of the discussions between the two agencies over shuttle design ever since, but apparently had had little influence on the assessment of the shuttle by the OMB civilian space staff. However, those potentialities were indeed known to and of interest to the top levels in the White House, including Richard Nixon. Ehrlichman in a 1983 interview suggested that "what the military could do with the larger bay in terms of the use of satellites" and the fact that "the space shuttle would have the capability of capturing satellites or recovering them" had "a strong influence on me" and "weighed into my attitude toward the larger shuttle. And I feel it is valid to say it also weighed into Nixon's" attitude.[11] What is not clear was how, and when, Nixon, Ehrlichman, and perhaps also Flanigan, Shultz, and Weinberger, were made aware of the national security potentials of the shuttle; because the issues involved were highly classified, any relevant documents are not contained in accessible archives. But as final decisions on shuttle size were reached at the end of December, the president's interest in national security uses of shuttle capability were known to his other senior associates and very likely influenced their willingness to go forward with NASA's full capability space shuttle.

It is not clear whether the Fletcher–Packard memorandum was ever sent; a final copy does not appear in NASA's files. But the memo stands as an example of the arguments that NASA was using in its effort to insure DOD support of the shuttle program. Fletcher and Smart did meet with Foster and several of his associates on December 3. But no formal statement of DOD views on the shuttle sent to the president in December 1971 was located in research for this study, and there is no record of a meeting with the president to discuss this issue.[12]

Engaging the National Security Council

While Nixon's most senior domestic policy advisers, Ehrlichman and Shultz, had become engaged in discussions of NASA's future, that was not the case with respect to national security adviser Henry Kissinger. Kissinger had gotten involved in evaluating post-Apollo space cooperation with Europe and the Soviet Union, but had not had much exposure to the broader issue of future U.S. space activities. Fletcher set out to remedy this situation, first by talking with Brigadier General Alexander Haig, Kissinger's deputy on the National Security Council staff. Fletcher reported to Low that "in suggesting that the National Security Council become more involved with NASA affairs, Al Haig needed absolutely no persuasion. He has, for the last year and a half, been convinced of this and so has Henry [Kissinger], but they have been so busy they haven't really tried to work the problem." Haig had suggested "that someone who regularly meets with the President ought to be intimately familiar with NASA affairs" and that, if Kissinger were to play that role, "some mechanism has to be set up whereby Henry is regularly informed on what the major issues are in NASA." Fletcher told Haig "that perhaps the principal issue before the President now was the space shuttle," and gave Haig a copy of the November 22 "best case" memorandum on the shuttle rationale, while observing that "it is doubtful whether he is going to have the time to read" the document.[13]

NASA's somewhat belated attempts to engage Kissinger as an advocate for the national security and foreign benefits of full capability shuttle and a strong civilian space program were intended as a corrective to the reality that from the start of the Nixon administration the future of the space program had been treated as an issue of domestic policy and thus had been evaluated in terms of employment effects, technological benefits, and budget priority. Whether NASA would have fared better in its post-Apollo aspirations if the Nixon White House had from the start seen the space program as a foreign policy and national security effort, as had been the case during the Kennedy administration, is an interesting but unanswerable question.

NASA and OMB Conflict Escalates

On December 12, George Low reported that "during the past two weeks we met with Don Rice, Tom Whitehead, Jonathan Rose, Ed David separately, and finally with Rice, David and Flanigan together, to discuss the kind of space shuttle that should be developed." Low once again stated that "the basic issue on the space shuttle concerns whether or not the shuttle should capture a majority of the payloads that will be flown in the 1980's."[14]

White House Support

NASA's efforts to gain support for its shuttle concept seemed to be paying off, at least in the view of Tom Whitehead. Whitehead wrote Flanigan on December 2, noting that he and Flanigan "had succeeded when we first came into office in averting NASA's high flying plans for space stations and Mars trips, and in bringing the budget down to a more realistic level consistent with the President's wishes." But, added Whitehead, it had not been their intention "to continue to erode NASA's budget indefinitely, but to induce them to come up with a sound, forward-looking evolutionary space program for the coming decade." Whitehead observed that "over the last few months, OMB and NASA have been bickering, principally about the space shuttle." He thought that Fletcher had "done what I believe to be an outstanding job of devising a space shuttle concept that is consistent with reasonable budget levels and sensible technology, and still builds for the future." Whitehead was aware of the alternative shuttle concepts then under discussion, and tended "to believe that the larger shuttle is the more prudent course, but the differences are so small that the choice should reasonably be left to NASA's discretion." He suspected that "OMB will try to push fairly hard for the smaller version. NASA might buy this as a last choice, but the impact on their morale and that of the aerospace industry would be unnecessarily negative."[15]

Attached to Whitehead's memorandum was a chart prepared by Bill Anders that summarized on one page the various shuttle alternatives that had been examined in the preceding months. Anders characterized the fully reusable shuttle that had been NASA's original hope as "Fat Albert" and the small glider that had been proposed during the Flax committee deliberations as "Weird Harold." The chart compared the then-current NASA and OMB shuttle configurations, noting that there were "relatively small (15–20%) payload differences and with reasonably broad consensus that we are talking about the right animal now, there would seem to be little further gain by delaying publicized commitment."[16]

The OMB Shuttle

On December 10, NASA received its FY1973 budget allowance from OMB, with the important exception that the budgets for the space shuttle and possible interim Earth-orbital missions were not specified; NASA was told that those budgets figures would be provided later. NASA was satisfied with the OMB allowances for the rest of its program, and told OMB Deputy Director Weinberger that it did not plan to ask for any reconsideration of the OMB-proposed budget levels.[17]

The positive feeling did not last long. Low recorded that "on Saturday, December 11, Fletcher and I met with Rice, David, and Flanigan and were told by Rice in that meeting that the President had decided to go ahead with the shuttle provided it was a smaller orbiter with a 10 × 30' payload bay, carrying a 30,000 pound payload." The rationale offered for arriving at this position was that "the shuttle would primarily be used for manned space flight missions and that this kind of shuttle was a major step beyond [Apollo] command and service modules." Considerable, rather heated, discussion followed; finally, Fletcher "indicated he could not accept this kind of edict and that he wanted to see the President."[18]

At this meeting, Rice gave Fletcher and Low a two-page document outlining the characteristics of the smaller shuttle that OMB was claiming that President Nixon had approved. This claim was not quite valid; Nixon had indeed approved the OMB proposal to work with NASA to develop a smaller, less expensive shuttle design, but in neither the OMB December 2 decision memo nor the discussion at the December 3 budget meeting had the president approved specific shuttle design characteristics. Ehrlichman, who was present at the meeting, suggested that "there was some explanation to him [Nixon] of what the differences were. They were not in great detail, I am sure, because those things just never were, not at that level." Rather, what OMB presented to NASA was its own preferred shuttle performance characteristics, which had been prepared with significant input from external sources. Presenting specific shuttle requirements as a presidential decision was an example of the tendency noted by Cap Weinberger of "the OMB staff acting on their own" with respect to the shuttle in a way that "may or may not have represented the policy of the appointed heads" of OMB, much less that of the president.[19]

The conservative philosophy behind the OMB-preferred shuttle was that it should "replace the current CSM [command and service module] capability for manned flight with increments of capability only to the extent they are both cost-effective and within overall fiscal feasibility." OMB argued that "a small, versatile system is more likely to be used and exploited and less likely to encounter development delays and cost overruns." With respect to orbiter size, OMB suggested that NASA should:

- "Exploit ability to dock payloads in orbit for near earth and synchronous missions (one flight carries payload and second payload carries tug)."
- "Rely on the ingenuity of payload designers to fit payloads into smaller compartments than currently projected."
- OMB argued that a "bay size of 10' × 30' with 30,000 # [pound] payload due East would add sufficient capability beyond manned flight to capture most payloads."

With respect to the "fiscal constraints" affecting shuttle development, OMB set demanding targets:

- "$4B maximum for DDT&E [design, development, test, and evaluation] including development vehicles";
- "Other investment costs (facilities and additional vehicles) should be held to a maximum of $.5 B";
- "Recurring costs per flight of $5 M";
- "Peak NASA budget level $3.2 B in FY73$ [Fiscal Year 1973 dollars]."

As the December 11 meeting broke up with OMB and NASA at loggerheads, NASA agreed "to do further analysis of the 10' × 30' payload so that we would have some good facts at hand and then we will have to decide whether the small shuttle makes any sense at all or whether we will have to fight for a larger one."[20]

Origins of the OMB Shuttle

The detailed performance and budget requirements for a smaller shuttle that OMB presented to NASA did not originate from the OMB staff, none of whom were aerospace engineers. Rice had sought outside advice on shuttle configuration and capability. He noted "some of my information came from the Defense Department, but not very much of it." He added "some of it came from industry. There were clearly some people in industry who were concerned that NASA was going to lead them down the road of another C-5A or F-111 debacle and that they would end up with nothing." The two programs Rice cited were DOD aircraft development efforts during the 1960s characterized by major cost overruns. Rice noted that "there was some interest at least among some people in the aerospace industry [in] having whatever was done be a program that was politically survivable." That interest translated into an attitude of "let's have it be less rather than more so it doesn't turn out to cost so much and it is less likely to overrun and you can keep it." Given this feeling, Rice worked with an aerospace firm, almost certainly one of NASA's shuttle study contractors, to help his staff spell out the characteristics of a shuttle concept that both made technical sense and could be developed at an acceptable cost. Weinberger was aware of what Rice was doing, saying "Don Rice asked me if he could go out and get some other people, or he told me that he had."[21]

NASA Reaction

The NASA leadership was angered by the idea that OMB, rather than NASA, should define the technical characteristics of the shuttle; that anger intensified when NASA discovered that some of the OMB requirements originated from a NASA shuttle study contractor feeding information to OMB on a confidential basis. Senior NASA official Willis Shapley characterized Rice's seeking advice from an aerospace firm as "dirty pool in the budget wars." He added "the one thing that really grated people wrong was that they [OMB] began getting engineering and technical information...and confronting our technical people with technical judgments." Low reported that Rice "claimed that the basis for the cost estimates he had were from a contractor whose name he could not divulge." Low called Rice and asked him "specifically to let me know who the contractor was so that we could verify his numbers or see where we, NASA, were going wrong. He refused to do so." Even in internal OMB correspondence related to the shuttle decision, the source of Rice's information was referred to only as "your contractor source." However, Low concluded, "based on the information we have and the questions Rice has asked, it is quite clear that he obtained his information from North American Rockwell."* Low reported that NASA Administrator Fletcher "objected to OMB designing the shuttle strongly"; Fletcher agreed, saying that in his dealings with Rice "the only thing I did resent was his trying to design shuttles of his own."[22]

Fletcher and Low felt that they had little choice but to begin evaluating a shuttle meeting OMB guidelines. On December 13, Low told Dale Myers that "the Office of Management and Budget has set forth certain concepts and assumptions concerning the Space Shuttle program" and that he and

---

* The author has not been able to find independent evidence supporting Low's conclusion that North American Rockwell was providing information to OMB counter to what NASA was advocating. This conclusion seems a bit questionable, given that in late 1971 the head of North American Rockwell, Willlard "Al" Rockwell, was visiting the White House to lobby for shuttle approval. But perhaps Rockwell was not aware of the fact that people at the working level within his company were cooperating with OMB.

Fletcher had "made a commitment to provide our assessment of the OMB assumptions and guidelines by December 31, 1971." He added that the evaluation of the OMB shuttle should compare it to the configurations NASA still was considering and should assume the use of the new shuttle engine and a reusable pressure-fed booster or, as an alternate, solid rocket motors. Low noted that "the initial development cost is of primary concern...Is it possible to start out with one of the smaller (or lighter) payload versions, and then grow to a full capability orbiter later on?"[23] There was no emphasis on minimizing the costs of operating the shuttle once it was in service.

Fletcher met with Flanigan on December 14 to protest the OMB directive. When Fletcher asked whether the OMB shuttle characteristics and the $4 billion development cost limit in fact were based on a presidential decision, Flanigan, who had not been at the December 3 meeting, said that he did not think so and that it was more likely that Shultz or Weinberger had thrown out the $4 billion figure and Nixon had said "it's worth a try." Flanigan also said that he would check whether there was a presidential decision document that included the shuttle characteristics specified in the OMB guidelines. He did check, and on December 17 told Fletcher that his (Flanigan's) "understanding of the request put to you regarding a smaller shuttle was correct. None of the figures in the paper given to you [by OMB] are set in concrete." Rather, he said, "they should be viewed as a new way to approach the problem, against which an initial estimate will have to be made in a couple of weeks." Flanigan added "there is no written directive from the President on this subject." In his December 14 meeting with Fletcher and later conversations, Flanigan advised against taking the NASA–OMB disagreement to the president for decision; he knew that Nixon tried to avoid refereeing such confrontations.[24]

New Technology Opportunities Effort Collapses

One issue that had been in the background through much of 1971 had been the possibility that President Nixon would decide to broaden NASA's mandate to include large-scale efforts to apply technology to the solution of various social problems outside of the aeronautics and space arena. Spearheading the effort to develop such "new technology opportunities" in the White House had been William Magruder, former head of the supersonic transport program. By November, Magruder had come up with a proposal to establish within the Executive Office of the President a new unit with some 300 staff members (many more than staff working for OMB or OST) as an interim step to coordinate planning a major technology initiatives effort, with the possibility that after sufficient planning was completed NASA might be asked to take on some or all of the new activities.

However, there were emerging problems with the Magruder effort. When Low in late November asked Magruder when the NASA people Magruder had requested to help staff the new office should report for duty, "it became quite apparent that he did not yet have clearance to move out with this so-called interim organization." At about the same time, Rice told Low that "there was a great deal of controversy within the White House as to whether or not Magruder ought to establish this organization." Rice thought that "nothing much will happen as a result of the New Technological Initiatives" and that "there was no sense in going ahead with the massive Magruder exercise." Rice was correct; during December both OMB and OST raised strong objections to the Magruder plan, and it was stillborn, with only a few modest efforts in stimulating technological innovation eventually approved. Low thought that "a strong NASA association" with the Magruder effort "would have done us a great deal of harm."[25]

The collapse of the Magruder exercise was not explicitly linked to increasing the chances of NASA's getting approval for a full capability shuttle and its other post-Apollo ambitions. The Domestic Council's Ed Harper said "I was at all the relevant meetings and the two programs [New Technology Opportunities and the space shuttle] were never discussed in terms of a trade-off." Weinberger suggested that "there was no connection between the two...The shuttle was already there on a separate track." When asked whether the collapse of the Magruder effort got linked to the shuttle decision at the president's level, Ehrlichman replied "I don't think so."[26]

The Showdown Looms

During the second half of December the White House prepared for a final decision on the space shuttle configuration. On December 16 the OMB space unit prepared a memorandum for the president discussing the "capabilities, size, and cost of the space shuttle" as "the one key Presidential issue remaining in the NASA FY 1973 Budget." The memorandum made the case for the OMB shuttle approach, and noted that "the difference between the employment impact of the two versions of the Shuttle on 11/72 [an indirect way of saying 'on the Presidential election'] is negligible. Announcement of a favorable decision for either version would be gratefully received by the aerospace industry." The memo recognized that the larger shuttle could "transport certain intelligence satellites and a relatively few large astronomy satellites," but that "achieving the extra capability of the larger version is not of near-term importance." It suggested that "it is important to maintain the Titan III for national security. Dropping the dependable Titan III would place too much reliance upon a new and unproven system for vital national security missions." Approving "the lower cost Shuttle would preserve the option to build bigger versions in the 1980's if really required. There is a high probability that this will not be the case." The memo recommended that "policy guidance be given to NASA that (a) the total investment cost of the Shuttle (including facilities and vehicles) is not to exceed $5 billion, (b) the recurring cost per Shuttle launch is not to exceed $6 million, and (c) the peak NASA budget during the rest of the 1970's is not to exceed $3.2 billion (in 1971 dollars)." These cost constraints had been modified slightly upward compared to the December 11 OMB shuttle paper. With this guidance, "NASA and its industrial contractors [should] proceed at once to begin to define the best system that can be developed within the overall fiscal constraints."[27]

Don Rice forwarded the draft presidential memo to Cap Weinberger together with a cover note that revealed some of the tactics that OMB was employing in its dealings with NASA. Rice noted that "the fact that the Shuttle decision is still open is our most significant bargaining point with NASA" with respect to the agency's future. He suggested, "as part of the decision on the Space Shuttle, an understanding be reached with Dr. Fletcher about the need for the closure of a manned space flight center after Apollo and Skylab are completed." In order to receive approval for the shuttle, NASA would have to agree in several years to reduce its institutional base, a particular OMB objective. But no action on this closure "would be initiated or announced" until after the November presidential election. Rice closed his note to Weinberger by suggesting that "it would seem unwise to approve... NASA's request for a large Space Shuttle."[28]

This draft memorandum was not forwarded to Richard Nixon; the space shuttle issue was instead addressed by his senior advisers. NASA was scheduled to meet with OMB on December 29 to make its final recommendation with respect to the shuttle. In preparation, on December 28 there was a meeting in the Indian Treaty Room of the Old Executive Office Building at which Don Rice discussed the various shuttle configurations with senior White House staff such as Ehrlichman, Shultz, Weinberger, Flanigan, and international economics counselor Peter Peterson. Bill Anders held models of the different configurations as Rice spoke. Later on the same day, Ehrlichman met separately with Ed David and Peter Flanigan to discuss the shuttle decision. During one of these meetings, Ehrlichman called Anders, asking which shuttle configuration would produce the most aerospace jobs in southern California. Anders replied "you don't need to be a rocket scientist to know that the bigger the shuttle, the more the jobs." Ehrlichman replied "OK, that will be the one" which would be approved. By the end of December, when the final decision on the shuttle design was to be made, there was thus a good understanding within the senior levels of the White House of the issues at stake.[29] It was clearly time for a decision.

Chapter 14

A "Space Clipper"

As a decision on the shuttle neared in the final days of 1971, Jim Fletcher and George Low continued to interact with the relevant White House and Executive Office officials. They told Cap Weinberger on December 22 that NASA was "not yet in the position to respond to Don Rice's request of December 11." NASA's human space flight element was still resisting serious analysis of the OMB-suggested shuttle design. The Weinberger meeting "was followed by another series of phone calls from Jonathan Rose in Peter Flanigan's office, who is primarily concerned with employment in Southern California." On December 23, Fletcher and Low had lunch with Bill Anders, his assistant David Elliott, Tom Whitehead, and Jonathan Rose. These individuals "were all trying to be very helpful and particularly wanted to bring the [shuttle] issue to a proper decision." On the basis of their White House discussions, Fletcher and Low learned "that there indeed was a Presidential decision to go ahead with the Shuttle; that the issue of size was not really raised as a major one with the President; but that David and Rice, and to a lesser extent, Flanigan, felt that the 15 × 60′ 65,000 pound shuttle proposed by NASA was really too big."[1] Based on messages such as these, the NASA leadership by the end of December was increasingly skeptical that it could get White House approval for a full capability shuttle, and was searching for an acceptable compromise.

What Shuttle to Recommend?

On December 27, Low met with those at NASA headquarters involved in the shuttle program "to discuss the various options of payload size and shape, payload carrying capability, booster options, etc." On the next day, he held individual meetings with his senior associates to get their frank assessment of the best course to pursue. The decision coming out of these meetings was to accept a slightly less ambitious shuttle design; Low reported that "as a result of these meetings, we decided that we should proceed with a Shuttle that has a 14 × 45′ payload [bay] and a payload carrying capability of 45,000 pounds. We further decided that we should hold open the option of a liquid vs. a solid booster for another two months." Jim Fletcher had not been involved in the December 27–28 meetings, but quickly accepted their conclusions. Low observed "from NASA's point of view, and not necessarily out in the open, the size and weight we picked could do most NASA missions and some of the DOD missions, but particularly would have the growth capability to the full size Shuttle should such a decision be made at a later date."[2]

Getting Close

On the afternoon of December 29, 1971, Fletcher and Low met with George Shultz, Cap Weinberger, and Don Rice from OMB, Peter Flanigan and Jonathan Rose from the White House, and science adviser Ed David to present NASA's proposal of how best to proceed with respect to the space shuttle. A decision was needed soon; the president's budget message was due to go to the printer in the first week of January, and it would have to contain some indication of the fate of the space shuttle program.

Fletcher on the morning of the meeting sent to Weinberger a letter reflecting the decisions reached within NASA in the past few days. The letter said: "We have concluded that the full capability 15 × 60' 65,000# payload shuttle still represents a 'best buy' and in ordinary times should be developed. However, in recognition of the extremely severe near-term budgetary problems, we are recommending a somewhat smaller vehicle—one with a 14 × 45'—45,000# payload capability, at a somewhat reduced overall cost." The letter added "this is the smallest vehicle we can still consider to be useful for manned flight as well as a variety of unmanned payloads." NASA gave highest priority to retaining a shuttle configuration that was large and powerful enough to eventually launch components of a space station, and the 14 × 45 foot shuttle it was now recommending had that capability, even though it would not be able to launch the largest intelligence satellites or astronomical observatories.

The Fletcher letter also reported NASA's assessment of the shuttle design suggested by OMB, saying that "we have not been able to meet" the objectives of a development cost of less than $4 billion and a cost per flight of less than $5 million. NASA noted that the 30-foot payload bay length suggested by OMB "eliminates nearly all DOD payloads, some important space science payloads, most application payloads, all planetary payloads, and useful manned modules." Attached to the letter was a table (reproduced on next page) showing the results of NASA's evaluation of various shuttle configurations.

The letter said that "the question of a liquid as opposed to a solid booster is not yet completely settled—there are some open technical questions" and "the differences in operating costs [for the two boosters] have not yet been determined with accuracy." For these reasons, NASA recommended that the choice among booster options should be deferred for two months to allow additional study.

NASA also asked for a "funding contingency," saying that "it is our intention to manage the program to bring it in" at the costs spelled out in the

Various Shuttle Options Presented by NASA to the White House, December 29, 1971

Payload bay size (foot)	10 × 30	12 × 40	14 × 45	14 × 50	15 × 65
Payload weight (pounds)	30,000	30,000	45,000	65,000	65,000
Development cost (billions)	4.7	4.9	5.0	5.2	5.5
Operating cost (millions)	6.6	7.0	7.5	7.6	7.7
Payload costs ($/pound)	220	223	167	115	118

Fletcher letter. NASA added "nevertheless, we believe that we should include a contingency against future cost growths due to technical problems...We believe a 20% contingency would be appropriate...Approval of a $5 billion program [for the 14 × 45' orbiter] would thus constitute a commitment by NASA to make every effort to produce the desired system for under $5 billion, but in no case more than $6 billion."

Finally, the letter argued that it was time for "a decision to proceed with full shuttle development" to be made. "Further delays would not produce significant new results," and "additional delays would have many unsettling effects...There is a great deal to be gained, and nothing to be lost, by making a decision to proceed now."[3]

Going into the meeting, Fletcher and Low were uncertain of its outcome; they even agreed in advance that they could accept a shuttle as small as one with a 14 × 40' payload bay and 40,000 pound lift capability, but that anything smaller "would require a Presidential decision." At the meeting, "the principal negative guy, once again, was Don Rice who indicated that he did not believe NASA's figures or the figures presented to us or to him by our contractors." However, "during the meeting Shultz looked at the facts and figures and decided that really the only thing that makes any sense, as NASA had said all along, is the 15 × 60'—65,000 lb. Shuttle capability." Fletcher recalled that "at the end of the meeting, George said, well, it's a pretty easy decision. We'll go for the 60-foot one. We had George saying that and no one arguing with him."[4]

Low noted that "no decision was made in the meeting," but added that "Fletcher and I were fairly confident that our recommendation of the 14 × 45' 45,000 lb. Shuttle would be accepted as a minimum and that even the full capability [shuttle] might still be accepted." A second senior-level meeting was scheduled for Monday, January 3, 1972, after the New Year's weekend, to make the final decision.[5]

Last Minute Objections Raised

Not surprisingly, given the deep skepticism of OMB and OST with respect to the wisdom of going ahead with a large space shuttle, last ditch opposition to approving NASA's shuttle plans continued. David wrote Shultz on December 30, saying that he was "disturbed by the prospect that a decision will be made" to approve the full-sized shuttle. It was David's view that "the large space program implicit in the large shuttle decision is not consistent with the best interests of this nation." David was joined in opposition by Rice and his OMB space staff, who remained dubious regarding the economic justification for a large shuttle, saying that "all of the decisions regarding the Shuttle should be made in the full awareness that the Shuttle is not a cost-effective system—whether at 15' x 60' or any other size." Thus, what should be approved was "the smallest Shuttle which can offer improvements over our current methods of operation. Any size Shuttle will provide manned space flight and national prestige."[6]

Weinberger offered Rice one last chance to make the case against a large shuttle. On December 30, Weinberger called Fletcher, asking for another look at a shuttle with a 14 × 45 foot payload bay but limited to lifting only 30,000 pounds to orbit. This suggestion came from Rice. Low reported that "Fletcher came close to telling Weinberger to go to hell." Fletcher then called Shultz "and had another lengthy conversation with him." Shultz was also "unwilling to make a decision and recommended that we should not yet go to see the President but take one more look at the request made by Rice and presumably David." The next day, Friday, December 31, Fletcher and Low "held a telephone conversation with David and Rice without really getting any new information. Rice did all of the talking and David was very quiet." Low added

> Rice said that he would still like to see us go ahead with the 12 × 40' 30,000 lb. Shuttle, but that they're willing to give in on size...He wanted us to crank up studies over the weekend to answer all of his questions. I pointed out that our people had already scattered for New Year's...I called Rice back later and asked him for a piece of paper to spell out in writing, once and for all, all that he wanted us to do. He indicated that the piece of paper would be a good idea but that he would not commit that it would, once and for all, ask all his questions.[7]

OMB sent eight questions to NASA, including "if future budgets for NASA were constrained to $3.2–$3.3 B, would you still want to do the large Shuttle?" and "why should a relatively few space station modules for the mid-1980's determine the size and weight capabilities of the Shuttle?" Other questions dealt with more specific technical issues. By the following Monday, NASA had developed answers to most of OMB's questions. With respect to the first query, NASA said "the answer is yes"; with respect to the second, NASA provided a detailed list of the payloads other than space station modules that required a weight-lifting capability of over 30,000 pounds.[8]

Reflecting on the need to respond to OMB's questions, a frustrated George Low complained that "there is nobody [likely referring to Shultz and Weinberger] in the White House willing to make any decisions. Everybody feels that the issue of Shuttle size is too small [not important enough] an issue to take to the President... but they're also unwilling to let the Administrator of NASA to make that decision." The result was that "they let their various staffs continue to... ask nickel and dime sized questions without ever calling a halt to that procedure and say it's about time we made up our mind and let's proceed." Looking back at the decision process several years later, Low added "the single most significant factor affecting the space shuttle decision was that there was no top-level leadership in the White House. President Nixon was unwilling to deal with his agency heads and dealt solely with his staff. This placed a great deal of decision-making responsibility with the OMB, and by definition the OMB is far more interested in short-range budgetary problems than in the long-range future of the nation." Low's criticism was not completely fair; both David and Rice couched their opposition to the large shuttle in terms of its longer term impact on the U.S. space program, not just on shorter-term budget issues.[9]

In advance of the late afternoon meeting on January 3, 1972, at which a decision how to proceed with the shuttle had to be made, David and Rice continued their opposition to the choice of a large shuttle. By this point, they recognized that some sort of announcement of presidential approval of shuttle development was a *fait accompli*, but were still arguing for having President Nixon make that announcement only in principle, without deciding on a specific shuttle design. David on January 3 sent another memorandum to Shultz, this time arguing that "it would be desirable to defer a decision on the configuration, while announcing the Administration's intention of proceeding with development of a new, reusable space vehicle for man and other payloads that will use advanced technology." David suggested a three-month delay in selecting a shuttle configuration; during that time, NASA studies would "complete detailed examination of the lower cost alternatives to the full Shuttle capability." Rice supported David's argument, telling Shultz "Dr. David's proposals... appear to make a great deal of sense," and that "the Administration should carefully examine the alternatives before committing itself to a very costly and potentially unpopular large new space program." Rice noted that "after urging by OMB and OST, NASA has only in December started looking at possible smaller and less costly alternatives—compared to about two years of study for the bigger system." With the deadline for printing the president's budget message fast approaching, Rice suggested that the budget message should include only "a general announcement of a decision to proceed with development of a new system for lower cost delivery of man and other payloads into space."[10] Both Rice and David resisted calling that new system a "shuttle."

David and Rice may have exceeded their appropriate staff roles in trying to change the minds of their political leaders by arguing in support of their strong conviction that approving the NASA shuttle was not in the country's interests. For example, Rice had sought outside help from the aerospace industry in developing OMB's alternate shuttle design and used shuttle approval as a bargaining chip in attempting to get NASA to downsize its institutional base. He gave little weight in his opposition to issues such as aerospace unemployment and the political impact of shuttle approval on the 1972 presidential election. In the judgment of one close observer of the decision process, Peter Flanigan's assistant Jonathan Rose, this behavior went beyond acceptable bounds. Rose observed that

> Ed David and Don Rice may well have been right that there existed a different cost curve than the one that NASA was able to find for a shuttle with a smaller bay and lighter payload. I am quite clear that only their pressure forced the shuttle modifications which produced the massive savings from the August shuttle [the two-stage design] to the December shuttle. They were however unable to prove their case when it came to another billion dollars in potential savings if we delayed for several months more. While NASA may not historically have effectively studied the smaller shuttle, I became convinced that Jim Fletcher had in the time given to him done the best he could. In the last analysis, that is all one can ask of an honest agency head. He should not be brutalized on a continuing basis by the budget process or by the White House staff when such pressure appears to reach the point of diminishing return.
>
> The essence of judgment is to know when to stop. I simply think that Don Rice failed us here. He viewed the political situation as well as the plight of the contractors very lightly. He was far more interested in pursuing the marginal cost savings which his staff led him to believe were possible. This in turn finally led him to some highly shoddy tactics in *ex parte* lobbying.
>
> I believe we reached the best possible result under the circumstances. In a non-election year I might have seen the equation differently and been willing to wait several extra months to see if Rice was right. But I believe you have to play the ball from where it lies, and this after all is 1972.[11]

It is arguable whether Rice and David "brutalized" NASA in opposing a full capability shuttle or whether they behaved responsibly in making sure the reasons for that opposition were fully understood by the political decision makers. What is clear is that they did state their case with vehemence, that short-term considerations related to aerospace employment in advance of the 1972 presidential election played a crucial role in the final decision to approve the full capability shuttle, that Rice and David were on the losing side of the argument, and that, with the benefit of hindsight, they were fully justified in their opposition.

Finally, a Decision

In preparation for the January 3 White House meeting, the NASA leaders prepared a letter reporting on their conclusions following the harried weekend of answering OMB's questions. The letter reported that "the previous conclusion that the full capability 15 × 60—65,000# shuttle makes the most sense has been reaffirmed and we now urge—even more strongly—that this configuration be adopted." It said that "the OMB proposed option of a 14 × 45—30,000# shuttle is not acceptable because it will not handle manned space station modules, manned sortie flights, or manned resupply missions in a standard space station orbit." In addition, "this shuttle would not handle 28 *different* science, applications and planetary payloads." Once again, NASA asked for an "Administrator's contingency" of 20 percent of the estimated development cost to accommodate "future cost growths due to technical problems."[12]

Before their meeting, Fletcher and Low stopped by the offices of the Space Council across Pennsylvania Avenue from the White House to discuss with Bill Anders, who had become an ally in their conflict with OMB and OST, "what they were going to say and what they thought the state of play was. Clearly they thought everything was still under scrutiny and study and it wasn't close to a decision." Then they went to Shultz's White House office; the 6:00 p.m. meeting was attended by Shultz, Weinberger, Rice, David, Flanigan, and Nixon Congressional liaison Clark MacGregor. David briefly restated his opposition to going ahead with the NASA-recommended 14 × 45 foot shuttle, but Shultz quickly overruled both David and Rice and told Fletcher and Low that they could proceed with their plans for the full capability 15 × 60 foot, 65,000 pound shuttle. At some point between December 29 and January 3, Shultz had telephoned fellow economist Oskar Morgenstern to discuss the Mathematica study of shuttle economics that Morgenstern's firm had carried out; Morgenstern assured him that the shuttle was a reasonable program in economic terms. (One report even had Shultz making the call to Morgenstern during the January 3 meeting, but this seems unlikely, given the short duration of the meeting.) With that assurance, aware of the impact of the shuttle on aerospace employment, and also apparently aware of President Nixon's interest in the national security missions enabled by the full capability shuttle, Shultz had decided before the meeting to approve NASA's full capability shuttle configuration. Within a few minutes, Fletcher and Low were back in the Space Council office, "kind of elated," to report "we didn't have to say a word; we were just told that the decision was to go ahead" with the full capability shuttle that NASA had been advocating all along. When the two NASA leaders returned to NASA headquarters and reported the outcome of their meeting to human space flight chief Dale Myers, he was "amazed."[13]

The next day, to be sure that his understanding of what had been decided was correct and to get that understanding on the record, Fletcher wrote Weinberger "to document the decision reached yesterday concerning the space shuttle." As Fletcher understood it: "NASA will proceed with the development of the space shuttle. The shuttle orbiter will have a 15 × 60-foot payload bay, and a 65,000-pound payload capability. It will be boosted either by a pressure-fed liquid recoverable booster or by solid rocket motors. NASA will make a decision between these two booster options before requests for proposals are issued in the spring of 1972." In addition, "NASA and industry will also continue to study, for the next several weeks, a somewhat smaller version of the orbiter...The main purpose of studying this smaller shuttle is to determine whether or not significant savings in operational costs can be realized, with [already existing] solid rocket motors, at this smaller size."[14]

A Related Issue

Even as NASA was receiving White House approval to proceed with the large space shuttle, Fletcher and Low were concerned about whether the shuttle program would gain Congressional approval; that was one of the reasons that Clark MacGregor was at the January 3 meeting. At the same time the shuttle was being approved, the White House had finally decided to cancel the NERVA nuclear rocket engine project after keeping it on life support for the previous several years. The NASA leaders' concern was that "without NERVA we will not have the political support in the Senate that we need for the Space Shuttle and other programs." Low's assessment was "that were we to cancel NERVA we have a 50/50 chance of completely losing all support by Howard Cannon [D-NV]." Fletcher agreed with Low, telling Shultz that other than Cannon, "there are no other spokesmen, on the Democratic side of our [Senate Space] Committee, that would or could carry the NASA bill through the Senate. Therefore, without a meaningful nuclear propulsion program, we are taking the very major risk of losing the space shuttle, as well as other pieces of the NASA program, in the Senate." Low even suggested that "the NERVA situation is to my mind more complicated and more difficult than the Shuttle question."[15]

The final outcome was to cancel NERVA, to allow NASA to carry out a study effort to define a smaller nuclear propulsion system, and to include in the president's budget request with respect to nuclear propulsion language intended to be palatable to Senators Cannon and Clinton Anderson (D-NM), another strong supporter of NERVA. Anderson was actually the chairman of the Senate Committee on Space and Astronautics, but he was old and ill, and not able to lead the Senate debate on the NASA budget. These moves may have been essential in assuring eventual Senate support of the space shuttle.

Was Richard Nixon Involved?

President Nixon had decided on December 3 to approve some form of a space shuttle program. Whether or not Richard Nixon was consulted later in December or over the New Year's weekend, as the decision to approve the full capability shuttle was made, is not clear. There is suggestive evidence to support either possibility.

Prior to the December 29 meeting at which Shultz gave the first indication that he would support the large shuttle, there had been general agreement among the White House staff that the issue of shuttle payload bay size and weightlifting capability was too detailed and too technical to bring before the president. There were no meetings with the president to make final decisions on any agency budget appeals, with the exception of the Department of Defense, in the days just before or after Christmas. Nixon was at his Key Biscayne, Florida, residence from December 27–31. There is no record on Nixon's official schedule of a phone call from anyone involved in the December 29 White House meeting with NASA to discuss its outcome with the president. On December 30, Shultz was still urging Fletcher not to insist on seeing the president with respect to the shuttle decision. On December 31, Nixon did try, unsuccessfully, to telephone Al Rockwell, but this appears to have been just one of many "Happy New Year" calls Nixon placed that day to people with whom he had a personal relationship. Shultz and Weinberger did meet with Nixon on the morning of January 3 after the president returned to Washington, but that meeting was to discuss the overall shape of the FY1973 budget; neither the space shuttle nor any other specific program was discussed.[16] All of this evidence tends to suggest that Nixon was not consulted as the final decision on shuttle size was being made.

There is some counterevidence, however, that Nixon might indeed have influenced the final decision on shuttle configuration, whether on that final weekend or before. John Ehrlichman suggested in a 1983 interview that he was sure it was "Nixon's decision on the thing [shuttle], because the way these things would come to him would be with alternate levels, and then a brief description of the differences—if you go to this level, you get that; if you go to this level, you get that plus this. There wasn't anybody during that time who made those final decisions except Nixon...Defense, space, certain kinds of domestic problems, he was the final arbitrator." However, Ehrlichman may well have been referring to the November 24 and December 3 meetings at which initial decisions on the space shuttle had been made. As noted in chapter 13, Ehrlichman and Nixon were attracted by the national security uses of the shuttle. Nixon would mention those uses in his January 5 meeting with NASA's Fletcher and Low, indicating that he was already aware of them. As the final decision on shuttle configuration was being made in March 1972, Cap Weinberger would reiterate to Fletcher "the President's strong interest in retaining the military capability" as a factor in the "decision on the larger size" shuttle, and Fletcher would say that "the President's expressed desire to make the shuttle a useful vehicle for military space operations could not be fulfilled with the smaller shuttle." When, and by whom, Nixon was briefed on the national security uses of the full capability shuttle is not clear. At least Ehrlichman, and probably also Shultz and Weinberger, were aware of Nixon's interest in national security applications of the shuttle's capabilities as they made the final decision on shuttle size. That they took the step of actually consulting with the president at that point appears unlikely. While there is no doubt that Richard Nixon gave the green light to developing the space shuttle and that he had expressed interest in the shuttle being able to carry out a wide range of national security missions, it is probable that George Shultz and Cap Weinberger, possibly after consulting Ehrlichman and Flanigan, were the individuals who made the final decision on approving the full capability shuttle. They made that decision on their own, not on the basis of a specific presidential directive.[17]

Transforming the Space Frontier

From a White House perspective, the December 29 meeting on the space shuttle had resulted in a definitive enough decision that there would be a space shuttle program to begin preparing for a presidential announcement of his approval of the shuttle. As those preparations began, there were two open questions: what should the presidential statement say and whether the space shuttle program should be given a distinctive name, just as prior U.S. human space flight programs had been christened Mercury, Gemini, and Apollo.

With respect to the first issue, NASA's Fletcher had been alerted by the White House during the New Year's weekend of the possibility of a presidential announcement. The decision to make that announcement was "firmed up" during the January 3 meeting in Shultz's office, and Flanigan asked NASA to prepare a draft statement. Even before this request, Fletcher and Jonathan Rose of Flanigan's office had also asked Bill Anders to start working on the presidential statement. Even though Fletcher and Low also prepared draft statements, it was the Anders draft that was the primary basis of the final presidential statement. With respect to the second issue, naming the space shuttle program, the decision was left to Richard Nixon himself.[18]

Presidential Announcement Scheduled

Nixon on November 24 had indicated that he should announce his approval of shuttle development "out in California where you are going to put it." Fortuitously, Nixon was scheduled to fly to California on the evening of January 3 in advance of a January 6 meeting at the Western White House in San Clemente with Japanese Prime Minister Eisaku Sato. His presence there provided the opportunity for an early announcement of shuttle approval.

There was some initial confusion at the White House about what actually was being announced. On December 30, Nixon's political advisor Charles "Chuck" Colson initiated a proposal that President Nixon should visit the Rocketdyne plant in Canoga Park, California, to "announce the initiation of research on engines for the space shuttle." Colson had not realized that the announcement would deal with the shuttle overall, not just its new engine. He also was seemingly unaware that the engine contract award was being protested by Rocketdyne's competitor, Pratt & Whitney, and thus it would have been inappropriate for the president to visit the Rocketdyne plant. When these realities were recognized, the Colson recommendation was quickly withdrawn in favor of a December 31 proposal by Peter Flanigan that Nixon meet with Fletcher and Low on January 4 (soon changed to January 5) "to discuss the decision to go ahead with the shuttle program which will insure the continuation and expansion of thousands of additional jobs in the space industry. This announcement is particularly significant to Southern California." Once again, the employment impact of starting the shuttle was identified as of high importance. Flanigan's schedule proposal noted that "at a recent budget session in Key Biscayne [likely referring to the December 3 meeting, since there were no other budget sessions on the president's published calendar while Nixon was in Key Biscayne between Christmas and New Year's Day] the decision was made to go ahead with the space shuttle program. Some of the mechanics in implementing the program have still to this moment not been completely resolved, but will be on Monday, January 3."[19]

A New Name for the Space Shuttle?

On December 29, Flanigan had asked NASA to suggest a new name for the program. Fletcher replied on December 30, telling Flanigan that the names he was proposing were "drawn from a much longer list previously generated by our Public Affairs department and by the people working on the shuttle itself, with the addition of several contributed by George Low and me." That longer list included suggestions such as: *Mayflower, Starship, Spaceliner, Star Frigate, Caravel, Star Packet, Star Freighter, Rocket Clipper, Star Ferry, Space Tram, Star Schooner,* and *Space Schooner,* all of which were rejected. Fletcher's preferred names were *Skyclipper, Skyship, Pegasus,* and *Hermes.* He gave second priority to *Space Clipper, Astroplane, Skylark,* and *Dragonfly.*[20]

In a January 4 memorandum to the president, Flanigan told Nixon that the name "space shuttle" does "not have the lift or importance that the project deserves. The word 'shuttle' has a connotation of second class travel and lacks excitement." Flanigan added "assuming that you wish to choose a name other than 'space shuttle,'" the suggested names were

1. *Space Clipper*—Generally agreed upon by NASA, Shultz, Safire, Moore, Davis and me, this name would describe the overall project. Individual vehicles might have individual names, the first being Yankee Clipper;
2. *Pegasus*—Preferred by the classicists, such as Jim Fletcher;
3. *Starlighter*—Dick Moore's favorite.

Commenting on the choice of names, speechwriter Bill Safire had suggested *Space Clipper, The Yankee Clipper, Rocket Ship #1,* and *Space Ship #1.* Safire was attracted to *The Yankee Clipper* name "because of its historic and patriotic association," which had been used to describe "a fleet of ships designed for speed and passengers rather than cargo and helped make the American merchant fleet preeminent in the early 19th century." (Safire's history of clipper ships was not quite accurate.) He added that "the name would be criticized as nationalistic, but I think that heat would be good." Safire advised against the name *Pegasus* "because it would soon be named Peggy and parodied with the old song title 'Peg of My Heart.'"[21]

Preparing President Nixon for the Shuttle Meeting

As President Nixon prepared for his meeting on the space shuttle decision, he was reminded of the overall situation with respect to California employment. Rose, in a January 3 memorandum forwarded to the president through Flanigan and Weinberger, reported that "a combination of actions set in motion by OMB, the Domestic Council, and this project [the White House California Employment Project] should produce at least 100,000 incremental jobs by November 1972," in time for the presidential election. One element of this job creation effort, Rose reported, was "a 'go' signal on the NASA shuttle (1600 California jobs and a tremendous lift for aerospace industry)."[22]

It was standard practice in the Nixon White House to provide Nixon with detailed briefing material in advance of a scheduled meeting; this was the case with respect to his meeting with Fletcher and Low. Late on the afternoon of January 4, the Nixon aide who managed presidential meetings, Alex Butterfield, gave Nixon a briefing paper that had been prepared by Flanigan, including suggested talking points and a draft of the statement that would be issued to the press after the meeting. Butterfield noted that the statement reflected the selection of "*Space Clipper*" as the name for the shuttle, but that "John Ehrlichman and others have expressed some [unspecified] reservations with regard to this particular name." Butterfield also gave the president as part of the briefing package Flanigan's January 4 memorandum that listed three alternate names for the shuttle.

The briefing paper indicated that the president's meeting with the NASA leaders was scheduled to last 15 minutes and its purpose was "to indicate your involvement in the decision to proceed with the development of a space shuttle." This was another sign that Nixon had not been previously involved as the final decisions on shuttle configuration were made. The paper reminded Nixon that "you have decided that NASA will continue a man in space program, the next step of which is the design and manufacture of a space shuttle. (Dr. Fletcher will show you a model.)" It noted that "there has been considerable debate between NASA and OMB as to the proper size of the shuttle, with OMB driving for a substantial cost saving, but NASA getting the size it wants." Also, "this program will greatly stimulate the aerospace industry." Flanigan suggested that Nixon might "wish to ask Fletcher to describe the various scientific, earth applications and military missions for which the shuttle can be used" and that Nixon "should tell Fletcher the name you have chosen for the shuttle system."[23]

Richard Nixon Meets the Space Shuttle

John Ehrlichman joined the president for the meeting with Fletcher and Low. Fletcher had suggested that Peter Flanigan also be at the meeting, given his important role in the shuttle decision, but Flanigan was not present. As the two NASA officials waited to enter the president's office with a shuttle model, Ehrlichman asked whether it was the NASA shuttle or the OMB shuttle. Low's reply was "it is the United States' shuttle."[24]

Ehrlichman took detailed notes during the meeting; there was no taping system in Nixon's San Clemente office. Several days later, George Low

Press photographers and reporters capture the moment as NASA's Jim Fletcher and George Low show President Nixon a model of the space shuttle in the president's San Clemente, California, office on January 5, 1972. The top of John Ehrlichman's head is in the foreground. (Photograph WHPO 8172–4, courtesy of Richard Nixon Presidential Library & Museum)

also prepared a "memorandum for the record" regarding the meeting and discussed it in one of his "personal notes." Ehrlichman also prepared a "Memorandum for the President's File" summarizing the meeting. So there is a good record of what transpired as the meeting stretched from its scheduled 15 minutes to over half an hour. Low reported that "it soon became apparent that he [Nixon] was interested in the shuttle and in the space program as a whole and wanted to spend more time with us. The discussion was warm, friendly, and productive."

First, reporters and press photographers were briefly present to see Fletcher present the shuttle model to President Nixon. After they left, the first order of business was whether to adopt *Space Clipper* as the new name for the shuttle program. Nixon decided to defer the decision to a later time; this led to a rapid modification of the planned presidential statement to remove any mention of the *Space Clipper* designation. (The name of course was never changed—space shuttle it would remain.) Nixon asked if the shuttle was really worth a $7 billion investment. Fletcher and Low replied in the affirmative. Fletcher said that the shuttle was a necessary step to future space exploration, that it was too expensive to explore and do other things in space using existing launchers, that the shuttle was useful for military purposes such as a "sudden need" and interception and inspection of others' satellites, and that it was part of the "new frontiers of the mind" with "unpredictable" impacts. Fletcher also mentioned speculative future uses of the shuttle such as facilitating solar power from space and nuclear waste disposal; Nixon's reaction was that "these kinds of things tend to happen much more quickly than we now expect and that we should not hesitate to talk about them now." Nixon observed that the shuttle would "open up entirely new fields" and was not a "$7 billion toy," since it would "cut operations costs by a factor of 10." He added that even if the shuttle "were not a good investment, we would have to do it anyway, because space flight is here to stay. Men are flying in space now and will continue to fly in space, and we'd best be a part of it." The president was very interested in the status of planning for a docking between U.S. and Soviet spacecraft, and suggested that Ehrlichman ask Henry Kissinger to be sure to add a discussion of that possibility to the draft agenda for the May 1972 U.S.-Soviet summit meeting in Moscow.

Ehrlichman's brief summary of the meeting said: "After the press and photographers left the NASA representatives explained the Shuttle to the President and the President asked questions about the Russian rendezvous, the Sky Lab, the use of solar power, the recent AEC [Atomic Energy Commission] proposal for the disposal of waste in space and other technical matters." Low recorded that Nixon told him and Fletcher that NASA "should stress civilian applications but not to the exclusion of military applications." However, Ehrlichman's notes say that Nixon's guidance was to "downplay" the shuttle's military aspects, particularly in the context of future international cooperation. Nixon stressed that from the start of his presidency he had "an interest in international peaceful applications of space programs." Low records Nixon as saying that "he was disappointed that we had been unable to fly foreign astronauts on Apollo...He understood that foreign astronauts of all nations could fly on the shuttle and appeared to be particularly interested in Eastern European participation in the flight program." Nixon was "not only interested in flying foreign astronauts, but also in other types of meaningful participation." Fletcher told Nixon that the shuttle program would have a "big job impact," with 3,500 jobs in 1972, 14,000 by 1973, and 50,000 at its peak. (Commenting on a draft of Low's memorandum for the record regarding the meeting, Fletcher noted that the president "wanted to be sure that aerospace employment was mentioned, particularly on [the] West Coast." But Fletcher thought that because of the political sensitivity of Nixon's indicating in the meeting that the shuttle prime contract should go to a California company, Low should not mention this interest in his memorandum.) Nixon stressed that it was his view that the United States needed to be "No. 1" in all fields of space activity. "Like the new world," he said, "someone will explore." And it was important for the United States to be in the vanguard.[25]

Ehrlichman commented on "Nixon's fascination with the [shuttle] model. He held it and, in fact, I wasn't sure Fletcher was going to be able to get it away from him" when the meeting was over. Actually, Fletcher and Low left the model behind for possible display in Nixon's White House office.[26]

After the meeting was concluded, the White House press office issued the presidential statement, quickly revised to delete any mention of *Space*

Clipper. In contrast to John Kennedy's high-profile speech before a joint session of Congress announcing his decision to go to the Moon, Richard Nixon did not speak to the press about his shuttle decision. In the statement, which based on a draft prepared by Bill Anders, Nixon declared "I have decided was today that the United States should proceed at once with the development of an entirely new type of space transportation system designed to help transform the space frontier of the 1970's into familiar territory." The statement added "the space shuttle will give us routine access to space by sharply reducing costs in dollars and preparation time... Most of the new system will be recovered and used again and again—up to 100 times. The resulting economies may bring operating costs down to as low as one-tenth of those for present launch vehicles." The shuttle would "take the place of all present launch vehicles except the very smallest and the very largest." It suggested that "we can have the shuttle in manned flight by 1978, and operational a short time later." The space shuttle, the statement suggested, "will revolutionize transportation into near space by routinizing it. It will take the astronomical costs out of astronautics."[27]

There were a number of loose ends to tie up over the next two months before NASA would be ready to announce the final configuration of the space shuttle and invite aerospace firms to bid on a contract to develop it. In particular, the choice of how the shuttle orbiter would be boosted off the launch pad had not been made; both liquid-fueled and solid-fueled boosters remained in contention. But with his January 5, 1972, statement, President Richard Nixon had formally approved the space shuttle program; the shuttle would be the centerpiece of U.S. human space flight activities for the next four decades.

Finale

As soon as NASA headquarters in Washington received confirmation that the presidential statement had been issued in San Clemente, Charles Donlan, director of the space shuttle program, sent a message to shuttle program manager Robert Thompson at the Manned Spacecraft Center, saying "NASA will proceed with the development of the space shuttle. The shuttle orbiter is expected to have a 15 × 60 foot payload bay, and a 65,000 pound payload capability. It will be boosted either by a pressure fed liquid recoverable booster or by solid rocket motors." The message contained detailed instructions to guide the next phase of shuttle studies.[1]

Reaction to the president's announcement was mixed. *The New York Times* quickly editorialized that the shuttle was an "investment in the future" and that Nixon's decision to approve the shuttle was "wise." Predictably, given their 1971 attempt to cut funding for shuttle studies, Senators William Proxmire (D-WI) and Walter Mondale (D-MN) announced that they would lead the Senate opposition to shuttle approval; in addition, Senator Edward Muskie (D-ME), at that point the likely opponent for Richard Nixon in the 1972 presidential election, also said that he opposed the program as a "boondoggle." Talking with his political operative Chuck Colson on January 9, Richard Nixon was pleased to hear that Muskie's opposition to the shuttle "may have blown his chances in Florida completely." Nixon noted that "in Florida and California this [approving the shuttle] is a big deal. It will save the aerospace industry."[2] Whatever else his approval of the space shuttle meant to Richard Nixon, he saw it as an asset in terms of his reelection prospects.

Unresolved Issues

There were two major issues with respect to the space shuttle left unresolved as the January 5 statement was issued. One was what means would be used to boost the shuttle orbiter off the launch pad. The other was the character of the budgetary commitment to the shuttle program being made by the Nixon administration. The first of these was resolved by early March; the second persisted over the next few years.

Which Booster?

NASA after January 5 began a rapid examination of four alternatives for lifting the shuttle orbiter off the launch pad: parallel burn solid rocket motors, series burn solid rocket motors, parallel burn recoverable pressure-fed boosters, or a single series burn pressure-fed booster. The two parallel burn configurations had their origin in the studies carried out by McDonnell Douglas and Grumman and resembled the thrust-assisted orbiter shuttle (TAOS) design suggested in Mathematica's October memorandum.

The preference of NASA engineers as intensive booster studies began in January 1972 was one of the pressure-fed alternatives. The pressure-fed design was an invention of NASA's engineers at the Marshall Space Flight Center; the German members of the engineering group who had been brought to the United States after World War II had career-long experience with liquid-fueled boosters. A division of labor between Houston, which would be in charge of developing the new shuttle orbiter, and Huntsville, which would be in charge of developing the new pressure-fed boosters in addition to the shuttle main engine and external propellant tank, made institutional sense. But the booster studies soon showed that developing the new pressure-fed booster would be more difficult than it had appeared at first glance and thus carried the possibility of higher costs and more technical risk than had been anticipated. The OMB kept pressure on NASA to select the booster option that had the least chance of cost overruns. Since there was extensive Air Force experience with solid-fueled rockets, OMB leaned in that direction. OMB was concerned whether "NASA could overcome its instinctive dislike" of solid rocket motors. But Don Rice's "contractor source" told OMB that NASA headquarters was "insisting on an honest comparison."[3]

By early March NASA headquarters had made its choice—to go with solid rocket motors fired at liftoff in parallel with the orbiter's engines to boost the shuttle off the launch pad. The "principal reason for going to the solids was the low technical risk and the approximate one half billion to one billion [dollar] savings" in development costs.

The final space shuttle configuration. (NASA photograph)

In addition, NASA had discovered that it might be possible to recover, refurbish, and reuse the casings of the solid rocket motors; this "tilted the scales because they made the operational costs reasonable." On March 13, Don Rice told OMB's George Shultz and Cap Weinberger that "we recommend acceptance" of the NASA choice; that recommendation was accepted, and the basic space shuttle configuration that would become so familiar over thirty years of shuttle flights was given a green light for development.[4]

What Budget Commitment?

In his December 29 letter to Cap Weinberger recommending how to proceed with respect to the space shuttle, and again in its January 3 letter responding to OMB questions, James Fletcher had asked for a

"Administrator's contingency" of 20 percent to guard against unexpected costs during shuttle development, saying "approval of a $5 billion program would thus constitute a commitment by NASA to make every effort to produce the desired system for under $5 billion, but in no case more than $6 billion." That funding reserve apparently was not discussed at the January 3 meeting at which Shultz gave quick approval to the full capability shuttle. Fletcher intended to bring up the issue with the president when he and Low met with Nixon on January 5, but forgot to do so. He called Bill Anders the next day, asking Anders to intercede with the White House to make sure that the budget reserve was part of NASA's understanding with OMB. When Anders called John Ehrlichman, he was told to relay the message to Fletcher that NASA would have to "eat" any cost overruns. In a February 16 letter from Shultz to Fletcher in which Shultz "recapitulated our understanding of the decisions that have been made to date on the space shuttle," there was no mention of a funding reserve; indeed Shultz told Fletcher that we "fully expect NASA to develop a shuttle system within the $5.5 billion of research and development costs, should we subsequently agree on the choice of pressure-fed boosters, or within an appropriately smaller amount should the choice be solid rocket motors."[5]

In the same letter, Shultz told NASA that it was "our specific understanding that NASA's peak annual spending during the period of development of the shuttle will not exceed $3.2 billion of outlays in the dollars of the FY 1973 budget." NASA up to that point had been arguing that the offer that NASA had made to develop the shuttle within the framework of a constant overall NASA budget, adjusted for inflation, was based on FY1971 dollars and on a FY 1973 new obligational authority level of $3.379 billion, rather than the $3.2 billion outlays level. The question of the budget baseline also had not been explicitly discussed at the January 3 approval meeting. When Fletcher and Low realized that their understanding was at variance with OMB's intent, a difference that could lead to more than a billion dollar shortfall in the funds available for shuttle development, they tried for the next month to convince OMB to agree to a constant NASA budget based on the $3.379 billion level for FY 1973 and shuttle cost estimates based on FY1971 dollars. They did not succeed. Indeed, in fall 1972, Weinberger, by then the director of OMB, refused to honor the constant budget agreement even at the $3.2 billion level, leading James Fletcher to conclude that "a commitment from OMB is worthless."[6]

These two differences of understanding between NASA and OMB with respect to the funding available for shuttle development meant that the program had to be managed under very tight financial constraints. When the almost inevitable technical problems arose, there was no margin in the shuttle budget to deal with them; as a result, there were cost overruns and schedule delays that in the late 1970s almost led to President Jimmy Carter canceling the shuttle program.

A "Go" for Shuttle Development

On March 14, Don Rice gave NASA oral approval for developing the NASA-recommended shuttle configuration, an orbiter with a 15 × 60 foot payload bay and using solid rocket motors to assist in its launch. The next day, NASA issued a press release saying: "NASA announced today that the Space Shuttle booster stage will be powered by solid rocket motors in a parallel burn configuration. The booster stage will be recoverable. Requests for proposals for design and development of the Space Shuttle are expected to be issued to industry about March 17." NASA estimated shuttle development costs would be $5.15 billion and the cost per flight would be $10.5 million. A contract for shuttle development would be issued in summer 1972, with the initial orbital test flights with a crew aboard to occur in 1978. The NASA release stated "the complete Shuttle system is to be operational before 1980."[7]

This announcement brought down the curtain on the drama that had begun more than three years earlier. President Richard Nixon and his associates, with the decision to develop the space shuttle, had finally given an answer to the question "What do you do next, after the Moon?" That answer defined much of the U.S. civilian space program for the next 40 years. John Kennedy's 1961 decision to go to the Moon led to the Apollo program, which lasted only from 1961 to 1975; Richard Nixon's decision to build the U.S. post-Apollo space program around the space shuttle had a far more lasting impact.

Epilogue

Richard Nixon and the American Space Program

President Richard Nixon and his associates between 1969 and 1972 made three major decisions with lasting consequences for the U.S. space program.[1] The preceding chapters have chronicled the making of those decisions. This summary chapter will assess their character and discuss their impact on the U.S. space program over the more than four decades since they were made.

The three principal Nixon administration space policy decisions were:

- *To treat the space program, not as a special, high-priority government activity as had been the case during Apollo, but rather as part of the "day in and day out" activities of government, with its budget determined "within a rigorous system of national priorities."** The Nixon administration formalized NASA's need to compete with other government agencies through the political and budgetary processes for priority, and then assigned a relatively low priority to space activities in that competition.
- *To lower U.S. ambitions in space by not setting another challenging space goal and thus ending for the foreseeable future human space flights beyond low Earth orbit.* As Assistant to the President Peter Flanigan remarked at the time, there was in the White House in 1969 and early 1970 "a feeling that the country had had enough excitement [in space] for now"; there was no inclination on the part of Richard Nixon to propose another Kennedy-like space goal for the post-Apollo period or even to indicate in any but the most general terms that the United States would continue to work toward human exploration of the solar system.
- *To build the post-Apollo program around the space shuttle without linking the shuttle to a long-term strategy for its use.* The shuttle was seen as a new capability for carrying out the space program of the 1980s and beyond. Those directly involved in shuttle planning saw it as a first step

---

* Citations to material quoted in previous chapters will not be repeated here.

toward a comprehensive space exploitation and exploration capability. However, NASA did not clearly present this perspective to Richard Nixon and his associates as the space agency sought shuttle approval, and the Nixon administration did not couple its approval to a strategic perspective on long term space program goals. As historian Walter McDougall later observed, "Apollo was a matter of going to the moon and building whatever technology could get us there; the Space Shuttle was a matter of building a technology and going wherever it could take us."[2]

The first two of these decisions were made early in the Nixon administration, in the context of the White House quickly rejecting the ambitious post-Apollo space program proposed in the 1969 Space Task Group report. While these decisions were resisted by NASA, there was little controversy among Richard Nixon and his advisers in making them; their collective intent from the start of their time in the White House was to follow Apollo with a much more modest space effort. In contrast, the decision to develop the space shuttle was the end product of a contentious three-year decision process, with NASA pushing for approval of a technologically ambitious shuttle design and White House budget and technical advisers opposing such an undertaking and proposing more modest approaches to keeping human space flight a part of post-Apollo activities. Richard Nixon and his senior advisers gave little weight to economic and technical arguments, seeing the shuttle program primarily in a political context. The NASA position prevailed, with four-decade consequences for the U.S. space program.

The Space Program and National Priorities

Richard Nixon made it clear to his associates that he did not want the post-Apollo space effort appear to take money away from government programs on Earth. As the March 1970 statement outlining his space policy was being prepared, Nixon stressed that it should avoid "positive statements on space" being "invidiously" compared to his attitude toward "problems in poverty and social problems here on earth." He did not want to be seen as "taking money away from social programs and the needs of the people here [on Earth] to fund spectacular crash programs out in space." Nixon was a careful reader of opinion polls and other indications of public sentiment, and his generalized sense that space achievements were part of "exploring the unknown" did not override his sense that an ambitious space program was not something that would gain political support. He was not interested in leading the nation to accept an ambitious post-Apollo space effort.

This perspective was applied in an ad hoc fashion to budget decisions on the NASA program in December 1969 and January 1970, and space did not fare well as it competed for funding with other Nixon administration priorities in the overall context of an imperative to balance the federal budget. In parallel with the chaotic Fiscal Year 1971 budget process, there was a move to formalize the approach the Nixon administration would take to setting the priority of post-Apollo space efforts. The result was what has been characterized in this study as the "Nixon space doctrine," clearly stated in the March 7, 1970, presidential statement on space. Characterizing this statement as a "doctrine" may be rather overstating the reality; it was more an after-the-fact rationalization of the perspectives that led Nixon and his associates to reject the recommendations of the Space Task Group and even in the aftermath of the Apollo success to continue to reduce the NASA budget. But in the sense that the framework for space decision making set out in the Nixon statement has in its essence been accepted by most presidents since, it can be said to deserve being called a "doctrine."

The Nixon space doctrine had two elements. The first was to change the status of the space program from an effort formally assigned the highest national priority, as had been the case during Apollo, to just one of many "normal" government activities. In the language of the space statement: "We must think of them[space activities] as part of a continuing process—one which will go on day in and day out, year in and year out—and not as a series of separate leaps, each requiring a massive concentration of energy and will and accomplished on a crash timetable." Space was to become "a normal and regular part of our national life."

The second element of the doctrine was to declare that the space program from 1970 forward would have to compete with other discretionary government activities for priority and corresponding budgetary support. The space statement said: "Space expenditures must take their proper place within a rigorous system of national priorities. What we do in space from here on in...must therefore be planned in conjunction with all of the other undertakings which are also important to us."

The Impact of the Nixon Space Doctrine

The proposition that the space program should not be based on "a series of separate leaps, each requiring a massive concentration of energy and will and accomplished on a crash timetable," has had a continuing impact on presidential decisions on the space program. President Jimmy Carter in 1978 approved a space policy statement that explicitly echoed the Nixon declaration; it said "our space policy will become more evolutionary rather than centering on a single, massive engineering feat."[3] Even though most presidents since Richard Nixon have proposed some type of major new space development and in most cases provided a timetable for its achievement, in none of those proposals was the undertaking to be carried out on a "crash" basis, and certainly none were accompanied by a "massive concentration of energy and will," not to mention adequate financial resources.

The Nixon decision that "space expenditures must take their proper place within a rigorous system of national priorities" has had an even more lasting impact on the U.S. space program. At the peak of the Apollo buildup in 1966, the NASA budget comprised nearly 4.4 percent of federal spending overall and 19 percent of discretionary nondefense federal spending. (The

NASA share of the federal budget is most frequently cited in terms of a percentage of the overall budget. Given the inexorable growth of the portion of the budget devoted to mandatory entitlements, it seems more useful to discuss the NASA budget in terms of its share of the discretionary nondefense budget, since it is in that realm that space spending competes with other discretionary government programs.) As President Lyndon B. Johnson refused to approve any of NASA's post-Apollo proposals, NASA's budget share quickly began to decline from its Apollo high point; by the time Richard Nixon became president the NASA budget had dropped to just above 8 percent of discretionary nondefense spending. The early Nixon space decisions continued this trend; in Fiscal Year 1973, the budget in which space shuttle approval was first reflected, the NASA discretionary budget share was approximately 6 percent and on a downward trajectory. While it was under Lyndon Johnson rather than any of his successors that the biggest percentage reduction in NASA's budget share occurred, that reduction came from deferring a decision on what to do in space after Apollo, not on the basis of a specific decision to lower the space program's priority. By contrast, Richard Nixon consciously made that crucial decision—to reduce NASA's priority rather than assign it new, expensive programs and thus continuing rather than reversing the decline in NASA's budget share. The NASA portion of discretionary nondefense spending vacillated between 6 and 4 percent between 1977 and 2002 and between 4 and 3 percent since. By any measure, the space program has not done well in competition for budget resources; in fact, compared to other government programs, it has declined in priority over the past 40 years.[4]

The consequences of this declining share of the overall discretionary budget have been clear to most observers. For example, the Columbia Accident Investigation Board in 2003 observed that "NASA has had to participate in the give and take of the normal political process in order to obtain the resources needed to carry out its programs." In this give and take, "NASA has usually failed to receive budget support consistent with its ambitions. The result... is an organization straining to do too much with too little."[5]

Increasing the NASA Budget Share?

The reaction to this situation on the part of the space flight community has been predictable—continuing advocacy that the NASA budget share should be increased. A 1990 space program review led by aerospace industry executive Norm Augustine suggested that "a reinvigorated space program will require real growth in the NASA budget of approximately ten percent per year (through the year 2000), reaching a peak spending level of about $30 billion per year (in constant 1990 dollars) by about the year 2000."[6] A NASA FY2000 budget of $30 billion in 1990 dollars would have been the equivalent of a budget of almost $40 billion in 2000 dollars; the actual NASA budget for Fiscal Year 2000 was $13.6 billion.[7] Almost two decades later, a similar review of NASA's human space flight program, again led by Augustine, reached only a slightly less ambitious conclusion, observing that "NASA's budget should match its mission and goals," but then suggesting that "meaningful human exploration" would be possible only if the NASA budget were increased by up to $3 billion per year for several years.[8] Given that the proposed NASA budget at the time the review was taking place was $18.7 billion, this was a call for an over 15 percent increase in NASA's annual resources. More recently, astrophysicist and science spokesperson Neil DeGrasse Tyson has gained widespread attention by his advocacy of doubling the NASA budget, bringing it back to 1 percent of overall Federal spending. Such an action, suggests Tyson, would "give NASA enough money to do everything everyone has wanted NASA to do over all these years *and* enable us to go back to the moon and on to Mars in a bold and audacious way."[9]

All of these recommendations and suggestions fly in the face of a reality set in motion by the Nixon space doctrine: When the priority of the space program is compared through the normal political process to the priority of other uses of government funds, the outcome is to allocate to the space program a relatively low share of federal discretionary spending, inadequate to support a vigorous and sustainable program of space exploration. This outcome has been consistent for over 40 years and is very unlikely to change anytime soon. A 2014 review of the U.S. human space flight program observed that "human spaceflight—among the longest of long-term endeavors—cannot be successful if held hostage to traditional short-term decision-making and budgetary processes." But the Nixon space doctrine declared that it was through those processes that space budget decisions should be made. The same report also noted that "it serves no purpose for advocates of human exploration to dismiss these realities [the lack of public interest in space and the attendant low priority given to increasing space spending] in an era in which the citizenry and national leaders are focused intensely on the unsustainability of the national debt, [and] the growth in entitlement spending...There is at least as great a chance that human spaceflight budgets will be below the recent flat line trend as above it."[10] The mismatch between the requirements of a successful program of human space exploration and the tenets of the Nixon space doctrine has been a central space policy reality since the doctrine was first stated in 1970.

Ending Exploration

Richard Nixon saw in the *Apollo 11* mission a unique opportunity. Project Apollo had been intended from its 1961 approval by President Kennedy to be a large-scale effort in "soft power," sending a peaceful but unmistakable signal to the world that the United States, not its Cold War rival the Soviet Union, possessed preeminent technological and organizational power.[11] Nixon agreed with this rationale for the lunar landing program, and in his first months as president made sure to identify himself and his foreign policy agenda with what he later would hyperbolically characterize as "the greatest week in the history of the world since the Creation." But Nixon's embrace of Project Apollo as a tool of American soft power was short-lived. Once the United States had won the race to the Moon, he perceived little foreign policy benefit from subsequent lunar landing missions or from approving a post-Apollo program focused on preparing for missions to Mars. Other considerations, primarily domestic in character, would determine the Nixon approach to space in the post-Apollo period.

Like many other Americans, Nixon quickly lost interest in continuing Apollo flights to the Moon. As early as December 1969, after the first two lunar landings, he remarked that he "did not see the need to go to the moon six more times." When the *Apollo 12* crew visited the White House that month, mission commander Pete Conrad came away "disappointed and disillusioned." He reported that Nixon evidenced an "apparent lack of interest in the space program." Nixon did become emotionally engaged with the fate of the *Apollo 13* crew, but that near-fatal experience only added risk avoidance to lack of interest as part of Nixon's attitude toward lunar missions. For the *Apollo 15* mission in July 1971, Nixon slept through the launch, even though the White House felt it should announce that he had followed the event closely. By that time Nixon was already urging his associates to find ways of canceling the last two Apollo missions, *Apollo 16* and *17*. By April 1970, the iconic "Earthrise" photograph taken during the *Apollo 8* mission that had been hanging on the Oval Office wall throughout 1969 was removed, a symbolic action reflecting the president's lack of commitment to continuing space exploration.

As *Apollo 17* lifted off the lunar surface on December 14, 1972, President Nixon issued a statement saying "this may be the last time in this century that men will walk on the Moon." As the statement was read to the *Apollo 17* crew as they circled to Moon before heading back to Earth, astronaut Harrison "Jack" Schmitt was furious, thinking "that was the stupidest thing a President could have said...Why say that to all the young people in the world...It was just pure loss of will."[12] By his space decisions, Nixon made sure that his forecast would become reality. As of this writing, humans have not traveled beyond the immediate vicinity of Earth for 42 years, and no such journey is planned before 2021, almost 50 years after the last Americans left the Moon.

Nixon coupled his lack of personal interest in continuing Apollo flights to a political judgment with respect to the space program—that the American public was not interested in supporting an expensive, exploration-oriented space program. As he met with NASA Administrator Tom Paine in January 1970 to explain his decision to reject the Space Task Group–recommended post-Apollo program, Nixon told Paine "the polls and the people to whom he talked indicated to him that the mood of the people was for cuts in space."

In May 1961, John Kennedy had paid little attention to poll results showing that the majority of the U.S. public opposed spending the sums of money needed to send Americans to the Moon; Kennedy proposed Apollo as a top-down leadership initiative based on geopolitical considerations. In

Richard Nixon's interest in Apollo missions was not long-lasting. As he met in December 1969 with his assistant Peter Flanigan (at the front of Nixon's desk) and science adviser Lee DuBridge, the famous *Apollo 8* "Earthrise" photograph was hanging on the Oval Office wall. (left image) By April 1970, the photograph was gone. (right image)[13] (National Archives photo WHPO 2598–15 (left) and WHPO photo 4518–6 (right), the latter courtesy of the Richard Nixon Presidential Library and Museum)

contrast, Richard Nixon saw no persuasive foreign policy or national security reason to lead a reluctant nation and its representatives in Congress toward accepting an ambitious post-Apollo space program, particularly one aimed at developing the capabilities needed for early trips to Mars. Staff assistant Clay Thomas "Tom" Whitehead, who among the White House staff had the most level-headed approach to future space activities, commented that "no compelling reason to push space was ever presented to the White House by NASA or anyone else."[14]

The immediate consequence of Richard Nixon's decision not to continue space exploration was suspending production of the Saturn V Moon rocket and approving a NASA budget outlook that forced the agency's leadership to cancel two planned Apollo missions in order to have funds available for future projects. During the 1960s NASA had developed the Saturn V and its related ground facilities on the expectation that the vehicle would remain in service for many years and would be the enabler of a continuing exploration-oriented space effort. These hopes were dashed by Richard Nixon's initial space decisions; those decisions meant that the United States was voluntarily giving up for the foreseeable future the results from its multibillion dollar investment in exploratory capabilities and transforming the unused Saturn V launchers into very impressive museum exhibits.

There is one sense in which Richard Nixon's decision to reject continued space exploration might seem somewhat surprising. Nixon often included space activity as an important aspect of his frequently repeated call for "exploring the unknown," an activity that he believed was essential if the United States was to remain a "great nation." For example, in February 1971 he told a group of astronauts "in the history of great nations, once a nation gives up in the competition to explore the unknown, or once it accepts a position of inferiority, it ceases to be a great nation." In a June

1972 conversation with the *Apollo 16* crew, Nixon equated exploring the unknown with concepts as varying as "science, breakthroughs in education, breakthroughs in technology, breakthroughs in transportation," adding "space—that's the unknown. What's out there?"[15] Nixon did communicate to his associates that he *was* interested in eventual human journeys to Mars, and even mused about the possibility of finding life on a moon of Jupiter, but he saw those activities in the far future, not as objectives related to the decisions he would make during his time in the White House. Nor did Nixon cast his decision to approve the space shuttle in the context of its being an initial step in a decades-long effort to explore destinations beyond the immediate vicinity of Earth. NASA in its input to the Space Task Group had portrayed the space shuttle as part of a coherent long-range strategy ultimately leading to outposts on the Moon and journeys to Mars, and even in 1971 retained elements of that thinking in its technical planning, but that perspective was not considered as part of Nixon's decision to approve the shuttle.

To Nixon, "exploration" was not a sharply defined concept, and his repeated calls for "exploring the unknown" seem to have been little more than what a historian would call a "trope"—an overused rhetorical device offered in the place of substantive thought. Nixon was notoriously poor at conversation with any but those in his inner circle, and falling back on repetitive rhetoric was often his way of dealing with discussions of policy issues with those outside that inner circle. The lack of logic in Nixon's attitude with respect to space activities was on display as he told one of his Congressional relations staff in 1971 that "the United States should not drop out of any competition in a breakthrough in knowledge—exploring the unknown. That's one of the reasons I support the space program." Without pausing, he then said "I don't give a damn about space. I am not one of those space cadets."

Exploring the space frontier was in reality not part of Richard Nixon's strategic vision for America, and thus his repeated call for "exploring the unknown" had little connection to his actual decisions on space policy and budgets in the post-Apollo period. By rejecting the recommendations of the Space Task Group, the Nixon administration attempted to reduce U.S. space ambitions to match the budget it deemed appropriate to allocate to NASA in the post-Apollo period. However, that lowering of ambitions did not happen, either then or since. The exploratory vision still persists; a 2009 blue-ribbon review of the U.S. human space flight program concluded that "Mars is the ultimate destination for human exploration of the inner solar system" and that "human [space] exploration... should advance us as a civilization towards our ultimate goal: charting a path for human expansion into the solar system." Discussing the persistence of this vision, Howard McCurdy suggests "expectations invariably fail, but the underlying vision rarely dies. Rather, people update the vision. The dream moves on."[16]

One can argue that Richard Nixon made a major policy mistake in mandating that the space program should be treated as just one of many government programs competing for limited resources. Certainly the belief that this judgment was ill-conceived is the long-held position of space advocates. But it is also possible that Nixon's decision-that U.S. space ambitions should be adjusted to the funds made available through the normal policy process-was a valid reading of public preferences, and there were no countervailing reasons for him to reject those preferences. If this is the case, then the Nixon administration in its space decisions was correctly reflecting the view of the majority of the U.S. public. There is no evidence that this situation has changed over the past 40 years; the most recent review of the U.S. space exploration program notes "lukewarm public support" for a program of human space exploration and the absence of "a committed, passionate minority large and influential enough" to provide a political basis for such a program.[17]

What has actually happened since Richard Nixon made his decisions to end lunar exploration, not to set a new exploratory goal, and to remove the space program's special priority is neither reduced ambitions nor increased budgets; instead, for more than 40 years there has been a continuing mismatch between space ambitions and the resources provided to achieve them. This outcome is close to the worst possible recipe for space program success; *a central part of Richard Nixon's space heritage is thus a U.S. civilian space program continually "straining to do too much with too little."*

Richard Nixon and the Space Shuttle

Although the Nixon decisions to treat the space program as just one of many government activities and to defer human space exploration for the indefinite future have had lasting impacts, the space shuttle program stands as Richard Nixon's most recognized space legacy. Thus any assessment of that legacy must give particular weight to the shuttle's influence on the evolution of the American space program.

Once NASA decided in 1970 to focus on developing the space shuttle as its major post-Apollo effort, there were many designs considered and a number of alternatives to going ahead with a shuttle suggested. During 1971, there was a somewhat confused sorting out of these various possibilities, but as the debate over developing a shuttle reached its final stage, there was little doubt that the White House would approve some version of a shuttle rather than pursue an alternative course. Other options, such as deferring a shuttle decision and carrying out an interim program of human space flight using surplus Apollo hardware or developing an unpowered space glider or a new crew-carrying capsule, had fallen by the wayside. The key decision to be made was thus what kind of shuttle, to carry out what missions, and with what rationale, would be approved.

The options for choice were clearly understood as the decision process reached its climax. As George Low observed in December 1971, "the basic issue on the space shuttle concerns whether or not the shuttle should capture a majority of the payloads that will be flown in the 1980's" and "whether the shuttle should be small or large and whether it should provide for routine operations or one or two flights per year." These two alternative approaches were embodied in two competing shuttle designs, called here the "NASA shuttle" and the "OMB shuttle." The NASA shuttle—the design ultimately selected—was the end product of more than two years of study by NASA and its aerospace contractors; that study effort had been guided by a combination of national security and NASA's own requirements and the OMB pressure to make the shuttle "cost effective." The NASA shuttle was a full capability vehicle incorporating advanced propulsion, thermal protection, and electronic systems technologies. It would have a 15 × 60 foot payload bay, be able to carry a 65,000 pound payload to a 100 nautical mile orbit due east from the Kennedy Space Center, launch or return a 40,000 pound payload from a polar orbit, and be capable of 1100 nautical miles of cross-range maneuvering. With these capabilities, the NASA shuttle would be able to carry out all planned and potential U.S. civilian, national security, and commercial missions. NASA claimed that it could be launched on a routine basis and at significantly lower cost than any alternative launch system and that it would provide valuable new capabilities for space operations. Such a shuttle, NASA claimed, could be developed for a budget of between $5 and $6 billion.

The staffs of the White House Office of Management and Budget (OMB) and Office of Science and Technology (OST) were deeply skeptical of the validity of these claims and indeed of the need for a system with the capabilities NASA was promising. Although many in the two staff offices were indeed skeptical of the value of human space flight itself, they recognized that no American president, and in particular not Richard Nixon, with his emotional view of NASA's astronauts, would choose to end the U.S. human space flight program. They therefore resonated to the advice given by Alexander Flax, chair of the ad hoc panel of the President's Science Advisory Committee set up to examine the NASA shuttle, that "serious consideration must be given to less costly programs which, while they provide considerably less advancement in space capability than the [NASA] shuttle, still continue to maintain options for continuing manned spaceflight activity, enlarge space operational capabilities, and allow for further progress in space technology." This perspective led to OMB recommending to President Nixon in early December 1971 that he direct "OMB and OST to work with NASA on the reorientation of the space program," with a central feature of that reorientation a "smaller, reduced cost" shuttle design. Nixon approved this recommendation on December 3, 1971, and a few days later OMB presented NASA with its concept for a less ambitious shuttle, with a 10 × 30 foot payload bay and 30,000 pound payload lifting capability, to be developed at a budget of no more than $4 billion.

The question of which of these two alternatives to approve was debated through most of December 1971. Even as the final choice to approve the NASA full capability shuttle was being made over the New Year's weekend, Don Rice, assistant director of OMB, and science adviser Ed David were still arguing strongly against that step. While Rice focused his opposition on the excessive cost of the NASA shuttle, David took a broader view, arguing that "the large space program implicit in the large shuttle decision is not consistent with the best interests of the nation." The opposition of Rice and David was well-founded and subsequently validated, but they were overruled by their White House bosses and ultimately by President Nixon. Nixon and his associates gave less weight to cost and technical issues than to other political and policy considerations as they decided to approve NASA's preferred shuttle design.

Why Did the Nixon Administration Select the NASA Shuttle?

On November 24, 1971, Richard Nixon had indicated his preference for a "more modest option" with respect to the space shuttle, and again on December 3 had said "yes" to the OMB proposal to work with NASA to examine a smaller, less expensive shuttle design. Yet when the final decision on which shuttle design would be approved was communicated, and most likely made, by George Shultz and Cap Weinberger, it was NASA's full capability shuttle that won the day. The evidence suggests that there were three primary reasons for that choice.

One reason was the importance to the White House of a continuing program of human space flight as a symbol of U.S. space leadership. Cap Weinberger in August 1971 told President Nixon that not moving ahead with some major post-Apollo space effort would be a "mistake," arguing that such a decision "would be confirming, in some respects, a belief that I fear is gaining credence at home and abroad: that our best years are behind us, that we are turning inward, reducing our defense commitments, and voluntarily starting to give up our super-power status, and our desire to maintain our world superiority." Nixon had written on Weinberger's memorandum "I agree with Cap." In making its best case for shuttle approval, NASA had argued "man has learned to fly in space, and man will continue to fly in space. This is fact. And, given this fact, the United States cannot forego its responsibility—to itself and to the free world—to have a part in manned space flight...For the U.S. not to be in space, while others do have men in space, is unthinkable, and a position which America cannot accept." As he met with Fletcher and Low on January 5, 1972, President Nixon asked the NASA leaders if they thought that the space shuttle was a good investment. They of course responded positively, but Nixon, echoing NASA's argument, then added "even if it were not a good investment, we would have to do it anyway, because space flight is here to stay. Men are flying in space now and will continue to fly in space, and we'd best be a part of it."[18] While, as OMB and OST were arguing, a smaller shuttle launched a few times a year would have kept the U.S. human space flight program alive and would have been a useful symbol of U.S. space leadership, clearly a large shuttle launched on a frequent basis would be a much more potent indication of that leadership.

A second reason for the decision to approve the full capability shuttle was President Nixon's interest in its national security uses. John Ehrlichman in a 1983 interview suggested that "what the military could do with the larger bay in terms of the use of satellites" and the fact that "the space shuttle would have the capability of capturing satellites or recovering them" had "a strong influence on me" and "weighed into my attitude toward the larger shuttle." He added "I feel it is valid to say it also weighed into" the president's evaluation of the shuttle. Nixon's interest in national security uses of the shuttle was well known as final decisions were made, and the president discussed them with Fletcher and Low at their January 5 meeting. As Fletcher and Low met with Cap Weinberger and Jon Rose on March 3, 1972, to discuss NASA's decision on using solid rocket motors to boost the shuttle orbiter off its launch pad, Weinberger reminded the NASA leaders that "the President's strong interest in retaining the military capability" was an important factor in "confirming our decision on the larger size" shuttle.[19]

The third and most immediate reason for the choice of the NASA shuttle was the anticipated short-term employment impact of that choice in parallel with Richard Nixon's 1972 reelection bid, particularly in Southern California. The evidence in support of this assertion is conclusive. Aerospace unemployment had emerged as an important political issue in early 1971, particularly after the congressional cancellation of the supersonic transport program. Meeting with Flanigan and Weinberger in May, the president discussed "what could be done about high unemployment areas with specific emphasis on California." Nixon "indicated a very great concern about the California area and the high level of unemployment among technically-trained individuals." The White House created a "California Employment Project" to address this issue in a systematic way. In June 1971, even before the shuttle decision process reached its final stage, Cap Weinberger told a California politician "I am sure that whatever is done will be largely based in California."

On November 24, as he made his initial decision on the space shuttle, President Nixon had commented to Ehrlichman: "Jobs—right, John? Do it in terms of jobs. It ought to be in California." Ehrlichman recalled that the issue of shuttle-related employment in Southern California "was a very important consideration in Nixon's mind...I can recall conversations about that, which were highly persuasive...You must not underemphasize that element, that employment element, in Nixon's decision [on the shuttle], the whole manned space program." When Bill Anders told Ehrlichman in late December that the NASA shuttle would create more short-term jobs than the OMB alternative, the reaction was "OK, that will be the one" chosen. In preparing President Nixon for his January 5 meeting with Fletcher and Low, Flanigan told him that "this program will greatly stimulate the aerospace industry." Nixon himself, discussing his shuttle approval with his political adviser Charles Colson a few days after it was announced, noted that "in Florida and California this [approving the shuttle] is a big deal. It will save the aerospace industry." A few months later, he told a delegation from New York lobbying for the shuttle contract to go to Grumman's Long Island facility that "I'll take the heat for putting the money there [on the space shuttle contract] rather than the ghettos and all that sort of thing, but then by God we'd better at least get a little credit for it. This is jobs. I mean that is really what is at stake here, jobs."[20]

Evaluating the Space Shuttle Decision

It was this combination of short and longer-term considerations—the creation of jobs in California before the 1972 election, the interest in potential national security uses of the space shuttle, and the desire to continue a human space flight program that would demonstrate U.S. space leadership to the world and be a source of national pride at home—that led to Richard Nixon's approval of NASA's full capability shuttle. Other factors, such as the ability of the shuttle to operate routinely and at greatly reduced costs, were not greatly influential as Nixon and his top advisers made that choice, even though they became the publicly offered justifications for shuttle development. So an evaluation of the shuttle's impact on the American space program must begin with an assessment of the shuttle program in terms of those objectives that were the proximate reasons for the choice of the NASA-preferred shuttle.

The Shuttle and Human Space Flight

In its 135 flights between April 1981 and July 2011, the space shuttle was undoubtedly the public face of the U.S. space program, communicating to the nation and the world an image of U.S. technological capability and American leadership. The shuttle orbiters carried 355 different people into orbit, including 306 men and 49 women, with many making multiple flights; two U.S. astronauts each flew on seven shuttle missions. The relatively nonstressful conditions of launching aboard the shuttle opened up the experience of space flight to scientists and engineers, and also to a few politicians, teachers, and industry representatives, not just to test pilots. Astronauts from 16 countries flew aboard the shuttle, thereby fulfilling Richard Nixon's "pet idea" of flying non-U.S. people on a U.S. spacecraft. (In fact, while Nixon wanted only the symbolism connected with flying non-Americans on a NASA spacecraft, his interest opened the door to intimate international participation in the U.S. human space flight program, leading to the European Spacelab effort and the Canadian robotic arm on the shuttle, the U.S.-Soviet *Apollo–Soyuz* Test Project, and eventually to the 15-partner International Space Station (ISS).) Of the shuttle's 135 missions, 37 were dedicated to assembling and outfitting the ISS; maintaining the capability to launch space station elements had been a NASA "bottom line" in the final stages of the shuttle debate. Demonstrating the unique capabilities offered by the shuttle, other missions launched, repaired, and recovered satellites, most notably the Hubble Space Telescope, sent probes to the Sun, Venus, and Jupiter, launched other telescopes to observe the universe, and hosted on-orbit research. There were nine shuttle dockings with the Soviet/Russian space station *Mir*. Unfortunately, two missions ended in catastrophe; in each, seven crew members lost their lives.

The shuttle was and continues to be a source of considerable pride for U.S. citizens. Images of a shuttle launch are global symbols of American accomplishment and technological leadership, and even after they have been retired from service the three remaining shuttle orbiters—*Discovery, Atlantis, and Endeavour*—are objects with high public appeal. In terms of its political impact and of offering unmatched capabilities for space operations, the space shuttle was a resounding success. The space shuttle met the objective of keeping Americans flying in space as a source of national prestige and pride; the capabilities offered by the shuttle made the United States the unquestioned leader in human space flight.

National Security Uses of the Space Shuttle

Another of the influences on the choice of the full capability shuttle was President Nixon's interest in its ability to launch the most advanced intelligence satellites and to carry out innovative national security missions. Those missions included the shuttle launching on demand during a political or military crisis, conducting a single-orbit satellite deployment or rendezvous, or inspecting or even destroying a potentially hostile satellite.

While the president himself may have been attracted by such national security uses, the reality was that support for the shuttle within the military and intelligence community was at best tepid, both at the time the shuttle decision was made and afterwards. Deputy Secretary of Defense David Packard's 1971 flexibility on shuttle requirements is suggestive of an ambivalent Department of Defense attitude toward the vehicle, and the effort in late 1971 to get a joint NASA–DOD statement to the president in support of the shuttle apparently did not bear fruit. During 1972 Congressional hearings on the shuttle program, DOD and Air Force testimony was supportive but guarded in character; the military took the position that the DOD would commit to depending on the shuttle only after its capabilities and constant availability had been fully demonstrated. During the mid-1970s top-level Department of Defense support for the shuttle ebbed and flowed. At lower levels of the national security community, there was strong opposition to phasing out expendable launch vehicles until the shuttle was demonstrated to be completely reliable. The DOD did agree to pay the costs of a west coast launch site for the shuttle at Vandenberg Air Force Base, since that location was primarily needed for national security launches into polar orbit. In addition, DOD agreed to be responsible for funding the "space tug" to move payloads from the shuttle to higher orbits and for covering the costs of separate launch control centers in Houston and Colorado Springs for managing classified shuttle missions. With the urging of Hans Mark, first as Undersecretary and then as Secretary of the Air Force from 1977 to 1981, and for much of that time also director of the National Reconnaissance Office (NRO), some national security satellites were redesigned to take advantage of the shuttle's attributes. When in 1979 President Jimmy Carter considered canceling the shuttle program because of its cost overruns, it was the national security uses of the shuttle, particularly in terms of launching the photo-reconnaissance satellites needed to verify arms control agreements, that convinced the president to continue the program. Once the Reagan administration took office in 1981, an early action was to confirm as national policy that the shuttle would be "the primary space launch system for both United States military and civilian government missions."[21]

This policy declaration represented the high point of the notion of using the shuttle for national security missions. Within the first years of the Reagan administration, Air Force and NRO resistance to total dependence on the shuttle escalated into a conflict that required a presidential decision to resolve. The consequences of total U.S. dependence on the shuttle had been predicted. In the midst of the shuttle debate in 1971, the OMB had warned "for national security purposes, we may not want all our eggs in one basket." The Air Force and NRO in 1984 won the right to develop an expendable launch vehicle as a backup to the shuttle for the largest national security payloads; this turned out to be the Titan IV booster. In the aftermath of the January 1986 *Challenger* accident, most national security payloads were removed from the shuttle and expendable launch vehicle production lines were activated; the nearly complete multibillion dollar West Coast launch site for the shuttle was mothballed.

Only ten dedicated national security missions, eight of which were classified, were launched aboard the space shuttle, including eight missions after the 1986 *Challenger* accident; the payloads for most of those missions had been uniquely designed for shuttle launch. Some of the capabilities relevant to national security uses, such as satellite repair, recovery, and refueling, were demonstrated on other early shuttle missions. But as a national security system, the shuttle had no continuing utility. One historian of national security space activities cites a Department of Defense 1992 report that set the cost of redesigning military and reconnaissance spacecraft first to launch on the shuttle and then reconfiguring them again to launch on the expensive Titan IV expendable launch vehicles after the *Challenger* accident as "in excess of $20 billion."[22] None of the ten national security shuttle missions required the cross-range capability that had been an original DOD demand, and none of the innovative missions described in the 1969 DOD/NASA space shuttle report that had influenced Richard Nixon's support of the NASA shuttle were ever attempted. Rather than provide new capabilities used by the national security community, the shuttle turned out to be a multibillion dollar drain on the national security space budget.

The Space Shuttle and Aerospace Employment

The space shuttle prime contract was awarded in mid-1972 to North American Rockwell, a company with its space operations based in Southern California. This award meant that the projected California employment impacts, both in advance of the 1972 presidential election and subsequently, were achieved. Although Rockwell barely beat out New York–based Grumman Aerospace for the contract award, there has been no evidence discovered in the course of research for this study that Richard Nixon's expressed wish to put a large share of shuttle work in California and his personal relationship with Willard "Al" Rockwell, the head of North American Rockwell, translated into an overt White House attempt to influence NASA as it selected the shuttle prime contractor. But NASA certainly was fully aware of the president's interest as that decision was made.

Basing shuttle approval on its job-creating impact set an unfortunate precedent for many subsequent space decisions. (In 1961, the politically driven decision to locate the Manned Spacecraft Center in Houston, Texas, as a new NASA facility for the Apollo program was a forerunner of this precedent.) From 1972 on, the employment and institutional impacts of various space program choices have been an important, sometimes overriding, factor in reaching a decision on how to proceed. This is especially the case since most decisions on large space projects since 1970 have been made through the normal political process, where such parochial considerations play a significant role. The widely accepted view of the civilian space effort as a "jobs program" had its origins in the Nixon administration's decision to "save the aerospace industry" by approving development of the full capability shuttle. The job-maintenance or job-creation impact of various space program options continues as a strong influence on twenty-first century decisions.

In terms of the proximate reasons for its White House approval, then, the space shuttle program must thus be judged a mixture of success and disappointment. In particular, the shuttle during its three decades of operation served the nation well as a focus for U.S. space leadership and the resultant prestige and pride. In terms of its role in U.S. military and intelligence efforts in space, some of the still classified national security missions launched aboard the shuttle are likely to have produced useful results, but overall the space shuttle program turned out to be a very expensive detour for the national security space program. The shuttle program's success in producing aerospace jobs in advance of the 1972 election and in the longer term helping revitalize the aerospace industry has been a mixed blessing; it achieved Richard Nixon's short-term political objectives while creating the image of the space program as a "public works" effort.

What about the Other Reasons for Shuttle Approval?

There were other reasons put forward for going ahead with the full capability shuttle, and the success or failure of the shuttle in satisfying those reasons must be included in an overall assessment of the shuttle as a Nixon space legacy. Although secondary factors in the presidential-level decision to approve NASA's full capability shuttle, the claims that NASA had made from 1969 on—that the shuttle could be launched on a routine basis and that it would significantly lower the costs of space operations—became the publicly offered reasons for developing the shuttle. The space shuttle failed to match those claims. As then-NASA Administrator Mike Griffin commented in 2010, "what the shuttle does is stunning, but it is stunningly less than what was predicted."[23]

In his prepared statement announcing shuttle approval, President Nixon said that the shuttle would "revolutionize transportation into near space by routinizing it." At the time the shuttle decision was made, the intent was to launch the shuttle 40 to 60 times per year once it became fully operational. This of course never happened. The most launches of the shuttle in a single calendar year ended up being nine in 1985, with the average annual launch rate over the 30-year lifetime of the program being 4.3 launches per year. The shuttle could not be launched on demand; rather, it took a lengthy and labor-intensive process to prepare each shuttle mission for launch. Rather than a vehicle capable of frequent and routine operation, the shuttle turned out to be, in the words of the Columbia Accident Investigation Board, a "complex and risky system" and a first generation "developmental vehicle" which required great care to operate safely.[24]

The Nixon January statement also said that the space shuttle would "take the astronomical costs out of astronautics." In March 1972, as the final shuttle design was announced, NASA Administrator Jim Fletcher claimed that once the shuttle became operational, its incremental cost per launch would be $10.5 million (almost $60 million in 2014 dollars). But NASA in 2012 estimated the cost of each shuttle launch, depending on how it was calculated, as between $814 million and $1.266 billion, up to 20 times higher than the 1972 estimate. NASA in December 1971 had said that the full-capability shuttle would carry payload into orbit at a cost of $118 per pound ($691 in 2014 dollars); in Congressional testimony, NASA listed the cost of using the shuttle to deliver cargo to the space station as $21,268 per pound in 2011 dollars.[25] By any measure, the shuttle did not "take the astronomical costs out of astronautics."

One may ask why NASA's 1972 estimates of shuttle operating costs were so far off the mark, while the estimates of shuttle development costs made at the same time were close to the actual amount. NASA cost estimators had significant experience in forecasting the costs to develop systems such as the Mercury, Gemini, and Apollo spacecraft, various launch vehicles, and scientific spacecraft. Based on that experience, they were able to estimate the cost of designing, developing, and testing the final NASA shuttle design with enough precision to allow Fletcher and Low to promise the White House that the cost was likely to be approximately $5 billion and certainly not more than $6 billion. The actual development cost was $5.5 billion in 1971 dollars, which were the basis for the NASA cost estimate.

None of those involved in the shuttle decision were as concerned about operating costs as they were about keeping the annual budget and total cost of shuttle development below politically acceptable levels. There was little White House or NASA leadership attention given to the quality of the cost-per-launch forecasts being put forward. NASA had little experience in estimating the costs of repetitive operation of a space system, and thus its estimates of shuttle operating costs were very uncertain. NASA, coming off its Apollo successes, was a technologically confident, perhaps even arrogant, organization, believing that it could incorporate advanced technology in such areas as propulsion, thermal protection, and on-board electronic systems into the shuttle design in ways that would make the vehicle able to be operated inexpensively and on a routine basis. The per flight cost estimate provided to the NASA leadership in late 1971 was based on assumptions of 50 flights per year, vehicle self-checkout, fully reusable thermal protection, and long engine life without major repair. While that estimate may have represented the best NASA could do in 1971, it ended up being based on the wrong technological assumptions and far off the mark. The shuttle required many hours of human labor for its checkout, a number of the shuttle's thermal protection tiles had to be repaired or replaced after each mission, and the shuttle's main engines required extensive refurbishment between uses. Shuttle program manager Robert Thompson characterized the NASA cost per flight estimate as an "optimistic guess." That "guess" was nowhere near the actual cost once the shuttle began flying; rather than the shuttle significantly lowering the cost of space launch, it became an extremely expensive system to operate. Because of the need for extensive checkout and refurbishment between flights, the shuttle that was developed had no chance of being operated at a 40 or more launches per year flight rate; in addition, the anticipated demand for that many shuttle launches a year never materialized.[26]

The Space Shuttle and the Space Station

From 1970 on, one of the performance requirements driving space shuttle design was NASA's intent at some future point to use the shuttle to launch elements of a space station. This was recognized by the OMB staff, who observed in 1971 that "in a sense, a commitment to a shuttle is an implicit commitment to a subsequent space station program." There is no evidence that this shuttle-station link was considered by the president and his senior advisers as the final shuttle decision was made, but the choice of the NASA shuttle design carried with it the virtual certainty that a future president would be asked to approve a shuttle-launched station.

That is precisely what happened. The shuttle's first flight was in April 1981; soon after that flight, President Ronald Reagan's nominees for NASA administrator and deputy administrator, James Beggs and Hans Mark, agreed that they "would try to persuade the new administration to adopt the construction of a permanently manned space station as the next major goal in space." The two announced their intent at their Congressional confirmation hearing in June 1981, in essence repeating Tom Paine's 1969 argument that the space station was "the next major evolutionary step in man's experimentation, conquest, and use of space." Beggs and Mark characterized the station as "the next logical step." It took almost three years for NASA to gain presidential approval; during his State of the Union address on January

25, 1984, Ronald Reagan announced that "I am directing NASA to develop a permanently manned space station and to do it within a decade."[27]

Discussing the long and troubled history of the space station project is beyond the scope of this study; the point here is that from its 1968 origin as the logistics vehicle for a Saturn V-launched space station, through the 1970 decision to switch to a shuttle-launched station and then to defer station development until the shuttle was flying, to the final July 2011 outfitting mission to what had become the International Space Station, there was an unbreakable link between the shuttle and the station. That bond meant that, unless the station program was terminated early, NASA had to keep the shuttle in service until station assembly and outfitting were completed. The high costs of the shuttle and station programs thus dominated the NASA human space flight budget for almost 40 years.

Was the Space Shuttle a "Policy Failure"?

In an article published soon after the 1986 *Challenger* accident, I suggested that "the space shuttle program must be assessed as a policy failure, at least in terms of meeting the objectives [lower cost and routine operation] that have been its articulated rationale since 1972." In deciding to approve the NASA shuttle, "too much attention was paid to the short term, while longer range considerations were inadequately considered... The shuttle decision stands as a powerful example of how not to make a national commitment to an undertaking on which many other significant projects depend."[28] Do these judgments still stand up, almost three decades later? Was the shuttle *program* itself a failure? Or was it the Nixon administration *decision* to approve the NASA full capability space shuttle that was the policy failure? I was not very clear in what I wrote in 1986, but it was my judgment then, and now, that the latter alternative is the case. As the preceding paragraphs have suggested, the record of the space shuttle program is a mixture of success and failure. But there were in 1971 better alternatives to approving development of the NASA full capability shuttle, and thus that approval is better described as a policy mistake, rather than a policy failure.

The Roots of the Policy Mistake

The policy mistake in the decision to develop the full capability space shuttle had deep roots in the history of the space shuttle program. The 2003 report of the Columbia Accident Investigation Board concluded that "the greatest compromise NASA made was not so much with any particular element of the technical design, but rather with the premise of the vehicle itself. NASA promised it could develop a Shuttle that would be launched almost on demand and would fly many missions each year." The report added "the increased complexity of a Shuttle designed to be all things to all people created inherently greater risks than if more realistic technical goals had been set from the start. Designing a reusable spacecraft that is cost-effective is a daunting engineering challenge; doing so on a tightly constrained budget is even more difficult."[29] That was the situation in which NASA found itself in 1970 and 1971, but NASA's leaders persisted in their advocacy of the full capability shuttle, even as some of them, particularly George Low, questioned the wisdom of that advocacy.

There were actually two policy mistakes associated with the shuttle decision. *The first and more fundamental mistake was the White House accepting as the basis for its shuttle decision NASA's claim that it could successfully go directly from the Apollo program, characterized by brute force launcher technology and crew-carrying capsules parachuting to an ocean landing, to developing a highly capable vehicle in terms of payload capacity, in-orbit operation, and maneuvering during entry, incorporating advanced technology in many of its systems, with a high degree of reusability, and able to land on a runway and quickly be readied for another launch, all at relatively modest cost compared to the alternatives.* The bullish vision of people such as Tom Paine and George Mueller pushed NASA to focus on an ambitious shuttle design incorporating advanced technology and capable of "airline type" operations. There was a significant degree of technological hubris in NASA's view of what would be achievable. After all, NASA engineers and managers had just succeeded in landing American astronauts on the Moon and were convinced that they could overcome the next set of technological challenges.

NASA and its contractors thus focused their attention during 1970 and the first half of 1971 on finding the best design meeting NASA and national security community requirements and employing cutting edge technology in areas such as propulsion, thermal protection, and onboard electronic systems. After May 1971 they had to carry out their design studies within an OMB-imposed budget ceiling in terms of both peak annual funding and the overall cost of the shuttle program. Although NASA recognized by mid-1971 that a two-stage fully reusable shuttle design was not feasible either financially or technologically, there was little emphasis on investigating less ambitious, less expensive, alternatives to an advanced technology shuttle orbiter with a variety of means for boosting it into space. There was essentially no attention given at the engineering level to concepts such as the glider favored by the Flax committee or the smaller shuttle proposed by OMB, or even to the Mark I, less technologically ambitious, shuttle proposed by NASA Headquarters.

In addition to designing a shuttle that could be developed within a constrained budget, NASA engineers were forced into demonstrating the shuttle's overall cost-effectiveness. In 1970, the Bureau of the Budget and then its successor OMB had insisted on proof that the shuttle development and operation would cost less than using expendable vehicles to launch U.S. space missions. NASA concluded that it had to satisfy that unprecedented OMB requirement. Demonstrating the shuttle's cost-effectiveness became perceived as a political necessity, and likely led to NASA's leaders and engineers deluding themselves about the costs of operating the shuttle on a frequent basis in order to make the economic case come out positively.

The design ultimately recommended was likely the best engineering solution to the demanding requirements NASA's technical staff was asked to meet. But that design created a first-generation experimental vehicle, not a shuttle capable of delivering the cost savings and routine operational benefits that NASA was promising. Basing the White House decision to approve the NASA shuttle on other factors while implicitly accepting NASA's optimistic claims with respect to the shuttle's operation was a policy mistake with long-lasting consequences.

Were There Alternative Choices Available?

There were some individuals both inside and outside of NASA who recognized the difficulties in pursuing the full capability shuttle as NASA's immediate post-Apollo project. NASA's top spacecraft designer, Max Faget, argued as the shuttle program was gaining momentum in early 1969 that a first step should be a relatively small vehicle, which he characterized as the space equivalent of the first-generation DC-3 commercial aircraft. Secretary of the Air Force Robert Seamans during the Space Task Group deliberations suggested that "it is not yet clear that we have the technology" for a reusable space transportation system that would produce major reductions in the cost of transporting payloads into space, and suggested "a program to study by experimental means including orbital tests" the feasibility of such a system. As NASA awarded shuttle design study contracts in 1970, veteran flight director and then-deputy director of NASA's Manned Spacecraft Center Chris Kraft warned "I don't think we should try and build an ultimate vehicle the first time...I think we've got to be extremely careful that we don't try to build a do-all vehicle. I don't think we ought to talk ourselves into the fact that the shuttle is to do every job in the space program."[30] The idea that there should be an interim step before deciding whether to develop a full-capability shuttle was present throughout the shuttle debate, but was never embraced by those directly in charge of the shuttle program, who remained convinced that they could design and develop the kind of shuttle approximating the advanced technology vehicle first suggested by George Mueller.

NASA's leaders themselves, as the final decision on the space shuttle approached, harbored reservations about the viability of the full capability concept. George Low in August 1971, as he assessed alternative courses of action, concluded that "we should drop the shuttle right now and come up with a different manned space flight program." He added that "this program should be based on an evolutionary space station development, leading from Skylab through a series of research and applications modules to a distant goal of a permanent space station. We should also set for ourselves a distant goal of a lunar base. The transportation system for this manned space flight program would consist of Apollo hardware for Skylab; a glider launched on an expendable booster for the research and application modules; and finally, the shuttle but delayed 5 to10 years beyond our present thinking." As Low and Jim Fletcher prepared the NASA budget request due at OMB on September 30, 1971, they "debated whether we should not forego the shuttle entirely and develop instead some alternate manned space flight program."

Fletcher and Low did not choose this option, getting little support for it from the NASA technical workforce and deciding that it was in NASA's institutional interests to seek immediate approval of an ambitious shuttle design. This was a fateful decision, since it polarized the shuttle debate; during the rest of 1971 NASA would make its case for approving the full-capability shuttle as a "best buy" rather than seek a compromise with shuttle skeptics in OMB and OST. The decision-making process functioned as it should, elevating two shuttle options for presidential choice. *But as the White House made the final shuttle decision, Nixon and his top advisers chose the wrong option. This was the second policy mistake connected to the space shuttle decision.*

That going ahead with the full-capability shuttle was a course of action fraught with the potential for future problems was clear to some of those examining shuttle choices. For example, Alexander Flax had reported to science adviser David in October that "most of the members of the Panel doubt that a viable program can be undertaken without a degree of national commitment over the long term analogous to that which sustained the Apollo program. Such a degree of political and public support may be attainable, but it is certainly not now apparent." Flax added "planning a program as large and as risky (with respect to both technology and cost) as the shuttle, with a long-term prospect of fixed ceiling budgets for the program and NASA as a whole does not bode well for the future." This was prescient advice, but it was not heeded.

The commitment to NASA's full-capability shuttle (which carried with it a future decision to develop a space station) created for more than four decades two very expensive "mortgages" on the NASA annual budget. Given that that budget was commanding a decreasing share of federal discretionary spending, the necessity of servicing these mortgages meant that there were limited funds available for other worthy space endeavors. As Bill Anders, a veteran of the shuttle decision process, recently commented, "the shuttle, like a cuckoo in the nest, pushed out many less sexy but higher pay-off science and commercial programs for lack of funds."[31]

It is of course impossible to know what might have happened if the White House had chosen the OMB shuttle option. But it does seem that pursuing a less ambitious shuttle design as an intermediate step in the evolution of U.S. space capability might well have made more technical sense and could have initiated an evolutionary U.S. space program that would have been a better fit to the resources that the political system has made available to NASA over the past four decades.

There is another piece of evidence suggesting that the 1972 decision to develop the full-capability orbiter was a policy mistake. The absence in the wake of the shuttle's 2011 retirement of any advocacy within the U.S. space community for replacing the NASA shuttle with a second generation system having similar or greater capabilities is striking. After 30 years of operating the shuttle, there is no current demand to replicate in one vehicle the capabilities that the shuttle provided.[32]

My assessment of the space shuttle *program* as a major element of the Nixon space legacy is a mixture of positives and negatives. It is a matter of judgment whether the former outweigh the latter. But I stand by my 1986 assessment that the *decision* to develop the full capability shuttle was indeed a "policy failure," better characterized as a "policy mistake," in that the consequences of that choice have had a strongly negative impact on the evolution of the U.S. space program. Jumping directly from Apollo to developing the full capability shuttle was "a leap too far." Rather than being a cost-effective system providing highly valued capabilities, the shuttle turned out to be an expensive and difficult to operate vehicle. Arguing that the shuttle enabled the United States to develop the International Space Station is somewhat circular, since it is not yet clear that the station will turn out to produce benefits worth its development cost, and most likely there would not have been a space station (or at least the station that was constructed) without the shuttle.

The Nixon Space Heritage

The space shuttle may be the most visible Nixon space legacy, but the consequences of the other Nixon administration decisions in the 1969–1972 period have also had pervasive and lasting impacts. A 2012 assessment of NASA's "strategic direction" observed that:

> The National Aeronautics and Space Administration (NASA) is at a transitional point in its history... The agency's budget... is under considerable stress, servicing increasingly expensive missions and a large, aging infrastructure established at the height of the Apollo program. Other than the long-range goal of sending humans to Mars, there is no strong, compelling national vision for the human spaceflight program, which is arguably the centerpiece of NASA's spectrum of mission areas. The lack of national consensus on NASA's most publicly visible mission, along with out-year budget uncertainty, has resulted in the lack of strategic focus necessary for national agencies operating in today's budgetary reality. As a result, NASA's distribution of resources may be out of sync with what it can achieve relative to what it has been asked to do.

This review concluded that "there is no national consensus on strategic goals and objectives for NASA."[33] This judgment was echoed in the most recent review of the human space flight program, which observed that "a national consensus on the long-term future of human spaceflight... remains elusive."[34]

To a significant degree this unsatisfactory condition of the U.S. human space flight program in the second decade of the twenty-first century is a heritage of the policy decisions made by Richard Nixon more than 40 years ago. Approving the space shuttle came without a meaningful national commitment to post-Apollo space program objectives—there was no "strategic focus." George Low in October 1970 had suggested that "with the shuttle the U.S. can have a continuing program of manned space flight...without a commitment to a major new manned mission goal." This proved to be a winning argument; by approving the space shuttle, a capability-justified means for carrying out a variety of space activities, Richard Nixon avoided having to define the long-term space objectives which the shuttle would serve. This lack of guiding goals for the U.S. space program has persisted for more than 40 years, causing many to characterize the program as "adrift." If this characterization is accepted, it was Richard Nixon that set NASA on that goal-less voyage.

Nixon and his closest advisers gave little attention to the longer term consequences of their decision to put the NASA full-capability space shuttle at the center of the post-Apollo space program. Those consequences were compounded by approving a shuttle design that from NASA's standpoint was a step toward an eventual space station. The consequences were exacerbated by setting out an approach to determining the NASA budget that was very likely to result in funding insufficient to support efficient development and operation of both the space shuttle and space station while also funding the space activities they were designed to serve. It was difficult to rally public and political support for the capability-driven approach inherent in the Nixon approach to the post-Apollo space program, and the lack of broad public support for the space program has persisted. The absence of a compelling exploration objective or other widely accepted goal has resulted for four decades in a human spaceflight program focused, for uncertain purposes, on developing and operating the shuttle and assembling the space station. Attempts in 1989 and 2004 to set an exploration goal to guide the space program have not taken root, and the fate of the current NASA exploration program is unclear.

There is no simple or immediate remedy to the current situation with respect to the U.S. space program. It will be very difficult to undo the consequences of flawed or mistaken policy decisions made more than four decades ago. Some suggest that the government should step aside and allow the U.S. private sector to take the leading role in the U.S. space program, including human exploration. My judgment is that such an approach is unrealistic; only governments can provide resources sufficient to lead the initial stages of a long-term exploration effort, although government–private sector partnerships (and international cooperation) should certainly be part of that effort. In my view, if the United States is to remain the leader in human space exploration it will take committed and continuing presidential leadership of the character provided so long ago by John F. Kennedy, once again singling out the space program for special priority and setting challenging goals, convincing a reluctant public and their representatives in Congress to accept those goals, and then, crucially, committing on a sustained basis the political, human, and financial resources needed to achieve them.[35] The alternatives are to continue to drift along, trying to do too much with too little, or, less likely, to lower U.S. ambitions in space to match the funding available. A comprehensive review of the U.S. space program in 2014 once again concluded, as had its 2009 and 2012 predecessors, that "the human spaceflight program conducted by the U.S. government today has no strong direction" and that "the long-term future of human spaceflight... is unclear."[36]

That situation is Richard Nixon's most fundamental space heritage.

Notes

Abbreviations Used

CTW	Papers of Clay Thomas Whitehead, Manuscript Division, Library of Congress, Washington, DC
GML	Papers of George M. Low, Folsom Library, Rensselaer Polytechnic Institute, Troy NY
LSN/NHRC	John Logsdon Source Notes, NASA Historical Reference Collection, NASA Headquarters, Washington, DC
NARA	National Archives and Record Administration, Greenbelt, MD
NHRC	NASA Historical Reference Collection, NASA Headquarters, Washington, DC
RNPL	Richard Nixon Presidential Library, Yorba Linda, CA
TOP	Papers of Thomas O. Paine, Manuscript Division, Library of Congress, Washington, DC

Overture

1. John M. Logsdon, *John F. Kennedy and the Race to the Moon* (New York: Palgrave Macmillan, 2010), 114.
2. Richard Nixon: "Statement Announcing Decision to Proceed with Development of the Space Shuttle," January 5, 1972. http://www.presidency.ucsb.edu/ws/?pid=3574. Online by Gerhard Peters and John T. Woolley, *The American Presidency Project*.
3. Richard Nixon: "Statement Following Lift-Off from Moon of the Apollo 17 Lunar Module," December 14, 1972. http://www.presidency.ucsb.edu/ws/?pid=3715. Online by Gerhard Peters and John T. Woolley, *The American Presidency Project*.

1 Richard Nixon and *Apollo 11*

1. Richard M. Nixon, *The Memoirs of Richard Nixon* (New York: Grosset & Dunlap, 1978), 361–362.
2. Richard Nixon, "Inaugural Address," January 20, 1969, http://www.presidency.ucsb.edu/ws?pid=1941. Online by Gerhard Peeters and John T. Woolley, *The American Presidency Project*.

3. Nixon, *Memoirs*, 428; William Safire, *Before the Fall: An Inside View of the Pre-Watergate White House* (New Brunswick, NJ: Transaction Publishers, 2005), 121, 145.
4. The term "soft power" was originated by Harvard professor Joseph Nye. See Joseph S. Nye, Jr., *The Paradox of American* Power (New York: Oxford University Press, 2002), 8–9. For a discussion of President Kennedy's motivations in initiating Project Apollo, see Logsdon, *John F. Kennedy and the Race to the Moon*. The quote from the Kennedy speech is on 113–114. For a very perceptive early analysis of the motivations underpinning the U.S. space program, see Vernon van Dyke, *Pride and Power: The Rationale of the Space Program* (Urbana: University of Illinois Press, 1964).
5. Interview with John Ehrlichman, May 6, 1983.
6. Frank Borman with Robert J. Serling, *Countdown: An Autobiography* (New York: William Morrow, 1988), 259–260.
7. Ibid, 226–227.
8. Richard Nixon: "Remarks Announcing a Goodwill Tour to Western Europe by Col. Frank Borman, USAF," January 30, 1969, http://www.presidency.ucsb.edu/ws/?pid=1997. Online by Gerhard Peters and John T. Woolley, *The American Presidency Project*; memorandum from Thomas Paine to Secretary of State William Rogers, January 24, 1969, Box 23, TOP; letter from Frank Borman to William Rogers, March 6, 1969 and letter from William Rogers to Frank Borman, March 27, 1969, both in Box 3017, Record Group 59, General Records of the Department of State, National Archives and Record Administration, Greenbelt, MD.
9. Borman, *Countdown*, 227, 260, 267–268.
10. Memorandum from Dwight Chapin to H. R. Haldeman, "Presidential Involvement in Apollo 11," May 28, 1969, Outer Space-3 File, RNPL.
11. Memorandum from John Ehrlichman to Peter Flanigan, May 29, 1969, Box 1, Federal Government 164 Files, RNPL.
12. Memorandum from H. R. Haldeman to Dwight Chapin, June 2, 1969; memorandum from Julian Scheer, NASA Assistant Administrator for Public Affairs, to Peter Flanagan (*sic*), June 9, 1969, both in Outer Space-3 File, RNPL; John Ehrlichman, "Notes of Meeting with President," June 10, 1969, Box 3, Papers of John Ehrlichman, RNPL.
13. Letter from President Richard Nixon to Archibald MacLeish, July 1, 1969, Outer Space-3 File, RNPL; Safire, *Before the Fall*, 147–148.
14. Memorandum from Dwight Chapin to H. R. Haldeman, "Apollo 11," June 16, 1969, Outer Space-3 File, RNPL; letter from Richard Nixon to Neil Armstrong, June 25, 1969, Box 1, Federal Government 164 Files, RNPL. Although Armstrong was a civilian who had resigned from the Navy in 1952, the letter was addressed to Commander Neil Armstrong, USN.
15. Memorandum from Dwight Chapin to H. R. Haldeman, "Status Report on Apollo 11 Project," July 1, 1969, Outer Space-3 File, RNPL; William Safire, "Of Nixon, Kennedy and Shooting the Moon," *The New York Times*, July 17, 1989, A17; memorandum from Jim Keogh (one of Nixon's speechwriters) to the President, June 16, 1969, Outer Space-3 File, RNPL; Safire, *Before the Fall*, 144–145.
16. Henry Kissinger, *White House Years* (Boston: Little, Brown and Company, 1979), 157.

17. Memorandum from Alexander Butterfield to Mr. Ehrlichman, July 7, 1969, Outer Space-3 File, RNPL; James R. Hansen, *First Man: The Life of Neil A. Armstrong* (New York: Simon & Schuster, 2005), 2–3; Michael Collins, *Carrying the Fire: An Astronaut's Journeys* (New York: Farrar, Straus and Giroux, 2009), 349; telegram from Neil Armstrong, Mike Collins, and Buzz Aldrin to the President, July 9, 1969, Papers of Spiro T. Agnew, University of Maryland Library, College Park, MD; telegram from the President to Commander Neil Armstrong and the Apollo XI Astronauts, July 15, 1969, Outer Space-3 File, RNPL.
18. I am grateful to former George Washington University graduate student Steve Wolfe for first bringing this episode to my attention. His article "Moonglow: Space Diplomacy in the Nixon Administration," *Quest*, 18:2 (2011) discusses the proposal. See also letter from Bill Moyers to Daniel Moynihan, June 4, 1969; memorandum from Daniel Moynihan to Bob Haldeman, June 6, 1969; and memorandum from Lee DuBridge to K. E. Cole, June 11, 1969, all in Outer Space-2 File, RNPL; memorandum from Arthur Burns to the Staff Secretary (Kenneth Cole), "Recommendation that the Apollo 11 Moon Shot be Commissioned the 'John F. Kennedy,'" June 10, 1969; memorandum from Herb Klein for the Staff Secretary, June 11, 1969, and memorandum from Stephen Bull to H. R. Haldeman, June 13, 1969, all in Outer Space-3 File, RNPL.
19. Richard Reeves, *President Nixon: Alone in the White House* (New York: Simon and Schuster, 2001), 40; Christopher Matthews, *Nixon & Kennedy: The Rivalry That Shaped Postwar America* (New York: Simon & Schuster, 1997), 279; "Nixoning the Moon," *The New York Times*, July 19, 1969; H. R. Haldeman, *The Haldeman Diaries: Inside the Nixon White House* (New York: G. P. Putnam's Sons, 1994), 72, 74; Action Memorandum 816 from the Staff Secretary to Pat Buchanan, July 21, 1969, Borman Alphabetical File, RNPL.
20. Craig Nelson, *Rocket Men: The Epic Story of the First Men on the Moon* (New York: Viking, 2009), 299.
21. Borman, *Countdown*, 237, 256; Borman's trip to the Soviet Union is described on 242–257 of his book; Wolfe, "Moonglow," 47.
22. Borman, *Countdown*, 240–241.
23. Memorandums from Frank Borman for the President and the First Lady, July 14, 1969, both in Outer Space-3 File, RNPL. I am grateful to Dr. Teasel Muir-Harmony for calling these documents to my attention.
24. Borman, *Countdown*, 239; memorandum from Frank Borman to the President, July 14, 1969, Outer Space-3 File, RNPL.
25. Haldeman, *Haldeman Diaries*, 71.
26. Memorandum from Jonathan Rose to Ken Cole, "Rain Plan for Apollo 11," Outer Space-3 File, RNPL; NASA, "Public Affairs Apollo Program Contingency Plan," undated, and memorandum from Bill Safire to H. R. Haldeman, "In Event of Moon Disaster," July 18, 1969, both in File 574, NHRC.
27. Richard Nixon: "Proclamation 3919—National Day of Participation Honoring the Apollo 11 Mission," July 16, 1969, http://www.presidency.ucsb.edu/ws/?pid=2128. Online by Gerhard Peters and John T. Woolley, *The American Presidency Project*; *The New York Times*, July 17, 1969, A1.
28. Hansen, *First Man*, 494.

29. Haldeman, *Haldeman Diaries*, 73. President Richard Nixon's Daily Diary is available at http://www.nixonlibrary.gov/virtuallibrary/documents/dailydiary.php.
30. The drafts prepared for President Nixon by Price and Safire can be found in Box 50, Speech File, President's Personal Papers, RNPL; Borman, *Countdown*, 238; Haldeman, *Haldeman Diaries*, 73; Safire, *Before the Fall*, 149. Safire says that Nixon called him after talking to the astronauts on the Moon, noting that he had worked in the "tranquility" theme. Richard Nixon: "Telephone Conversation with the Apollo 11 Astronauts on the Moon," July 20, 1969, http://www.presidency.ucsb.edu/ws/?pid=2133. Online by Gerhard Peters and John T. Woolley, *The American Presidency Project*.
31. Hansen, *First Man*, 506–507; Collins, *Carrying the Fire*, 434.
32. *The New York Times*, July 21, 1969, A1.
33. Collins, *Carrying the Fire*, 424–425.
34. Interview with Thomas Paine, August 12, 1970.
35. John Ehrlichman, *Witness to Power: The Nixon Years* (New York: Simon & Schuster, 1982), 297; John Ehrlichman, Notes of Meeting with President, undated but July 1969, Box 3, Papers of John Ehrlichman, RNPL.
36. Haldeman, *Haldeman Diaries*, 75.
37. Hansen, *First Man*, 554–556; Collins, *Carrying the Fire*, 444; Stephen Ambrose, *Nixon: The Triumph of a Politician*, Volume Two (New York: Simon and Schuster, 1989), 285; Richard Nixon: "Remarks to Apollo 11 Astronauts Aboard the U.S.S. Hornet Following Completion of Their Lunar Mission," July 24, 1969, http://www.presidency.ucsb.edu/ws/?pid=2138. Online by Gerhard Peters and John T. Woolley, *The American Presidency Project*.
38. Milt Putnam's account of the *Apollo 11* recovery can be found at www.navy-history.org/2012/02/navy-photographer-apollo-11-recovery/
39. Richard Nixon: "Remarks on Arrival at Manila, the Philippines," July 26, 1969, http://www.presidency.ucsb.edu/ws/?pid=214; Richard Nixon: "Remarks on Arrival at Bucharest, Romania," August 2, 1969, http://www.presidency.ucsb.edu/ws/?pid=2172. Online by Gerhard Peters and John T. Woolley, *The American Presidency Project*.
40. Drew Pearson and Jack Anderson, "Nixon Gambled on Moon Publicity," *The Washington Post*, July 29, 1969, B11.
41. Hansen, *First Man*, 566–567; Richard Nixon: "Remarks at a Dinner in Los Angeles Honoring the Apollo 11 Astronauts," August 13, 1969, http://www.presidency.ucsb.edu/ws/?pid=2202. Online by Gerhard Peters and John T. Woolley, *The American Presidency Project*; Haldeman, *Haldeman Diaries*, 80.
42. Memorandums from Peter Flanigan to Julian Scheer, August 6, 1969 and August 11, 1969 and memorandum from Peter Flanigan to Dr. Paine, August 15, 1969, all in Box 1, Federal Government 164 Files, RNPL; Haldeman, *Haldeman Diaries*, 81.
43. Memorandum from Julian Scheer to Peter Flanigan, with attachments, August 15, 1969, Box 392, National Security Files, RNPL.
44. Wolfe, "Moonglow," 43; memorandum from Peter Flanigan to Julian Scheer, August 23, 1969, Box 1, Federal Government 164 Files, RNPL.
45. Undated and handwritten memorandum, "President's changes in astronaut schedule," Box 393, National Security Files, RNPL; memorandum from

Peter Flanigan for Dr. Paine, "Apollo 11 World Tour," August 26, 1969, Box 392, National Security Files, RNPL; Wolfe, "Moonglow," 43.
46. Memorandum from L. Nicholas Ruwe to Henry Kissinger, "Apollo 11 World Tour," September 23, 1969, Box 1, Federal Government 164 Files, RNPL; Hansen, *First Man*, 576; Edwin E. "Buzz" Aldrin, Jr. with Wayne Warga, *Return to Earth* (New York: Random House, 1973), 55.
47. "Telephone Call to Neil Armstrong, Commander, Apollo XI Flight," Undated, Outer Space 3-1 File, RNPL.
48. For accounts of the "Giant Step" tour, including Aldrin's problems, see Aldrin, *Return to Earth*, 56-83, Hansen, *First Man*, 577-580, and Collins, *Carrying the Fire*, 456.
49. Richard Nixon: "Remarks Welcoming the Apollo 11 Astronauts Following Their Goodwill Tour," November 5, 1969, http://www.presidency.ucsb.edu/ws/?pid=2309. Online by Gerhard Peters and John T. Woolley, *The American Presidency Project*; Hansen, *First Man*, 579-580; Aldrin, *Return to Earth*, 84-86; Collins, *Carrying the Fire*, 457.

2 Setting the Post-Apollo Stage

1. Hugh Sidey, *John F. Kennedy, President* (New York: Atheneum, 1964), 98.
2. L. A. Minnich, Jr., "Legislative Leadership Meeting, Supplementary Notes," February 4, 1958, reprinted in John M. Logsdon et al., eds., *Exploring the Unknown: Selected Documents in the History of the U.S. Civil Space Program*, Volume I: Organizing for Exploration, NASA SP-4407 (Washington, DC: Government Printing Office, 1995), 631-632; Charles Townes et al., "Report of the Task Force on Space," January 8, 1969, ibid, 509.
3. Yanek Mieczkowski, *Eisenhower's Sputnik Moment: The Race for Space and World Prestige* (Ithaca, NY: Cornell University Press, 2013), 31, 254; Nixon, *Memoirs*, 428-429.
4. William Leavitt, "Space: Now the Nixon Years," *Space Digest*, December 1968, 73-74; Republican National Committee News Releases, "Statements by Richard Nixon," September 6, 1968, September 30, 1968, and November 2, 1968, all in LSN/NHRC.
5. Interview with Arnold Frutkin, September 15, 1970. Frutkin was NASA's long-time assistant administrator for international affairs, but in 1968 was temporarily working on the NASA-wide planning effort. Joan Hoff, "The Presidency, Congress, and the Deceleration of the U.S. Space Program in the 1970s," in Roger D. Launius and Howard E. McCurdy, eds., *Spaceflight and the Myth of Presidential Leadership* (Urbana, IL: University of Illinois Press, 1997), 92. Matthew D. Tribbe, *No Requiem for the Space Age: The Apollo Moon Landings and American Culture* (New York: Oxford University Press, 2014), 4.
6. Interview of Willis Shapley by John Mauer, April 12, 1989, Archives Division, National Air and Space Museum, Smithsonian Institution, Washington, DC.
7. Letter from James E. Webb to President John F. Kennedy, November 30, 1962, reprinted in Logsdon, *Exploring the Unknown*, Volume I, 461, 465. The definitive biography of James Webb is W. Henry Lambright, *Powering Apollo: James E. Webb of NASA* (Baltimore, MD: The Johns Hopkins University Press, 1998).

8. Logsdon, *Exploring the Unknown*, Volume I, 495; NASA, *Preliminary History of NASA, 1963–1969*, January 15, 1969, II-53. This document was prepared at the end of the Johnson presidency for submission to the Lyndon B. Johnson Presidential Library and can be found in the NHRC. Letter from James E. Webb to the President, October 2, 1968, File 16697, NHRC.
9. Interview of Willis Shapley by Martin Collins, September 27, 1994, Archives Division, National Air and Space Museum, Smithsonian Institution, Washington, DC; White House, "Transcript of James E. Webb Press Conference," September 16, 1968, File 16697, NHRC; *The Washington Post*, September 18, 1969, A24.
10. See Robert Seamans, *Aiming at Targets*, NASA SP-4106 (Washington, DC: Government Printing Office, 1996). Seamans's autobiography minimizes his conflict with James Webb, but by late 1967 Webb had lost trust in Seamans as his deputy.
11. NASA, "Summary of Dr. Paine's Remarks to the Management Council," May 14, File 1968, NHRC.
12. *Wall Street Journal*, November 11, 1968, 12.
13. Memorandum from Floyd Thompson to the Administrator with attached report, "Report of the Post-Apollo Advisory Group," July 20, 1968, 4, 10, LSN/NHRC.
14. George E. Mueller, "Honorary Fellowship Acceptance Speech," August 10, 1968, reprinted in John M. Logsdon et al., eds., *Exploring the Unknown: Selected Documents in the History of the U.S. Civil Space Program*, Volume IV: Accessing Space, NASA SP-4407 (Washington, DC: Government Printing Office, 1999), 202–203. For an uncritical biography of Mueller, see Arthur Slotkin, *Doing the Impossible: George E. Mueller & the Management of NASA's Human Space Flight Program* (London: Springer Praxis, 2012).
15. Memorandum from Associate Administrator to Deputy Administrator, "Evolution of NASA's Long Range Planning," February 9, 1970, LSN/NHRC; interview with Homer Newell, July 15, 1970.
16. Letter from Thomas Paine to Charles Zwick, October 14, 1968; Bureau of the Budget, "National Aeronautics and Space Administration – Highlights Summary," October 30, 1968, both in LSN/NHRC.
17. Letter from Thomas Paine to Charles Zwick, November 10, 1968, LSN/NHRC; Lyndon B. Johnson, *The Vantage Point: Perspectives of the Presidency, 1963–1969* (New York: Holt, Rinehart and Winston, 1971), 286.
18. William Normyle, "Alternatives Open on Post-Apollo," *Aviation Week and Space Technology*, January 13, 1969, 12.
19. Letter from Thomas Paine to Abe Silverstein, marked "For Eyes Only," January 14, 1969, LSN/NHRC. Similar letters were sent to the other invitees to the meeting.
20. Notes taken by Thomas Paine from "Meeting on Space Stations," January 27, 1969, File 4184, NHRC.
21. William Normyle, "NASA Aims at 100-Man Station," *Aviation Week and Space Technology*, February 24, 1969, 16; handwritten note from Tom Paine to Gene (Eugene Emme, NASA's chief historian), February 3, 1969, File 4148, NHRC.
22. Bryce Nelson, "Nixon Forms Advisory Panels on Science, Space, Health," *Science*, December 13, 1968, 1255.

23. Interview with Russell Drew, Office of Science and Technology, August 31, 1970; Drew was the primary OST staff person for space.
24. Townes et al., "Report of the Task Force on Space," in Logsdon, *Exploring the Unknown*, Volume I, 505–504.
25. Ibid, 512.
26. Letter from Thomas Paine to Lee DuBridge, May 6, 1969, LSN/NHRC.
27. Rowland Evans, Jr. and Robert D. Novak, *Richard Nixon in the White House: The Frustration of Power* (New York: Random House, 1971), 9, 11; Elisabeth Drew, *Richard M.* Nixon (New York: Henry Holt and Company, 2007), 23.
28. Ambrose, *Nixon*, 228; Reeves, *President Nixon*, 29; John Ehrlichman, *Witness to Power*, 78.
29. Haldeman, *The Haldeman Diaries*. The quoted passage is on page 15. A copy of the complete diary is available at RNPL.
30. Evans and Novak, *Nixon in the White House*, 46, 50.
31. Leonard Garment, *Crazy Rhythm: My Journey from Brooklyn, Jazz, and Wall Street to Nixon's White House, Watergate, and Beyond...* (New York: Random House, 1997), 152.
32. Ehrlichman, *Witness to Power*, 48.
33. James MacGregor Burns, *Running Alone: Presidential Leadership – JFK to Bush II* (New York: Basic Books, 2006), 98; Drew, *Richard M. Nixon*, 1–2.
34. Two memos from President Richard Nixon to Lee DuBridge, February 4, 1969, Box 1, Federal Government 164 files, RNPL.
35. Thomas Paine, Memorandum for the Record, "Telephone Conversations with Dr. DuBridge, Saturday, February 8, 1969 and events preceding them," February 11, 1969, Box 32, TOP.
36. Interviews with Russell Drew, August 31, 1970 and October 13, 1970; interviews with Thomas O. Paine, August 12, 1970 and September 3, 1970; interview with Willis H. Shapley, July 30, 1970.
37. T.O. Paine, Memorandum for the Record, "Telephone Conversation with Dr. DuBridge of February 10, 1969," Box 32, TOP.
38. Memorandum from Lee DuBridge to the President, February 6, 1969, File 3–1, Papers of Arthur Burns, RNPL; memorandum from Lee DuBridge to the President, February 10, 1969, Box 26, Papers of Edward E. David, RNPL. (All science adviser papers in the Nixon Library are filed under the name of DuBridge's successor, Edward E. David.); memorandum from Kenneth Cole to Bob Haldeman, February 12, 1969, Outer Space-3 File, RNPL; memorandum from the President for the Vice President, The Secretary of Defense, The Acting Administrator of the National Aeronautics and Space Administration, and The Science Adviser, February 13, 1969, reprinted in Logsdon, *Exploring the Unknown*, Volume I, 513.
39. Nixon, *Memoirs*, 300, 310–313.
40. Charles Townes et al., "Report of the Task Force on Space, January 8, 1969," reprinted in Logsdon, *Exploring the Unknown*, Volume I, 511; memorandum from Russell Drew to Dr. DuBridge, "Steering Committee for the Space Program," February 6, 1969, Box 35, Papers of Edward David, RNPL.
41. Interview with William Anders, January 6, 2013; letter from Spiro T. Agnew to Thomas Paine, June 6, 1969, Papers of Spiro T. Agnew, University of Maryland, College Park, MD.
42. Memorandum from H. R. Haldeman to Dwight Chapin, January 27, 1969, and letter from Patrick Haggerty to the Honorable Richard Nixon, February

4, 1969, Box 1, Federal Government 164 Files, RNPL; H. R. Haldeman, diary entry for January 28, 1969, not in printed version of diary but available at RNPL.
43. Thomas O'Toole, "President Must Soon Appoint Space Agency Administrator," *The Washington Post*, February 9, 1969, A11; H. R. Haldeman, diary entry for February 24, 1969, RNPL; Richard Nixon: "Remarks on Presenting the Robert H. Goddard Memorial Trophy to the Apollo 8 Astronauts," March 5, 1969. http://www.presidency.ucsb.edu/ws/?pid=2445. Online by Gerhard Peters and John T. Woolley, *The American Presidency Project*.

3   After the Moon, Mars?

1. Memorandum from T. O. Paine for the President, "NASA Activities," February 4, 1969, File 012578, NHRC.
2. Interview of Willis Shapley by Martin Collins, September 27, 1994, Archives Division, National Air and Space Museum, Smithsonian Institution, Washington, DC.
3. T. O. Paine, Memorandum for the Record, "First Meeting of the President's Task Group on the National Space Program," March 11, 1969, Box 23, TOP.
4. Letter from Thomas Paine to Robert Mayo, February 24, 1969, LSN/NHRC.
5. T. O. Paine, Memorandum for the President, "Problems and Opportunities in Manned Space Flight," February 26, 1969, reprinted in Logsdon, *Exploring the Unknown*, Volume I, 513–519.
6. Lewis M. Branscomb, "Comments on the Memorandum to the President from the Acting Administrator of NASA, dated February 22, 1969," attached to letter from Lewis Branscomb to Lee DuBridge, March 4, 1969, Box 35, Papers of Edward E. David, RNPL.
7. "Report of Space Task Group Staff Directors Committee on NASA's Request for Amendments to the FY1970 Budget" and T. O. Paine, Memorandum for the Record, "Space Task Group Meeting, March 22, 1969," both in Box 32, TOP.
8. Homer Newell, "Report of the Space Task Group Meeting of March 22, 1969," LSN/NHRC; T. O. Paine, Memorandum for the Record, "Space Task Group Meeting, March 22, 1969," Box 32, TOP.
9. Memorandum from Thomas Paine to Space Task Group Members, March 22, 1969, Folder 19910, NHRC.
10. Homer Newell, "Report of the Space Task Group Meeting of March 22, 1969," LSN/NHRC; T. O. Paine, Memorandum for the Record, "Space Task Group Meeting, March 22, 1969," Box 32, TOP.
11. Interview with Thomas Paine, August 12, 1970.
12. Memorandum from James Schlesinger to the Director, "Meetings of the Space Task Group and the Marine Council," May 27, 1969, LSN/NHRC. Two senators, Clinton Anderson of New Mexico, who was chair of the Senate Committee on Aeronautical and Space Sciences, and Carl Curtis of Nebraska, and eight members of the House of Representatives (George Miller, Olin Teague, Joseph Karth, Ken Hechler, James Fulton, Charles Mosher, Richard Roudebush, and Thomas Pelley) attended the meeting.

13. Handwritten comment by Vice President Agnew on a Memorandum from Lee DuBridge to the Vice President, "Recommendations Concerning Operations of the Space Task Group," April 16, 1969, Papers of Spiro T. Agnew, University of Maryland, College Park, MD.
14. Memorandum from J. R. W. (Jerome Wolff) to the Vice President, July 3, 1969, Papers of Spiro T. Agnew, University of Maryland, College Park, MD. Attending one or both of the meetings, in addition to Shirley Temple Black, were: Jamie Benitez, University of Puerto Rico; Angus Campbell, University of Michigan; C. Stark Draper, MIT; William Foster, former head of the Arms Control and Disarmament Agency; T. Keith Glennan, first NASA administrator; Peter Goldmark, CBS Laboratories; Najeeb Halaby, Pan American Airlines; Frederick Kappel, former head of AT&T; Foy Kohler, former ambassador to the Soviet Union; William Deming Lewis, Lehigh University; John Love, governor of Colorado; Richard Olgilvie, Governor of Illinois; Henry Rowen, Rand Corporation; Leon Schacter, Amalgamated Meat Cutter and Butchers Workman Union; Dan Seymour, J. Walter Thompson Company; and Frank Stanton, CBS. (Stanton had also been one of the nongovernment people consulted by Lyndon Johnson in 1961 as he prepared the recommendations to President Kennedy that led to Project Apollo.)
15. Remarks of Vice President Agnew, July 7, 1969, attached to Memorandum from Russell Drew to Members and Observers of the Space Task Group, July 29, 1969; no author, "Report of Discussion at Invited Contributors Meeting, July 7, 1969," July 18, 1969; memorandum from Milton Rosen to Thomas Paine, "Meeting of Invited Contributors to the President's Space Task Group," August 6, 1969, all in LSN/NHRC.
16. American Institute of Aeronautics and Astronautics, "The Post-Apollo Space Program: An AIAA View," May 20, 1969, LSN/NHRC. The quoted passages are on 3, 17, and 24.
17. Letter from Harry Hess, Chair, Space Science Board, to Lee DuBridge, June 23, 1969, LSN/NHRC.
18. Interviews with Thomas Paine, August 12, 1970, and September 3, 1970.
19. NASA Planning Steering Group, "Discussion Paper," June 5, 1969; presentation by DeMarquis Wyatt at Program Alternatives Meeting, June 24, 1969; NASA, "America's Next Decade in Space: A Draft Report for the Space Task Group," July 9, 1969, ii, iv, 19, 32, 34, all in LSN/NHRC.
20. NASA, "America's Next Decade in Space: A Draft Report for the Space Task Group," July 9, 1969, LSN/NHRC; interview with George Mueller, September 28, 1980.
21. Interviews with Demarquis Wyatt, NASA Assistant Administrator for Planning, July 23, 1970; Charles Donlan, Deputy Associate Administrator for Manned Space Flight, August 28, 1970; and William Lilly, NASA Assistant Administrator for Administration, August 21, 1970.
22. NASA, "An Integrated Space Program of Space Utilization and Exploration for the Decade 1970 to 1980," July 16, 1969 (the date of the Apollo 11 launch), LSN/NHRC.
23. Remarks by the Vice President, July 7, 1969, attached to memorandum from Russell Drew to Members and Observers of the Space Task Group, July 29, 1969, LSN/NHRC; Richard Witkins, "Agnew Proposes a Mars Landing," *The New York Times*, July 17, 1969, 1, 22; interview with Jerome Wolff, November 3, 1970. In a discussion with the author, *Apollo 10* astronaut Tom

Stafford suggested that he also had urged Agnew to articulate the Mars goal; Stafford was the vice president's escort during the *Apollo 11* events.

24. For one of many discussions of the allure of Mars, see John Noble Wilford, *Mars Beckons: The Mysteries, the Challenges, the Expectations of Our Next Great Adventure in Space* (New York: Viking, 1991). See also Howard E. McCurdy, *Space and the American Imagination*, 2nd edition (Baltimore: The Johns Hopkins Press, 2011).
25. Interviews with Thomas Paine, August 12 and September 3, 1970.
26. Milton Rosen, Memorandum for the File, July 22, 1969, LSN/NHRC, and interview, August 12, 1970; interviews with Thomas Paine, August 12, 1970 and September 3, 1970; interview with George Mueller, September 28, 1980.
27. Personal communication to author from Milton Rosen, November 22, 1974.
28. See Michael J. Neufeld, *Von Braun: Dreamer of Space, Engineer of War* (New York: Alfred A. Knopf, 2007) for the definitive biography of Wernher von Braun. Von Braun's presentation to the Space Task Group is discussed on 434–437.
29. Interview with Wernher von Braun, August 25, 1970; Neufeld, *von Braun*, 437.
30. Memorandum from STA (Spiro T. Agnew) to Lee DuBridge, "Space Task Group," July 25, 1969, Box 41, Papers of Edward E. David, RNPL.
31. A copy of von Braun's presentation is in File 2662, NHRC; see also David S. F. Portree, *Humans to Mars: Fifty Years of Mission Planning, 1950–2000*, NASA SP-2001-4521, available at history.nasa.gov/monograph21.pdf.
32. NASA, Presentation to Space Task Group on "A Program of Space Exploration in the 1970's and the 1980's," August 4, 1969. Part III, LSN/NHRC. The same presentation was given to the Senate Committee on Aeronautical and Space Sciences the next day, August 5. The Senate presentation is contained in Committee on Aeronautical and Space Sciences, "Future NASA Space Programs," Hearing, August 5, 1969, copy in File 580, NHRC.
33. Senators Mansfield and Kennedy are quoted in *The New York Times*, July 17, 1969, 22; Anderson's remarks are in the *Congressional Record*, July 29, 1969, S8739. Miller is quoted in *Aviation Week and Space Technology*, August 18, 1969, 16; "After Apollo-Mars?" *The New York Times*, July 18, 1969; results of the Gallup Poll appeared in the *Congressional Record*, August 13, 1969, H7361.
34. Interview with Thomas O. Paine, August 12, 1970.
35. A detailed description of the August 4 meeting is contained in the Department of Energy, *Journal of Glenn T. Seaborg*, Volume 19, PUB-625. Seaborg, the long-time chairman of the Atomic Energy Commission, was an observer of the Space Task Group's proceedings. Copies of his journals were graciously provided to the author by Dr. James Dewar. Letter from Robert Seamans to the Vice President, August 4, 1969, reprinted in Logsdon, *Exploring the Unknown*, Volume I, 520–522; interview with Robert Seamans, September 2, 1970.
36. *Journal of Glenn T. Seaborg*, 542–544. Interviews with Robert Seamans, September 2, 1970, and with Robert Mayo, December 29, 1970.
37. Memorandum from Russell Drew for Space Task Group Members and Observers, "Staff Director's Committee Report," July 26, 1969, File 581, NHRC.

38. Journal of Glenn T. Seaborg, 544; Russell Drew, "Notes for Staff Directors Committee," August 9, 1969, File 8199, NHRC.
39. NASA, Material Submitted to Staff Directors Committee, August 14, 1969, LSN/NHRC.
40. Whitehead's papers, including those related to his work on NASA issues, are located in the Manuscript Division of the Library of Congress and are also available online. See www.claytwhitehead.com.
41. Memorandum from Clay T. Whitehead to Mr. [Peter] Flanigan, June 25, 1969, Box 44, White House Central Files, RNPL.
42. Memorandum from Mr. [Donald] Crabill to the Files, "1971 budget options for NASA," August 25, 1969, LSN/NHRC.
43. Ibid.
44. Memorandum from Homer Newell to Thomas Paine, "Comments on the Report of the Space Task Group," August 29, 1969, LSN/NHRC.
45. This account of the September 3 STG meeting is drawn from the detailed account in Department of Energy, *Journal of Glenn T. Seaborg*, Volume 20, PUB-625, 16–21.
46. Letter from Robert Seamans to Russ Drew, September 5, 1969, File 581, NHRC.
47. Space Task Group Report, marked "Draft," *The Post-Apollo Space Program: Directions for the Future,* September 8, 1969, LSN/NHRC.
48. Ehrlichman, *Witness to Power,* 144–145.
49. This account of the September 11 meeting is drawn from Department of Energy, *Journal of Glenn T. Seaborg*, Volume 20, 69–73.
50. The quoted material in the preceding paragraphs is taken from Space Task Group Report to the President, "The Post-Apollo Space Program: Directions for the Future," September 1969, which is reprinted in Logsdon, *Exploring the Unknown,* Volume I, 523–543.
51. Letter from Vice President Spiro T. Agnew to the President, September 15, 1969, included in Department of Energy, *Journal of Glenn T. Seaborg*, Volume 20, 157–160.
52. Note from Milton Klein, Atomic Energy Commission (who attended the STG report briefing to the president in Glenn Seaborg's absence) for Chairman Seaborg, September 17, 1969, included in Department of Energy, *Journal of Glenn T. Seaborg*, Volume 20, 104–106.
53. Office of the White House Press Secretary, "Transcript of Press Conference," September 17, 1969, NHRC; letter from T. O. Paine to the President, September 19, 1969, Box 3, Federal Government Task Group Files, RNPL.
54. "A Spaceman's Sense of Balance," *The Washington Post,* September 19, 1969, A26; John Noble Wilford, "Soft Deadline for a Trip to Mars," *The New York Times,* September 21, 1969, E8; *Science,* October 10, 1969, 32.
55. Memorandum from Milton Rosen to Dr. Paine, September 10, 1969, File 581, NHRC.

4 Space and National Priorities

1. See Evans and Novak, *Nixon in the White House,* 184–203 for an account of the confused economic policy and budget process during the first year of the Nixon administration.
2. On NASA's attitude that it deserved high priority in the post-Apollo period, see Joan Hoff, "The Presidency, Congress, and the Deceleration of the U.S.

Space Program in the 1970s" in Launius and McCurdy, *Spaceflight and the Myth of Presidential Leadership*, 93. Interviews with Thomas O. Paine, August 12 and September 3, 1970.

3. Memorandum from Peter Flanigan to Robert Mayo, September 22, 1969, Outer Space-3 File, RNPL; memorandum from Peter Flanigan to David Derge, September 25, 1969, LSN/NHRC.
4. Interview with Willis Shapley by Martin Collins, September 27, 1994, Archives Division, National Air and Space Museum, Smithsonian Institution, Washington, DC.
5. Memorandum from John Ehrlichman to the Staff Secretary, "Director Mayo's Memo on Space Task Force Report, Log No. 1491," October 6, 1969, Box 3, Federal Government Task Force Files, RNPL.
6. Letter from George H. W. Bush to Bryce Harlow, March 28, 1969, Box 2, Federal Government 164 Files, RNPL; memorandum from Thomas Paine to the President, "NASA Deputy Administrator," September 19, 1969, LSN/NHRC; memorandum from Peter Flanigan to Bryce Harlow, John Ehrlichman, Bob Haldeman, Darrell Trent, Harry Flemming, and Herb Klein, October 7, 1969, Box 12, CTW.
7. George M. Low, "Personal Notes No. 1," January 1, 1970, GML. During his time as NASA deputy administrator and acting administrator, Low dictated detailed notes of events in which he was involved. These notes constitute an invaluable record of space policy and program decisions in the 1970–1977 period.
8. Ibid; memorandum from Peter Flanigan to the President, October 21, 1969, Box 44, White House Sensitive Files, RNPL; Low, "Personal Notes No. 1," GML.
9. Interviews with Bill Lilly, NASA Assistant Administrator for Administration, August 21, 1970, and Thomas Paine, September 3, 1970.
10. Letter from Robert Mayo to Thomas Paine, April 4, 1969, File 22858, NHRC; letter from Robert Mayo to Thomas Paine, July 9, 1969, LSN/NHRC; Bureau of the Budget, "Fiscal '71 Savings from a Decision to Reduce the Number of Manned Lunar Flights to One per Year," April 3, 1969, provided to author by Richard Speier.
11. Letter from Robert Mayo to Thomas Paine, July 28, 1969, LSN/NHRC; interview with Bureau of the Budget examiner Earl Rhode, August 28, 1970.
12. Bureau of the Budget Staff Paper, untitled and undated, but late August/early September 1969, Folder 5, Box 11, CTW; Donald Derman, Economics, Science, and Technology Division, Bureau of the Budget, Memorandum for the Director, "Space Task Group Report," September 12, 1969, LSN/NHRC.
13. Thomas Paine, Memorandum from Administrator to Heads of Program Offices, "Development of the FY1971 Agency Budget Proposals," September 22, 1969, LSN/NHRC; interview with Earl Rhode.
14. Letter from Thomas Paine to Robert Mayo, October 8, 1969, LSN/NHRC.
15. Bureau of the Budget, "National Aeronautics and Space Administration—Tentative Allowance—1971 Budget," November 13, 1969, LSN/NHRC.
16. Letter from Thomas Paine to Robert Mayo, November 18, 1969, LSN/NHRC.

17. Interviews with Bill Lilly and Earl Rhode.
18. Letter from Thomas Paine to Robert Mayo, December 5, 1969, File 1996, NHRC.
19. On Nixon's dislike of Robert Mayo, see Ehrlichman, *Witness to Power*, 92; on the budget decisions, Richard Speier, handwritten notes on NASA appeal strategy, November 24, 1969, provided to the author by Dr. Speier.
20. John S. Carroll, "Nixon Applauds 2nd Moon Venture," *Baltimore Sun*, November 15, 1969, A1.
21. Richard Nixon, "Remarks to NASA Personnel at the Kennedy Space Center," November 14, 1969, http//www.presidency.ucsb.edu/ws/pid=2322. Online by Gerhard Peters and John T. Woolley, *The American Presidency Project*.
22. Peter Flanigan, Memorandum to the President, December 6, 1969, reprinted in Logsdon, *Exploring the Unknown*, Vol. I, 546.
23. Elizabeth Covert, Secretary to Dr. Paine, Memorandum for the Record, "Phone Conversation, Dr. Paine and Mr. Peter Flanigan, Assistant to the President," December 8, 1969, LSN/NHRC.
24. Letter from Thomas Paine to the President, December 17, 1969, File 2578, NHRC.
25. Memorandum from John Brown III to John Erhlichman and Dr. Moynihan, December 6, 1969, LSN/NHRC. Brown was one of Haldeman's staff; that staff frequently relayed Richard Nixon's thoughts and orders to others in the White House.
26. George Low, "Personal Notes No. 1," GML.
27. Memorandum from Peter Flanigan to the Staff Secretary, "Log No. 2518 (NASA Budget)," December 19, 1969, Folder 1, Box 11, CTW.
28. John Ehrlichman, Meeting Notes, December 23 and December 26, 1969, Box 3, Papers of John Ehrlichman, RNPL; George Low, "Personal Notes No. 1," GML.
29. Interview with Clay Thomas Whitehead, October 19, 1970.
30. No author but almost certainly Clay Thomas Whitehead, untitled paper on the NASA FY1971 budget, December 2, 1969, Folder 6, Box 11, CTW.
31. Interview with Bill Lilly.
32. Whitehead, untitled paper on the NASA FY1971 budget, December 2, 1969, CTW.
33. Interviews with Bill Lilly and Earl Rhode; Low, "Personal Notes No. 1," GML; John Ehrlichman, "Meeting Notes," December 31, 1969, Box 3, Papers of John Ehrlichman, RNPL.
34. Letter from Thomas Paine to Robert Mayo, January 2, 1970, File 12578, NHRC; interview with Earl Rhode.
35. Memorandum from Clay T. Whitehead to Mr. Flanigan, January 5, 1970, Folder 1, Box 11, CTW; memorandum from Peter Flanigan to Thomas Paine and Robert Mayo, January 6, 1970, File 27995, NHRC.
36. Handwritten notes from Tom (Paine) to Peter (Flanigan), January 9 and January 12, 1969, TOP; edited version of press conference statement attached to Memorandum from Thomas Paine to Peter Flanigan, January 12, 1970, Folder 1, Box 13, CTW; George Low, "Personal Notes No. 5," January 17, 1970, GML; Richard Lyons, "50,000 NASA Jobs to Be Eliminated," *The New York Times*, January 14, 1970, A1.
37. Eileen Shanahan, "Tax Bill Signed: President Pledges Budget Balance," *The New York Times*, December 31, 1969, A1. The account of the origins of

"Operation Paring Knife" is based on Evans and Novak, *Nixon in the White House*, 202–203 and John Osborne, *The First Two Years of the Nixon Watch* (New York: Liveright, 1971), "The Second Year of the Nixon Watch," 15–19.
38. Evans and Novak, *Nixon in the White House*, 202–203.
39. Interview with Robert Mayo, December 29, 1970.
40. Letter from Thomas Paine to the President, January 15, 1970, LSN/NHRC.
41. This account of the events of January 15–16 is taken from George Low, "Personal Notes No. 5," January 17, 1970, GML and interviews with Thomas Paine, Robert Mayo, and with Peter Flanigan, November 9, 1970.
42. George Low, "Personal Notes No. 5," GML.
43. Peter Flanigan, Memorandum for the President, "Meeting with Dr. Thomas Paine, Administrator of NASA," January 21, 1970, Outer Space-1 File, RNPL.
44. Thomas Paine, Memorandum for the Record, "Meeting with the President, January 22, 1970," LSN/NHRC; memorandum from H. R. Haldeman to John Ehrlichman, January 23, 1970, Box 14, Papers of Edwin Harper, RNPL.
45. Memorandum from Spiro Agnew to Thomas O. Paine, January 30, 1970, Papers of Spiro T. Agnew, University of Maryland, College Park, MD.
46. Thomas Paine, Memorandum for the Record, "Meeting with the President, January 22, 1970," LSN/NHRC.

5   The Nixon Space Doctrine

1. Memorandum from Earl Rhode to Mr. (Jack) Young, "Outline for Dr. Schlesinger's speech to the Space Council," February 10, 1971, LSN/NHRC. Rhode was a NASA budget examiner in the Bureau of the Budget, and the Schlesinger speech was intended to spell out the philosophy behind the FY1971 budget decisions.
2. Memorandum from Milton Rosen to Dr. Paine, September 10, 1969, File 581, NHRC.
3. For a discussion of the Kennedy speech, see Logsdon, *John F. Kennedy and the Race to the Moon*.
4. Memorandum from Peter Flanigan to Tom Whitehead, October 6, 1969, Box 12, CTW; memorandum from Peter Flanigan to the Staff Secretary, "Log 1491," October 6, 1969, Box 3, Federal Government Task Force Files, RNPL; memorandum from Lee DuBridge to Hugh Sloan, "Request that the President Participate in the Dedication of the Lunar Science Institute," October 14, 1969, Outer Space-2 Files, RNPL.
5. Memorandum from Tom Whitehead to Peter Flanigan, November 17, 1969, with attached outline of "President's Statement on Our Next Decade in Space," File 012578, NHRC.
6. Ibid.
7. Memorandum from John R. Brown III to Peter Flanigan with attached essay by J. R. Bruckner, "Future of Space Program is Reaching a Critical Point," December 2, 1969, Folder 2, Box 13, CTW.
8. Memorandum from Peter Flanigan to Ken Cole, December 9, 1969; memorandum from Ken Cole to Jeb Magruder, December 12, 1969, both in Box 3, Federal Government Task Force Files, RNPL.

9. Memorandum from Raymond Price to Tom Whitehead, with attached draft outline dated 12/12/69, December 13, 1969, Folder 6, Box 11, CTW.
10. Memorandum from Tom Whitehead to Thomas Paine, December 16, 1969, and memorandum from Eva (Whitehead's secretary) to Marge (Flanigan's secretary), December 16, 1969, Box 12, CTW.
11. Action Memorandum 2530, "Proposed Presidential statement on space," December 18, 1969, with attached draft statement, Box 12, CTW.
12. Memorandum from Lee A. DuBridge to K. R. Cole, Jr., "Proposed Presidential Statement on Space," December 22, 1969, Box 35, Papers of Edward E. David, RNPL; record of telephone call for Tom Whitehead regarding release of the space statement, December 18, 1969, Folder 6, Box 11, CTW; letter from Thomas Paine to James Keogh, December 22, 1969, with attached "Suggested Statement on Space with NASA Proposed Revisions" and "Suggested [NASA] Draft: 'A New Space Program – Challenge and Opportunity,'" LSN/NHRC.
13. Memorandum from Ken Cole to Peter Flanigan, Bryce Harlow, and Wilf Rommel, December 30, 1969; memorandum from Ken Cole to Jim Keogh, January 5, 1969, both in Outer Space-1 Files, RNPL.
14. Thomas Paine, Memorandum for the Record, "Meeting with the President, January 22, 1970," LSN/NHRC.
15. Memorandum from Peter Flanigan to Ken Cole, January 24, 1970; memorandum from Peter Flanigan to Dr. Paine. January 28, 1970, both in Outer Space-1 Files, RNPL.
16. Letter from Thomas Paine to Peter Flanigan, January 25, 1970, LSN/NHRC; letters from Thomas Paine to Peter Flanigan, February 6, 1970, and February 16, 1970, Outer Space-1 Files, RNPL.
17. Memorandum from Lee Huebner to Herb Klein, February 10, 1970, Folder 1, Box 11, CTW.
18. Memorandum from Secretary of State William Rogers for the President, "International Space Cooperation," March 14, 1969, File 580, NHRC.
19. Richard Nixon: "Address Before the 24th Session of the General Assembly of the United Nations," September 18, 1969, http://www.presidency.ucsb.edu/ws/?pid=2236. Online by Gerhard Peters and John T. Woolley, *The American Presidency Project*.
20. Memorandum from Henry Kissinger to the President, "Reply to Letter from Senator Fulbright on Space Cooperation and Proposed Interagency Study," January 28, 1969, Box 392, National Security Files, RNPL; memorandum from Bud Wilkinson to John Ehrlichman, February 24, 1969, Box 1, Federal Government 164 Files, RNPL; letter from Frank Borman to William Rogers, March 6, 1969; letter from William Rogers to Frank Borman, March 27, 1969, Box 3017, Record Group 59, General Records of the Department of State, National Archives and Record Administration.
21. I am grateful to Frank Borman for providing me a copy of this memorandum. A copy can also be found in Outer Space-1 Files, RNPL.
22. Interview with Thomas Paine, August 12, 1970; letter from Thomas Paine to the President, August 22, 1969, File 84104, NHRC.
23. Memorandum from Peter Flanigan to the President, August 20, 1969, Outer Space-1 File, RNPL.

24. Letter from Thomas Paine to the President, November 7, 1969, LSN/NHRC. For a more detailed discussion of all aspects of post-Apollo space cooperation, see John Krige, Angela Long Callahan, and Ashok Maharaj, *NASA in the World: Fifty Years of International Collaboration in Space* (New York: Palgrave Macmillan, 2013).
25. Memorandum from the President to Peter Flanigan, November 24, 1969, Outer Space-1 Files, RNPL.
26. Memorandum from Colonel Frank Borman to Peter Flanigan, December 2, 1969, Outer Space-1 Files, RNPL.
27. Letter from Thomas Paine to Peter Flanigan, December 5, 1969, LSN/NHRC; memorandum from Peter Flanigan to Thomas Paine, December 9, 1969, Outer Space-1 Files, RNPL.
28. Memorandum from William Watts (a member of Haldeman's staff) to Robert Behr (a member of the National Security Council staff), "Dr. Paine Reports on International Space Cooperation," February 12, 1970, Folder 1, Box 11, CTW.
29. Memorandum from Willis Shapley to Clay T. Whitehead, February 26, 1970, Folder 1, Box 11, CTW.
30. Memorandum from Lee Huebner to Herb Klein, February 10, 1970, Folder 1, Box 11, CTW.
31. Memorandum from Ken Cole to Jim Keogh and Peter Flanigan, March 2, 1970, Box 3, Federal Government Task Forces Files; memorandum from Dwight Chapin to Ron Zeigler, Jerry Warren, Jeb Magruder, and Ken Cole, Box 1, Federal Government 164 Files, both at RNPL.
32. Memorandum from Willis Shapley to Tom Whitehead, March 5, 1970 with attached draft of space statement with NASA's suggested changes, Box 9, Papers of Peter Flanigan, RNPL; memorandum from Tom (Whitehead) to Marge (Flanigan's secretary), March 5, 1970, Box 10, CTW.
33. Memorandum from Ken Cole to John Ehrlichman, March 5, 1970; memorandum from John Ehrlichman for the President, March 5, 1970, both in Outer Space-1 Files, RNPL.
34. Memorandum from Peter Flanigan to John Ehrlichman, March 5, 1970, with attached Memorandum for the President, "Meeting with Dr. Thomas O. Paine, March 7, 1970," Box 9, Papers of Peter Flanigan, RNPL.
35. Memorandum from Peter Flanigan to Mr. Haldeman, February 2, 1970, Outer Space-1 Files, RNPL.
36. Richard Nixon: "Statement About the Future of the United States Space Program," March 7, 1970. http://www.presidency.ucsb.edu/ws/?pid=2903. Online by Gerhard Peters and John T. Woolley, *The American Presidency Project*.

6 The End of the Apollo Era

1. Office of The White House Press Secretary, "Press Conference of Dr. Thomas O. Paine, Administrator, National Aeronautics and Space Administration," March 7, 1970, Box 10, CTW.
2. George M. Low, "Personal Notes No. 25," June 21, 1970, GML.
3. Memorandum from Robert Lohman to Deputy Director, Space Station Task Force, "Shuttle-Sized Station Modules," April 1, 1970; teletype message from Douglas Lord, Deputy Director, Space Station Task Force to

Addressees, "Space Station Definition Studies Guidelines," May 12, 1970, both in File 18469, NHRC.
4. George M. Low, "Personal Notes No. 27," July 18, 1970, GML; interview with George Low, January 30, 1981.
5. Letter from George Low to George Shultz, September 30, 1970, File 9198, NHRC.
6. Andrew Chaikin, *A Man on the Moon: the Voyages of the Apollo Astronauts* (New York: Viking, 1993), 286. Chris Kraft, in 1970 director of flight operations at the Manned Spacecraft Center, in a January 3, 2014, telephone conversation with the author suggested that Chaikin's quote was not an accurate representation of Gilruth's views at the time, but there have been many discussions since 1970 of the hesitation of Gilruth, Low, and others associated with Apollo to continue to launch missions to the Moon because of their high risk.
7. Letter from T. O. Paine to John Findlay, August 5, 1970; letter from T. O. Paine to Charles Townes, August 13, 1970; letter from T. O. Paine to Peter Flanigan, August 11, 1970; memorandum from Peter Flanigan to Thomas Paine, August 13, 1970, all in Box 1, Federal Government 164 Files, RNPL.
8. Martin Sedlazek, Assistant Executive Secretary, NASA, Memorandum for the Record, "Apollo Flight Decision Meeting," August 31, 1970; letter from Charles Townes and John Findlay to Thomas Paine, August 24, 1970, both in File 87, GML.
9. Letter from Lee DuBridge to Tom Paine, August 28, 1970, File 87, and George M. Low, "Personal Notes No. 30," September 6, 1970, both in GML.
10. Letter from T. O. Paine to the President, September 1, 1970, File 87, GML.
11. Elizabeth Covert, Secretary to Dr. Paine, "Memorandum for the Record—Phone Conversation, Dr. Paine and Mr. Peter Flanigan, Assistant to the President," December 8, 1969, Folder 5, Box 34, TOP; George M. Low, "Personal Notes No. 25," June 21, 1970, GML.

Intermission

1. Interview with Peter Flanigan, November 9, 1970.
2. Interview with Clay Thomas Whitehead, October 19, 1970; memorandum from Tom Whitehead to Peter Flanigan, February 6, 1971, Folder 1, Box 13, CTW.

7 A New Cast of Characters

1. Richard Nixon, "Message to the Congress Transmitting Reorganization Plan 2 of 1970," March 12, 1970, http://www.presidency.ucsb.edu/ws/?pid=2907, Gerhard Peters and John T. Woolley, *The American Presidency Project*.
2. Memorandum from John Ehrlichman to Edward David, November 17, 1970, Box 54, Papers of John Ehrlichman, RNPL.
3. Richard Nixon "Message to the Congress Transmitting Reorganization Plan 2 of 1970," March 12, 1970, http://www.presidency.ucsb.edu/ws/?pid=2907, Gerhard Peters and John T. Woolley, *The American Presidency Project*.

4. Richard Nixon, "Message to the Congress Transmitting Reorganization Plan 1 of 1970 to Establish an Office of Telecommunications Policy," February 9, 1970, http://www.presidency.ucsb.edu/ws/?pid=2735, Gerhard Peters and John T. Woolley, *The American Presidency Project*.
5. "President's New Science Adviser," *The New York Times*, September 15, 1970, 26.
6. Memorandum for the President, February 13, 1970, Box 4, National Aeronautics and Space Council Files, RNPL.
7. On keeping Agnew out of policy issues, see Haldeman, *The Haldeman Diaries*, 247; interview with Bill Anders, January 6, 2013.
8. Memorandum from Ken Cole for George Shultz, August 24, 1970; memorandum from Dwight Ink to Arnold Weber, "National Aeronautics and Space Council," September 25, 1970; memorandum from Donald Rice to Jack Young, OMB, October 13, 1970; and memorandum from Arnold Weber to Kenneth Cole, "National Aeronautics and Space Council," October 29, 1970, all in Box 4, National Aeronautics and Space Council Files, RNPL.
9. Memorandum from John Ehrlichman to Arnold Weber, Office of Management and Budget, December 15, 1970, Box 4, National Aeronautics and Space Council Files, RNPL; Ehrlichman Meeting Notes, December 17, 1970, Box 5, Papers of John Ehrlichman, RNPL; letter from William Anders to John Ehrlichman, December 21, 1970, Box 4, Federal Government 6 Files, RNPL.
10. George Low, "Personal Notes No. 1," January 1, 1970 and "Personal Notes No. 4," January 13, 1970, GML; Neufeld, *Von Braun*, 440–445.
11. Memorandum from Thomas Paine to Addressees, "NASA Long-Range Planning Conference," May 25, 1970, File 19909, NHRC.
12. Transcript of Concluding Remarks by Dr. Paine, NASA Long Range Planning Conference, June 14, 1970, File 15994, NHRC.
13. Memorandum from Milton Rosen to Dr. Paine, "Comments on STG Report," September 10, 1969, File 581; James Daniels, Memorandum for the Record, "Meeting on Apollo/Skylab Program Options," June 15, 1970, File 8219; memorandum from Rene Berglund, "Space Station Task Group Staff Meeting of June 22, 1970," June 22, 1970, File 9199, all in NHRC.
14. Memorandum from Peter Flanigan to Bob Haldeman, July 22, 1970, Box 1, Federal Government 164 Files, RNPL.
15. Memorandum from Thomas Paine for the President, "Request for Appointment to Review Our Long-Range Future in Space," July 9, 1970, File 012578, NHRC.
16. George Low, "Personal Notes No. 28," August 8, 1970, GML; Richard Nixon: "Statement on the Anniversary of the First Manned Lunar Landing," July 20, 1970, http://www.presidency.ucsb.edu/ws/?pid=2587, Gerhard Peters and John T. Woolley, *The American Presidency Project*.
17. Memorandum from Dwight Chapin to H. R. Haldeman, "Dr. Paine," July 27, 1970, Box 1, Federal Government 164 Files, RNPL.
18. George M. Low, "Personal Note No. 28," August 8, 1970, GML.
19. Memorandum from Hugh Sloan via Dwight Chapin to the President, August 17, 1970, Box 1, Federal Government 164 Files, RNPL.
20. Interview of Willis Shapley by Martin Collins, September 27, 1994, Archives Division, National Air and Space Museum, Smithsonian Institution, Washington, DC; interview with Thomas Paine, August 12, 1970.

21. Interview with Thomas Paine, September 3, 1970; Homer Newell, *Beyond the Atmosphere: Early Years of Space Science*, NASA SP-4211 (Washington: Government Printing Office, 1980), 228–229; Neufeld, *Von Braun*, 446; Joan Hoff, "Deceleration of the U.S. Space Program in the 1970s," in Launius and McCurdy, *Spaceflight and the Myth of Presidential Leadership*, 100; interview with John Young, Office of Management and Budget, December 11, 1970; interview with Peter Flanigan, November 9, 1970; interview with John Ehrlichman, May 6, 1983.
22. Letter from Thomas Paine to the President, July 31, 1970, Box 9, Papers of Peter Flanigan, RNPL.
23. Memorandum from Peter Flanigan for the President, August 10, 1970, Box 9, Papers of Peter Flanigan, RNPL.
24. Memorandum from Peter Flanigan to David Packard, Deputy Secretary of Defense, September 11, 1970, Box 1, Federal Government 164 Files, RNPL.

8 The Space Shuttle Takes Center Stage

1. George M. Low, "Personal Notes No. 27," July 18, 1970, GML.
2. George M. Low, "Personal Notes No. 30," September 6, 1970, GML.
3. Edwin Harper, "Possible August 13, 1970 cut targets," undated, Box 11, Papers of Edwin Harper, RNPL; memorandum from John Whitaker to John Ehrlichman, "Political evaluation of cutback or elimination possibilities on June 26 Agency planning targets of OMB," July 7, 1970, Box 12, Papers of Edwin Harper, RNPL.
4. George M. Low, "Personal Notes No. 31," September 20, 1970, GML.
5. Safire, *Before the Fall*, 273.
6. Memorandum from Ed Harper to Robert Finch, "Key Election Issue: Federally Caused Unemployment," September 23, 1970, Box 1, Federal Government 164 Files, RNPL.
7. George M. Low, "Personal Notes No. 32," October 3, 1970, GML.
8. Letter from George Low to George Shultz, September 30, 1970, File 9198, NHRC.
9. Letter from George Low to William Pickering, October 7, 1970, File 19909, NHRC. For Low's 1961 report, see Logsdon, *John F. Kennedy and the Race to the Moon*, 51–52.
10. George M. Low, "Personal Notes No. 34," November 7, 1970, GML; letter from George Low to Ed David, October 30, 1970, File 01258, NHRC.
11. Letter from George Low to Henry Kissinger, October 30, 1970, File 8220, NHRC. This study does not give much attention to the discussions with respect to post-Apollo international cooperation in the 1969–1972 period. For a detailed account of those discussions, see Krige, Callahan, and Maharaj, *NASA in the World*.
12. George M. Low, "Personal Notes No. 34," November 7, 1970, GML.
13. Letter from George Low to Caspar Weinberger, October 28, 1970, File 1849, NHRC.
14. Letter from George Low to George Shultz, with attached paper "The Role of Manned Space Flight," November 2, 1970, File 70-4, GML.
15. Memorandum from Russell Drew to Edward David, "Space Strategy Paper," October 22, 1970, LSN/NHRC.

16. "List of Actions and Decisions, NASA Director's Review," November 3, 1970, attachment to Memorandum from Ed Harper to John Ehrlichman, "Key Budget Issues," undated, Box 12, Papers of Edwin Harper, RNPL.
17. George M. Low, "Personal Notes No. 34," November 7, 1970 and "Personal Notes No. 35," November 15, 1970, GML; interview with Bill Anders, January 6, 2013.
18. Office of Management and Budget, "National Aeronautics and Space Administration, 1972 Budget," undated, attachment to Memorandum from Ed Harper to John Ehrlichman, "Key Budget Issues," undated, Box 12, Papers of Edwin Harper, RNPL; memorandum from Edward David to Ed Harper, "Apollo 17," November 21, 1970, Box 61, Papers of Edwin Harper, RNPL.
19. Memorandum from John Ehrlichman to the President, "Unemployment Implications of Cuts in the NASA Budget," November 27, 1970, with attached memorandum from Will Kriegsman to John Ehrlichman, "NASA Budget," November 8, 1970, Box 8, Handwriting Files, RNPL.
20. Ed Harper, Memos for John Ehrlichman, "Employment Impact of NASA Budget Options," and "NASA Budget Options," December 1, 1970, Box 12, Papers of Edwin Harper, RNPL.
21. John Ehrlichman, Meeting Notes, December 1, 1970, Box 5, Papers of John Ehrlichman, RNPL.
22. George M. Low, "Personal Notes No. 37," December 20, 1970, GML.
23. Letter from George Low to the President, December 14, 1970, File 012578, NHRC; George Low, "Personal Notes No. 37," December 20, 1970, GML.
24. Kissinger, *White House Years*, 483; Haldeman, *Haldeman Diaries*, 150; Reeves, *President Nixon*, 189.
25. Kissinger, *White House Years*, 483.
26. Haldeman, *Haldeman Diary*, 150; Borman, *Countdown*, 259.
27. Memorandum from Dwight Chapin to H. R. Haldeman, "Apollo 13," April 15, 1970, Outer Space-3 Files, RNPL; Haldeman, *Haldeman Diary*, 150–151.
28. H.R. Haldeman, diary entry for April 17, 1970. The entry as included in the published *Haldeman Diaries* does not contain the whole text; it can be found in the complete version of Haldeman's diary at the RNPL. Reeves, *President Nixon*, 191–192. With respect to the plausibility of Reeve's description of Nixon drinking in his excitement at the successful conclusion of the Apollo 13 mission, there are many accounts of Nixon enjoying alcohol in the evening both as he sat alone reviewing his paperwork and on yacht cruises on the Potomac, but less evidence that Nixon on occasion also drank during the day.
29. Office of the White House Press Secretary, "Remarks of the President upon the Presentation of the Medal of Freedom to the Apollo 13 Mission Operations Team" and "Exchange of Remarks between the President and Captain James Arthur Lovell, Jr., USN, upon the Presentation of the Medal of Freedom to the Apollo 13 Astronauts," April 18, 1970, Box 61, Papers of Edwin Harper, RNPL.
30. Memorandum from Dwight Chapin to H.R. Haldeman, "Apollo 13 – White House Activity," April 21, 1970, Outer Space-3 Files, RNPL; Hugh Sidey, "Marshaling the Good Guys," *Life*, August 21, 1970, 28.

31. John Ehrlichman, Meeting Notes, December 28, 1970, Box 5, Papers of John Ehrlichman, RNPL.
32. George M. Low, "Personal Notes No. 38," January 3, 1971, GML; Clare Farley, Memorandum for the Record, "FY'72 Budget Decisions," January 20, 1971, File 19909, NHRC.
33. George M. Low, "Personal Notes No. 38," memorandum from Paul McCracken to the President, December 30, 1970, Box 8, Presidential Handwriting Files, RNPL.
34. Clare Farley, Memorandum for the Record, "FY'72 Budget Decisions," January 20, 1971, File 19909, NHRC; George Low, "Personal Notes No. 38."
35. Memorandum from Edward David to the President, December 31, 1970, File 012578, NHRC; John Ehrlichman, Meeting Notes, December 31, 1970, Box 5, Papers of John Ehrlichman, RNPL.
36. Letter from George Low to Donald Rice, December 11, 1970; memorandum from Daniel Taft to Mr. Rice, "1972 Budget Posture Statement for Space Shuttle," December 15, 1970; memorandum from William Niskanen to Donald Rice, "The Advanced Space Engine," December 26, 1970; memorandum from Daniel Taft to Mr. Rice, "FY 1972 budget language for the space shuttle or other lower-cost launch vehicle," January 8, 1971, all in LSN/NHRC.
37. George Low, "Personal Notes No. 39," January 9, 1971, GML; letter from George Shultz to George Low, February 19, 1971, File 19909, NHRC.
38. George Low, "Personal Notes No. 39."

9 National Security Requirements Drive Shuttle Design

1. Comprehensive studies of the technological evolution of the space shuttle's design include Dennis R. Jenkins, *Space Shuttle: The History of the National Space Transportation System* (Stillwater, MN: Voyageur Press, 2002) and Jenkins's forthcoming three-volume study of the shuttle with the same title; T. A. Heppenheimer, *The Space Shuttle Decision: NASA's Quest for a Reusable Space Vehicle*, NASA SP-4221 (Washington: Government Printing Office, 1999), reprinted as *The Space Shuttle Decision, 1965–1972* (Washington, DC: Smithsonian Institution Scholarly Press, 2002); Roger D. Launius, John Krige, and James L. Craig, eds., *Space Shuttle Legacy: How We Did It and What We Learned* (Reston, VA: American Institute of Aeronautics and Astronautics, 2013); Wayne Hale, executive editor, *Wings in Orbit: Scientific and Engineering Legacies of the Space Shuttle, 1971–2010*, NASA SP-201-3409, available at www.nasa.gov/centers/johnson/wingsinorbit/ and Scott Pace, "Engineering Design and Political Choice: the Space Shuttle 1969–1972," Master's Thesis, Massachusetts Institute of Technology, May 1982. For other discussions of shuttle technical issues, see the entries in Roger D. Launius and Aaron K. Gillette, *Toward a History of the Space Shuttle: An Annotated Bibliography,* NASA History Office, December 1992, available at http://history.nasa.gov/708235main_Shuttle_Bibliography_1-ebook.pdf and John Guilmartin and John Mauer, "A Shuttle Chronology, 1964–1975," Johnson Space Center, Contract 23309, NASA, December 1988, available through the Johnson Space Center History Collection at the University of Houston Clear Lake.

2. See the chapter by Roger Lanius, "Defining the Shuttle: The Spaceplane Tradition," in Launius, Krige, and Craig, *Space Shuttle Legacy*.
3. George E. Mueller, "Honorary Fellowship Acceptance Speech," August 10, 1968, reprinted in Logsdon, *Exploring the Unknown*, Volume IV, 202–203.
4. Jenkins, *Space Shuttle*, 78; Guilmartin and Mauer, "Shuttle Chronology," Volume I, entry for January 24, 1969.
5. Jerry Grey, *Enterprise* (New York: William Morrow and Company, 1979), 67–68.
6. Jenkins, *Space Shuttle*, 80.
7. NASA Space Shuttle Task Group, "NASA Space Shuttle Summary Report," May 19, 1969, 1, 32, LSN/NHRC.
8. For background on the National Reconnaissance Office and its activities, see Jeffrey Richelson, *America's Secret Eyes in Space: The U.S. Keyhole Satellite Program* (Cambridge, MA: Ballinger, 1990) and *The Wizards of Langley* (New York: Basic Books, 2002), and L. Parker Temple III, *Shades of Grey* (Reston, VA: American Institute of Aeronautics and Astronautics, 2004).
9. Department of Defense and NASA, "Joint DOD/NASA Study of Space Transportation Study of Space Transportation Systems—Summary Report," June 16, 1969, originally classified Secret but declassified by the Historical Research Division, Maxwell Air Force Base on December 21, 1999, under the authority of Executive Order 12958. I am grateful to Dennis Jenkins for providing me a copy of this document.
10. For information on the Corona program, which was declassified in 1995, see Dwayne Day, John Logsdon, and Brian Latell, eds., *Eye in the Sky: The Story of the Corona Spy Satellites* (Washington, DC: Smithsonian Institution Press, 1999). For information on Gambit, which was declassified in 2011, see James Outzen, ed., *Critical to US Security: The Gambit and Hexagon Satellite Reconnaissance Systems Compendium*, Center for the Study of National Reconnaissance, National Reconnaissance Office, January 2012.
11. DOD/NASA, "Summary Report," 1, 15–18, 11–12.
12. Ibid, 15, 2, 13.
13. Ibid, 8, 20.
14. Ibid, 61–62.
15. The Mark interview with NRO historian Gerald Haines took place on March 12, 1997. It remains classified, but a version with many words and passages redacted (blacked out) was declassified in 2012 and is available at www.nro.gov/foia/docs/Hans%20Mark.PDF For information on Hexagon, see Outzen, *Critical to National Security*. For a discussion of the planning for a successor to Hexagon, see Temple, *Shades of Grey*, Chapter 15. I am grateful to Dwayne Day and L. Parker Temple III for their comments on this section of the study.
16. Once the shuttle program was approved, NRO and its Hexagon contractor Lockheed did examine various interactions between the shuttle and Hexagon, including returning the satellite to Earth. See Dwayne Day, "The HEXAGON and the Space Shuttle," *The Space Review*, October 21, 2011, available at http://www.thespacereview.com/article/1960/1
17. Interview with Robert Seamans, June 24, 1971.
18. During 1970 and 1971, there were a number of organized attempts to convince NASA and the White House to locate the launch and landing site

for the shuttle in a Western state. Alternatives proposed included not only California but also Utah, New Mexico, or Colorado. This lobbying effort did not bear fruit, and is not discussed in this study.
19. Johnson Space Center, NASA, "Internal Note No. 73-FM-47, Shuttle Baseline Reference Missions—Mission 3A and Mission 3B," Volume III, March 26, 1973 and Revision 1 to this document, May 7, 1974. I am grateful to retired NASA engineer Jim Behling for providing the 1973 version of this document and to Dennis Jenkins for providing the 1974 revision.
20. Memorandum from Jacob Smart, NASA Assistant Administrator for DOD and Interagency Affairs to Administrator, "Security Implications in National Space Program," December 1, 1971, with attached draft letter to David Packard, File 19910, NHRC. It is not clear whether the letter was ever sent.
21. For a discussion of the Manned Orbiting Laboratory program and the DORIAN camera system, which was declassified in 2012, see Dwayne Day, "The Hour of the Wolf," *The Space Review,* July 16, 2012, available at http://www.thespacereview.com/article/2121/1#. Memorandum from Jacob Smart, NASA Assistant Administrator for DOD and Interagency Affairs to Administrator, "Security Implications in National Space Program," December 1, 1971, with attached draft letter to David Packard, File 19910, NHRC.
22. Guilmartin and Mauer, "Space Shuttle Chronology," Volume I, entry for June 13, 1969.
23. Jenkins, *Space Shuttle*, 80; Roy Day, Teletype to organizations involved in shuttle studies, "Decisions and Actions of the Space Shuttle Meetings of August 5, 6, 1969," File 16298, NHRC.
24. Memorandum from General John D. Ryan to Dr. Seamans, "Space Transportation System," September 15, 1969, originally classified "Confidential" but declassified, File 011709, NHRC.
25. Letter from Robert Seamans, Jr. to Spiro Agnew, August 4, 1969, reprinted in Logsdon, *Exploring the Unknown,* Volume I, 521; letter from Robert Seamans to Thomas Paine, November 7, 1969, File 08210, NHRC.
26. For a discussion of the Gemini experience, see Temple, *Shades of Grey,* 420–422.
27. NASA Management Instruction 1052.130, February 17, 1970, File 59971-1, NHRC.

10 A Time of Transitions

1. Memorandum from Fred Malek to Peter Flanigan, "References on Frank Jameson and James Fletcher," January 6, 1971, Box 9, Papers of Peter Flanigan, RNPL.
2. Memorandum from H. R. Haldeman to Mr. Flanigan, February 9, 1971 and memorandum from Peter Flanigan to H. R. Haldeman, February 15, 1971, both in Box 44, Confidential Files, White House Sensitive Files, RNPL.
3. Memorandum from Peter Flanigan to the President, "NASA Administrator," February 17, 1971, Box 2, Federal Government 164 Files, RNPL.
4. The Schorr incident is discussed in John Osborne, *The Third Year of the Nixon Watch* (New York: Liveright, 1972), 153. Nixon's reaction is in Conversation 6, Tape 465, March 10, 1971, RNPL. In addition to the availability of the tapes at the Nixon Presidential Library, there are several on-line locations for

accessing the tapes, including www.nixontapes.org and www.millercenter.org/presidentialrecordings.
5. A recent book by Douglas Brinkley and Luke Nichter, *The Nixon Tapes* (New York: Houghton Mifflin Harcourt, 2014) contains over 700 pages of Nixon tape transcriptions, but none directly relevant to this study.
6. Interview with James Fletcher, September 21, 1977.
7. Interview of Willis Shapley by Martin Collins, October 13, 1994, Archives Division, National Air and Space Museum, Smithsonian Institution, Washington, DC.
8. Ibid; George Low, "Personal Notes No. 43," March 7, 1971, and "Personal Notes No.45," April 10, 1971, GML.
9. Interview with George Low, January 31, 1981.
10. Memorandum from Peter Flanigan to Dwight Chapin, March 31, 1971, Box 2, Federal Government 164 Files, RNPL.
11. Speech by James C. Fletcher to Annual Conference of the Aerospace Industries Association, Williamsburg, VA, May 20, 1971, Papers of James Fletcher, Marriott Library, University of Utah, Salt Lake City, UT.
12. Conversation 7, Tape 464, March 9, 1971, RNPL.
13. Conversation 10, Tape 471, March 24, 1971, RNPL.
14. Interview with John Ehrlichman, May 6, 1983.
15. Conversations 11 and 15, Tape 498, May 13, 1971; conversation 16, Tape 500, May 18, 1971; conversation 13, Tape 253, May 26, 1971, RNPL.
16. Memorandum from John Ehrlichman to George Shultz, May 17, 1971, File 19909, NHRC; memorandum from Don Rice to Mr. Shultz, "Apollo Program," May 19, 1971 and memorandum from John Ehrlichman to the President, May 27, 1971, both in File 021578, NHRC.
17. Unsigned, "Memorandum for the President's File," February 23, 1971, Box 1, Federal Government 6–9 Files, RNPL; memorandum from "Peter Flanigan for the President's File," May 5, 1971, Box 88, President's Office Files, RNPL.
18. Haldeman, *Haldeman Diaries*, 331.
19. George Low, "Personal Notes No. 42," February 21, 1971, GML.
20. Conversation 24, Tape 490, May 6, 1971, RNPL; H. R. Haldeman, diary entry for May 11, 1971, RNPL. This entry is not included in the published version of the Haldeman diary, but appears in the full copy of the diary in the Nixon Library.
21. Conversations 1 and 9, Tape 497, May 11, 1971, RNPL.
22. Memorandum from John Ehrlichman to George Shultz, May 17, 1971, NHRC; memorandum from Bill Safire to H.R. Haldeman, May 17, 1971, Box 23, Federal Government 164 Files, RNPL.
23. Memorandum from J. C. F. (Jim Fletcher) for Dr. Low, "Conversation with George Shultz on May 26, 1971" and memorandum from George Low to the Administrator, "NASA as a Technology Agency," May 25, 1971, both in File 19909, NHRC.
24. Memorandum from George Low to Edgar Cortright, "Broadened Role for NASA," Folder 57, GML.
25. John Guilmartin and John Mauer, "A Shuttle Chronology, 1964–1975," Johnson Space Center, Contract 23309, NASA, December 1988, available through the Johnson Space Center History Collection at the University of Houston Clear Lake. The quoted passages are in Volume IV, V-1, V-5.
26. Ibid, entry for January 19–20, 1971.

27. Ibid, entries for June 25 and June 30, 1971.
28. George Low, "Personal Notes No. 41," February 7, 1971, and "Personal Notes No. 36," November 28, 1970, GML.
29. Memorandum from James Fletcher to George Low, "Luncheon Conversation with Peter Flanigan and Ed David, May 4," May 4, 1971, File 57, GML.
30. Memorandum from Frank Pagnotta, Executive Officer, Office of Science and Technology, to Edward David, May 6, 1971 and memorandum from Edward David to John Ehrlichman, "NASA Space Shuttle Program," May 7, 1971, Box 35, Papers of Edward David, RNPL.
31. Letter from James Fletcher to William Anders, June 23, 1971, File 012291, NHRC.
32. Letter from Donald Rice to James Fletcher, May 17, 1971, LSN/NHRC; interview with Don Rice, November 13, 1975; interview with Dale Myers, August 23, 1977.
33. George Low, "Personal Notes No. 47," May 18, 1971, GML.
34. Letter from Robert Mayo to Thomas Paine with attached enclosure "Analysis of Alternative Systems for Reducing the Cost of Payload in Orbit," March 18, 1970, LSN/NHRC. I am grateful to Richard Speier for providing a copy of this document. Interview with George Low, January 30, 1981.
35. Mayo to Paine letter, March 18, 1970. For a fuller discussion of the discount rate as applied to the space shuttle, see Heppenheimer, *Space Shuttle Decision*, 253–262.
36. Ibid, 264; Office of Management and Budget, "Documentation of the Space Shuttle Decision Process" in Logsdon, *Exploring the Unknown*, Volume IV, 223.
37. Ibid, 225–227; interview with Klaus Heiss, March 27, 1978.
38. Mathematica, Inc., "Economic Analysis of New Space Transportation Systems," Executive Summary, May 31, 1971, 0–9, 0–11, 0–60.
39. Email from the first shuttle program manager, Robert Thompson, June 2, 2014.
40. Mathematica, Inc., "Economic Analysis of New Space Transportation Systems," Executive Summary, May 31, 1971, 0–62 and 0–63.
41. Jenkins, *Space Shuttle*, 129, 131–132, 140–141.
42. Letter from Dale Myers to Grant Hansen, May 25, 1971, File 19909, NHRC.
43. Letter from Grant Hansen to Dale Myers, June 21, 1971, ibid.
44. George Low, "Personal Notes No.48," June 6, 1971, GML; Clare Farley, Memorandum for the Record, "NASA Alternative Program Plans," May 28, 1971, File 19909, NHRC.
45. Interview of James Fletcher by John Mauer, May 17, 1989, and interview of Charles Donlan by John Mauer, October 19, 1983, both available through the Johnson Space Center History Collection at the University of Houston Clear Lake, Clear Lake, TX; Robert Thompson, email communication to the author, February 14, 2014.
46. NASA Press Release 71–107, "Space Shuttle Studies," June 16, 1971, LSN/NHRC.
47. Letter from James Fletcher to James van Allen, July 12, 1971, LSN/NHRC.
48. Letter from Robert Seamans to James Fletcher, July 7, 1971, File 19909, NHRC.
49. Letter from Don Rice to James Fletcher, July 20, 1971, File 57, GML; interview with Donald Rice, November 13, 1975.

50. Claude Barfield, "Intense Debate, Cost Cutting Precede White House Decision to Back Shuttle," *National Journal*, August 12, 1972, 1294; interview with Klaus Heiss, March 27, 1978. The Barfield article is a very good summary of the shuttle decision process written close to the time that that process unfolded.

11 A Confused Path Forward

1. Memorandum from Tom Whitehead to Peter Flanigan, February 6, 1971, Folder 1, Box 13, CTW.
2. Conversation 23, Tape 455, February 22, 1971, RNPL.
3. See Chapter 5 in Krige, Callahan, and Maharaj, *NASA in the World* for an account of this period in post-Apollo collaboration.
4. For background on the *Apollo–Soyuz* Test Project, see Chapter 7 in ibid. and especially Edward Ezell and Linda Ezell, *The Partnership: A History of the Apollo-Soyuz Test Project*, NASA SP-4209 (Washington, DC: Government Printing Office, 1978).
5. Conversation 27, Tape 524, June 17, 1971, RNPL.
6. "Eyes Only" Memorandum from George Low to James Fletcher, "Items of Interest," August 12, 1971, File 70, GML; letter from Bill Anders to James Fletcher, August 30, 1971, File 102291, NHRC.
7. Memorandum from Clay T. Whitehead to Peter Flanigan, September 24, 1971, Outer Space-1 File, RNPL.
8. George Low, "Personal Notes No. 50," July 4, 1971, and "Personal Notes No. 51," July 18, 1971, GML.
9. George Low, "Personal Notes No. 51," and "Personal Notes No. 52," August 15, 1971, GML; letter from James Fletcher to George Low, August 31, 1971, File 57, GML.
10. George Low, "Personal Notes No. 55," October 2, 1971; draft paper on "Technology for Society," attached to letter from James Fletcher to William Magruder, September 28, 1971, Marriott Library, University of Utah, Salt Lake City, Utah. A final version of this paper was not located in either the Marriott Library or NASA files. "Eyes Only" Memorandum from James Fletcher to George Low, "Suggestions on Implementation of the 'Magruder Response,'" September 30, 1971, File 57, GML.
11. George Low, "Personal Notes No 45," April 10, 1971, GML. For background on Gene Fubini, see a biography written by his son, David G. Fubini, *Let Me Explain: Eugene G. Fubini's Life in Defense of America* (Santa Fe, NM: Sunstone Press, 2009).
12. Letter from Edward David to James Fletcher, July 26, 1971, File 19908, NHRC.
13. "Eyes Only" Memorandum from Deputy Administrator to Administrator, "Items of Interest," August 12, 1971, File 70, GML.
14. Dale Myers, Memorandum for the Record, "Woods Hole Review of the Shuttle Program," August 19, 1971, File 19908, NHRC; memorandum from Associate Administrator for Manned Space Flight to Administrator, "Flax Shuttle Committee Meeting at Woods Hole," August 19, 1971, File 59971, NHRC.
15. "Eyes Only" Memorandum from George Low to the Administrator, August 24, 1971, File 19908, NHRC.

16. John Ehrlichman, Notes from July 23 Meeting with the President, Paul McKracken, John Connally, Cap Weinberger, George Shultz, Kenneth Cole, and Edward Harper, Box 6, Papers of John Ehrlichman, RNPL.
17. Letter from Caspar Weinberger to James Fletcher, August 2, 1971, File 19909, NHRC.
18. George Low, "Personal Notes No. 52," GML; James Fletcher, "Notes from Meeting with Mr. Weinberger," August 5, 1971, File 19844, NHRC.
19. Low, "Personal Notes No. 52."
20. Ibid.
21. George Low, "Space Transportation System Planning," August 18, 1971, File 19844, NHRC; George Low, "Personal Notes No. 53," August 22, 1971, GML. Myers is quoted in Heppenheimer, *Space Shuttle Decision*, 368.
22. Memorandum from John Huntsman (one of Haldeman's staff) to George Shultz and Caspar Weinberger, "Space Shuttle," August 6, 1971, Box I:275, Papers of Caspar Weinberger, Library of Congress.
23. "Eyes Only" Memorandum from James Fletcher to Dr. Low, "Meeting with Ed David," August 24, 1971, File 012578, NHRC.
24. George Low, "Personal Notes No. 54," September 19, 1971, and "Personal Notes No. 55," October 2, 1971, GML.
25. Ibid.
26. Letter from James Fletcher to George Shultz, September 30, 1971, File 19909, NHRC.
27. Memorandum from Caspar Weinberger through George Shultz to the President, "Future of NASA," August 12, 1971, reprinted in Logsdon., *Exploring the Unknown*, Volume I, 547.
28. Memorandum from Jon Huntsman to George Shultz, "Future of NASA," September 13, 1971, LSN/NHRC.

12 Debating a Shuttle Decision

1. George Low, Memorandum for the Record, "NASA Priorities," October 15, 1971, File 35, GML.
2. George M. Low, "Personal Notes No. 56," October 17, 1971, GML; interview with Donald Rice, November 13, 1975; interview of James Fletcher by John Mauer, May 17, 1989; interview of Willis Shapley by John Mauer, April 12, 1989.
3. Office of Management and Budget, "Documentation of the Space Shuttle Decision Process," February 4, 1972, reprinted in Logsdon, *Exploring the Unknown*, Volume IV, 227, and Office of Management and Budget, "The U.S. Civilian Space Program: A Look at the Options," October 14, 1971, Summary, 1–2, 1–7, 1–8, Folder 32, CTW.
4. Office of Management and Budget, "The U.S. Civilian Space Program: A Look at the Options," 1–13, 1–17, 1–20.
5. Office of Management and Budget, "The U.S. Civilian Space Program: A Look at the Options," V-1,V-2.
6. Tape 596, Conversation 4, October 19, 1971, RNPL.
7. Memorandum from H. R. Haldeman to Mr. Weinberger, October 19, 1971, and memorandum from Caspar Weinberger to Bob Haldeman, October 21, 1971, both in File 19908, NHRC.

8. Interview with Caspar Weinberger, August 23, 1977; email communication from Bill Anders, March 10, 2014.
9. Interview with Caspar Weinberger.
10. George Low, "Personal Notes No. 57," October 31, 1971, GML.
11. "Eyes Only" Memorandum from James Fletcher to George Low, "Conversation with Ed David and Russ Drew Today," September 30, 1971, File 57, GML.
12. Letter from Alexander Flax to Edward David, with attached report, October 19, 1971, File 19908, NHRC.
13. Ibid.
14. Ibid.
15. George M. Low, "Personal Notes No. 56," October 17, 1971, GML; interview with Dale Myers, August 23, 1977.
16. Interview with Klaus Heiss, March 27, 1978.
17. Ibid.
18. Klaus Heiss and Oskar Morgenstern, "Factors for a Decision on a New Reusable Space Transportation System," October 28, 1971, File 8245, NHRC.
19. Heppenheimer, *Space Shuttle Decision*, 375; interview with Klaus Heiss; interview with Dale Myers; interview with James Fletcher, September 21, 1977.
20. George Low, "Personal Notes No. 57" and "Personal Notes No. 58," November 14, 1971, GML.
21. Memorandum from Acting Director, Space Shuttle Program (Charles Donlan) to Deputy Administrator, "Additional Space Shuttle Information," December 5, 1971, Box 79, GML.
22. "Eyes Only" Memorandum from J. C. F. (James C. Fletcher) to Dr. Low, "Luncheon Conversation with Dave Packard," October 20, 1971, LSN/NHRC.
23. Low, "Personal Note No. 57."
24. Interview with Klaus Heiss.
25. "Eyes Only" Memorandum from J. C. F (James Fletcher) to Dr. Low, "Luncheon with Tom Whitehead and Bill Anders," November 5, 1971, File 57, GML.
26. George Low, "Memorandum for the Record," November 15, 1971, quoted in Heppenheimer, *Space Shuttle Decision*, 382–383.
27. George Low, "Personal Notes No.58," GML.
28. "Eyes Only" Memorandum from George Low to Dr. Fletcher, "Meeting with Al Flax," November 15, 1971, File 70, GML.
29. George Low, "Personal Notes No. 59," November 28, 1971, GML.
30. George Low, "Personal Notes No. 59."
31. George Low, "Personal Notes No. 57."
32. This discussion is drawn from Neufeld, *Von Braun*, 446–457.
33. Letter from James Fletcher to Caspar Weinberger, November 3, 1971, File 70, GML; memorandum from Edward David to the President, "Apollo 16 and 17," November 29, 1971, Box 2, Federal Government 164 Files, RNPL; George Low, "Personal Notes No. 50."
34. George Low, "Personal Notes No. 57."
35. "Eyes Only—Sensitive" Memorandum from Deputy Administrator to Administrator, "Telephone Call from Bill Magruder," November 16, 1971, File 70, GML; George Low, "Personal Notes No. 59."

36. Memorandum from Jonathan Rose through Peter Flanigan for the President, August 28, 1971, and Memorandum from H. R. Haldeman to Cap Weinberger, September 9, 1971, Box I:272, Papers of Caspar Weinberger, Manuscript Division, Library of Congress, Washington, DC; interview with John Ehrlichman, May 6, 1983.
37. Letter from Newton Russell to H. R. Haldeman, June 2, 1971 and letter from Caspar Weinberger to Newton Russell, June 25, 1971, Box I:275, Papers of Caspar Weinberger.
38. George Low, "Personal Notes No. 58"; identical letters from James Fletcher to Peter Flanigan and Caspar Weinberger, November 3, 1971, File 8209, NHRC; NASA, "California Employment," November 19, 1971, Box 8, Papers of Peter Flanigan, RNPL.
39. Memorandum from George Low to Jonathan Rose, December 1, 1971, File 79, GML.
40. Interview with John Ehrlichman.
41. George Low, "Personal Notes No. 59"; letter from Peter Flanigan to W. F. Rockwell, Jr., December 21, 1971, Outer Space-2 Files, RNPL.
42. "Eyes Only—Sensitive" Memorandum from Deputy Administrator to Administrator, "Discussions with Bill Anders," November 16, 1971, File 70, GML.
43. Letter from George Low to Don Rice, with attachment, November 22, 1971, File 19844, NHRC.
44. Interview with Don Rice, November 13, 1975. Rice repeated the same point in a September 10, 2012 interview with the author.
45. Letter from George Low to the author, January 23, 1979.
46. Letter from James Fletcher to Jonathan Rose with attached paper on "The Space Shuttle," November 22, 1971, reprinted in Logsdon, *Exploring the Unknown*, Volume I, 555–558.
47. Letter from George Low to Caspar Weinberger, October 28, 1970, File 1849, NHRC.

13 Which Shuttle to Approve?

1. Interview with Donald Rice, September 10, 2012.
2. Draft Memorandum from Office of Management and Budget to the President, "The future direction of the U.S. civilian space program," November 11, 1971, Box 61, Papers of Edwin Harper, RNPL.
3. George Low, "Personal Notes No. 58," November 14, 1971, GML.
4. Memorandum from George Shultz to the President, "Meeting with Secretary Connally, John Ehrlichman, and George Shultz, November 24,1971, 10:00 a.m.," Federal Government 6–16 Papers, RNPL; memorandum from Ed Harper to John Ehrlichman, "Budget Problems," November 23, 1971, Box 13, Papers of Edwin Harper, RNPL.
5. Conversation 13, Tape 624, November 24, 1971, RNPL. For John Ehrlichman's notes on the November 24 meeting, see Box 6, Papers of John Ehrlichman, RNPL.
6. Memorandum from John Ehrlichman for the President's Files, "Report on Meeting with Secretary Connally, George Shultz, John Ehrlichman, November 24, 1971, 11:00 a.m.," Federal Government 6–16 Files, January 4, 1972, RNPL.

7. Memorandum from Office of Management and Budget for the President, "NASA Budget for FY 1973 and the Future Manned Space Program," December 2, 1971, File 19909, NHRC.
8. John Ehrlichman, Meeting Notes, Key Biscayne, December 3, 1971, 11:40 a.m., Box 6, Papers of John Ehrlichman, RNPL.
9. Memorandum from Deputy Administrator to Administrator, "Discussions with Johnny Foster," December 2, 1971, LSN/NHRC.
10. Memorandum from Assistant Administrator for DOD and Interagency Affairs to Administrator, "Security Implications in National Space Program," with attached draft memorandum to Deputy Secretary of Defense David Packard, File 19910, LSN/NHRC.
11. Interview with John Ehrlichman, May 6, 1983.
12. Assistant Administrator for DOD and Interagency Affairs, "Security Implications."
13. "Eyes Only" Memorandum from J. C. F. (James Fletcher) to Dr. Low, "Conversation with Al Haig," December 2, 1971, File 8209, NHRC.
14. George Low, "Personal Notes No. 60," December 12, 1971, GML.
15. Memorandum from Tom (Whitehead) to Jon (Rose), December 2, 1971 with attached Memorandum for Mr. Flanigan including shuttle configurations summary sheet prepared by Bill Anders, CTW.
16. Ibid.
17. Letter from James Fletcher to Caspar Weinberger, December 14, 1971, File 19909, NHRC.
18. George Low, "Personal Notes No. 60."
19. Interview with John Ehrlichman, May 6, 1983; letter from Caspar Weinberger to the author, March 13, 1979, LSN/NHRC.
20. Office of Management and Budget, "Space Shuttle Program," December 10, 1971, attached to Memorandum from Deputy Administrator to Associate Administrator for Manned Space Flight, "Shuttle Cost and Payload Analysis," December 13, 1971, Box 70, GML.
21. Interview with Donald Rice, November 13, 1975; interview with Caspar Weinberger, August 23, 1977.
22. Interview with Willis Shapley by John Mauer, October 26, 1984, JSC History Collection, University of Houston Clear Lake, Clear Lake, TX; interview of Willis Shapley by John Mauer, May 17, 1989, Archives Division, National Air and Space Museum, Smithsonian Institution, Washington, DC; George Low, "Personal Notes No. 61," January 2, 1972, GML; interview with George Low, January 30, 1981; interview with James Fletcher, September 21, 1977.
23. Memorandum from Deputy Administrator to Associate Administrator for Manned Space Flight, "Space Shuttle Cost and Payload Analysis," December 13, 1971, Box 70, GML.
24. Papers of James Fletcher, Box 1, Marriott Library, University of Utah, Salt Lake City, Utah; memorandum from Peter Flanigan to James Fletcher, December 17, 1971, Box 2, Federal Government 164 Files, RNPL.
25. George Low, "Personal Notes No. 60" and "Personal Notes No. 62," January 15, 1972, GML.
26. Harper is quoted in an article on the Magruder effort by Claude Barfield, *National Journal,* May 6, 1972; interview with Caspar Weinberger, August 23, 1977; interview with John Ehrlichman, May 6, 1983.

27. Office of Management and Budget, Memorandum for the President, "The Space Shuttle Decision," undated, Box I:275, Papers of Caspar W. Weinberger, Manuscript Division, Library of Congress, Washington, DC.
28. Unsigned memorandum (but from Don Rice) to Mr. Weinberger, "FY 1973 NASA Appeal," December 16, 1971, Box 1:275, Papers of Caspar W. Weinberger, Manuscript Division, Library of Congress, Washington, DC.
29. Emails from Bill Anders commenting on earlier drafts of this chapter, April 17, May 10, May 11, and May 28, 2014. I am grateful to researcher Alicia Fernandez for searching files at the Nixon Library to identify the date of the Indian Treaty Room meeting and pointing me to the Weinberger papers in the Library of Congress.

14 A "Space Clipper"

1. George Low, "Personal Notes No. 61," January 2, 1972, GML; email from Bill Anders commenting on a draft of this chapter, April 17, 2014.
2. George Low, "Personal Notes No. 61."
3. Letter from James Fletcher to Caspar Weinberger, December 29, 1971, reprinted in Logsdon, *Exploring the Unknown,* Volume IV, 245–249.
4. George Low, "Personal Notes No. 61"; interview with James Fletcher, September 21, 1977.
5. George Low, "Personal Notes No. 61."
6. Memorandum from Edward David to George Shultz, "Space Shuttle Decision," December 30, 1971, Box 44, Papers of Edward David, RNPL; memorandum from OMB Staff to Mr. Rice, "Analysis of NASA Position on the Shuttle," December 30, 1971, LSN/NHRC.
7. George Low, "Personal Notes No. 61."
8. NASA, "Space Shuttle Questions Provided by OMB on 12/31/71," January 3, 1972, LSN/NHRC.
9. George Low, "Personal Notes No. 61"; letter from George Low to author, January 29, 1979.
10. Memorandum from Edward David to George Shultz, "The Space Shuttle Decision," January 3, 1972, Box 44, Papers of Edward David, RNPL; memorandum from Donald Rice to the Director, "Dr. David's Memorandum on the Space Shuttle Decision," January 3, 1972, File 19844, NHRC.
11. Memorandum from Jon Rose to Peter Flanigan, "NASA," January 27, 1972, LSF/NHRC.
12. Letter from James Fletcher to Caspar Weinberger, January 3, 1972, File 19884, NHRC.
13. Interview with Space Council staff member David Elliott, April 22, 1977; George Low, "Personal Notes No. 62," January 15, 1972, GML; on the Shultz call to Morgenstern, see John Mauer's interviews with Bill Lilly, October 20, 1983, and Willis Shapley, October 26, 1984, JSC Historical Collection, University of Clear Lake, Clear Lake, TX; interview with Dale Myers, August 23, 1977.
14. Letter from James Fletcher to Caspar Weinberger, January 4, 1972, File 12459, NHRC. If a smaller orbiter could be launched using the 120-inch diameter solid rocket motors already in production for use by the Air Force Titan III booster, there was a possibility of avoiding the costs of developing and using a new, larger solid rocket motor to launch the full-capability orbiter.

15. George Low, "Personal Notes No. 61" and "Personal Notes No. 62"; letter marked "Very Sensitive-No Copies" from James Fletcher to George Shultz, December 30, 1971, Box 70, GML. For a detailed history of the NERVA program, which in many ways was the last vestige of NASA's ambitious post-Apollo plans, see James Dewar, *To the End of the Solar System: The Story of the Nuclear Rocket*, 2nd edition (Burlington, Ontario, Canada: Apogee Books, 2008).
16. This meeting was taped; it is conversation 29 on Tape 642, January 3, 1972, RNPL. The author reviewed the recording to make sure that the shuttle was not discussed. See also Caspar Weinberger, Memorandum for the President's File, Box 87, President's Office Files, January 3, 1972, RNPL.
17. Interview with John Ehrlichman, May 6, 1983; James Fletcher, Memorandum for the Record, "Meeting with Cap Weinberger and Jon Rose on the Shuttle, this date," March 3, 1972, File 57, GML; letter from James Fletcher to Caspar Weinberger, March 6, 1972, File 7915, NHRC. A 2013 attempt to contact George Shultz with respect to the space shuttle decision was not successful. His office reported that he remembered no details with respect to the decision.
18. George Low, "Personal Notes No. 62."
19. Schedule Proposal to Dwight Chapin from Bill Rhatican via Charles Colson, "Presidential Visit to Rocketdyne Division of North American Rockwell," December 30, 1971, Outer Space-4 File, RNPL; schedule Proposal from David Parker, "Meeting with NASA Director Fletcher," December 31, 1971, Box 2, Federal Government 164 Files, RNPL.
20. Handwritten list of names in File 79, GML; memorandum from James Fletcher to Peter Flanigan, "New Names for Shuttle," December 30, 1971, File 8207, NHRC.
21. Memorandum from Peter Flanigan to the President, "Name to describe program to develop the new space transportation system," January 4, 1972, and memorandum from Bill Safire to Peter Flanigan, January 4, 1972, both in Box 9, Outer Space-3 File, RNPL.
22. Memorandum from Jonathan Rose through Peter Flanigan and Caspar Weinberger to the President, "Status of California Employment Project," January 3, 1972, Box 16, Handwriting File, RNPL.
23. Note from Alex (Butterfield) to Mr. President, January 4, 1972, 7 p.m.; memorandum from Peter Flanigan to the President, "Meeting with Dr. James C. Fletcher, Wednesday, January 5, 1972, 10:00 A.M. (15 minutes)," undated but January 4, 1972, both in Box 9, Outer Space-3 File, RNPL.
24. Interview with George Low, January 30, 1981; interview with John Ehrlichman, May 6, 1983.
25. George Low, "Personal Notes No. 62"; John Ehrlichman. "Notes from January 5, 1972 Meeting," Box 7, Papers of John Ehrlichman, RNPL; George Low, Memorandum for the Record, "Meeting with the President on January 5, 1972," January 12, 1972, reprinted in Logsdon, *Exploring the Unknown*, Volume I, 558–559; John Ehrlichman, Memorandum for the President's File, "Report on meeting with NASA Administrator, James Fletcher; Deputy Administrator, George Low," January 5, 1972, Box 87, President's Office Files, RNPL; note from James Fletcher to George Low commenting on a draft of Low's memorandum, January 11, 1972, Box 149, GML.

26. Interview with John Ehrlichman, May 6, 1983.
27. Richard Nixon, "Statement Announcing Decision to Proceed with Development of the Space Shuttle," January 5, 1972. http://www.presidency.ucsb.edu/ws/?pid=3574. Gerhard Peters and John T. Woolley, *The American Presidency Project*.

Finale

1. Telegraphic message from Charles Donlan to R. F. Thompson, "Space Shuttle Study Redirection," January 5, 1972, LSN/NHRC.
2. Conversation 19, Tape 18, January 9, 1972, RNPL.
3. Memorandum from Daniel Taft to Mr. Rice, "Next Steps on the Space Shuttle," January 27, 1972, LSN/NHRC.
4. Memorandum from J.C.F. (James Fletcher) to George Low, "Summary of David/Flanigan meeting, this date," March 3, 1972, File 57, GML; memorandum from Don Rice to Mr. Shultz and Mr. Weinberger, "Final Decisions on the Space Shuttle," March 13, 1972, LSN/NHRC.
5. Letter from James Fletcher to Caspar Weinberger, December 29, 1971, reprinted in Logsdon, *Exploring the Unknown*, Volume IV, 245–249; letter from James Fletcher to Caspar Weinberger, January 3, 1972, File 19884, NHRC; emails from Bill Anders commenting on earlier drafts of this study, April 17, May 10, May 11, and May 28, 2014; letter from George Shultz to James Fletcher, February 16, 1972, File 78, GML.
6. Memorandum from Deputy Administrator to Administrator, "The 'Constant' Budget," March 23, 1972, File 57, GML; interview with James Fletcher, September 21, 1977.
7. NASA Press Release 72–61, "Space Shuttle Decisions," March 15, 1972, LSN/NHRC.

Epilogue   Richard Nixon and the American Space Program

1. It is important to note that on many issues of space policy and space programs, Richard Nixon personally did not get deeply involved; space was not often high on his policy agenda. This meant that Nixon usually delegated the substantive aspects of the three key decisions identified in this chapter to his senior policy and budget advisers, who presented their recommendations to Nixon for his approval. On occasion, Nixon did provide general guidance with respect to space decisions, but often it was the perspective of those advisers, most notably assistant to the president Peter Flanigan and his staff person Clay Thomas Whitehead and Caspar Weinberger and Don Rice of the Office of Management and Budget (OMB), that defined the Nixon administration position. Even so, as president, Richard Nixon was ultimately responsible for the decisions of his administration.
2. McDougall is quoted in Tribbe, *No Requiem for the Space Age,* 218.
3. Presidential Directive NSC-42, "Civil and Further National Space Policy," October 10, 1978, reproduced in Logsdon, *Exploring the Unknown*, Volume I, 576.
4. These data are derived from Table 1–5 in Committee on NASA's Strategic Direction, National Research Council, *NASA's Strategic Direction and the Need for a National Consensus* (Washington, DC: National Academies Press, 2012).

5. Columbia Accident Investigation Board, *Report,* August 2003, 210. The report is available at http://s3.amazonaws.com/akamai.netstorage/anon.nasa-global/CAIB/CAIB_lowres_full.pdf. The author was a member of the Board and thus shares responsibility for the content of its final report.
6. "Report of the Advisory Committee of the Future of the U.S. Space Program," December 1990, 24. The report is available at history.nasa.gov/augustine/racfup1.htm
7. Both here and later in this chapter, current-year dollars have been calculated using the Bureau of Labor Statistics inflation calculator, which can be found at www.bls.gov/data/inflation_calculator.htm
8. Review of U.S. Human Space Flight Plans Committee, *Seeking a Human Spaceflight Program Worthy of a Great Nation,* October 2009, 115, 96. The report is available at http://www.nasa.gov/pdf/396093main_HSF_Cmte_FinalReport.pdf.
9. The quote is taken from a Forbes magazine interview with Neil DeGrasse Tyson, March 13, 2012. For a full exposition of Tyson's views, see Neil DeGrasse Tyson and Avis Lang, *Space Chronicles: Facing the Ultimate Frontier* (New York: W.W. Norton, 2012).
10. Committee on Human Spaceflight, National Research Council, *Pathways to Exploration—Rationales and Approaches for a U.S. Program of Human Space Exploration* (Washington, DC: National Academies Press, 2014), 43, 42.
11. For a discussion of the use of Apollo as an instrument of public diplomacy, see Teasel Muir-Harmony, "Project Apollo, Cold War Diplomacy and the American Framing of Global Interdependence" (Ph.D. dissertation, Massachusetts Institute of Technology, 2014).
12. Schmitt is quoted in David Meerman Scott and Richard Jurek, *Marketing the Moon: The Selling of the Apollo Lunar Program* (Cambridge, MA: MIT Press, 2014), 111–112.
13. I am grateful for photo archivist Jon Fletcher at the Richard Nixon Presidential Library and Museum for checking Oval Office photographs during 1970 to confirm the removal of the "Earthrise" photograph and making available this September 1970 photograph. The "Earthrise" photograph had already been removed by April 27, as President Nixon swore in James Fletcher as NASA administrator.
14. Interview with Clay Thomas Whitehead, October 19, 1970.
15. Conversation 4, Tape 735, June 15, 1972, RNPL.
16. U.S. Human Spaceflight Plans Committee, *Seeking a Human Spaceflight Program,* 37, 33. Howard E. McCurdy, *Space and the American Imagination,* 2nd ed. (Baltimore: Johns Hopkins Press, 2011), 31.
17. Committee on Human Spaceflight, *Pathways to Exploration,* 42.
18. Logsdon, *Exploring the Unknown,* Volume I, 559.
19. James C. Fletcher, "Eyes Only" Memorandum for the Record, "Meeting with Cap Weinberger and Jon Rose on the shuttle, this date," March 3, 1972, Folder 57, GML.
20. Conversation 10, Tape 739, June 21, 1972, RNPL.
21. For discussions of the shuttle and the national security community, see Temple, *Shades of Gray,* Chapter 15 and Hans Mark, *The Space Station: A Personal Journey* (Durham, NC: Duke University Press, 1987), Chapter VII. For the Reagan administration policy statement making the space shuttle the

primary U.S. launch vehicle, see Logsdon, *Exploring the Unknown*, Volume IV, 333.

22. Footnote 13 in R. Cargill Hall, "National Space Policy and Its Interaction with the U.S. Military Space Program," in George C. Marshall Institute, *Military Space and National Policy: Record and Interpretation*, 2006, 24. This report is available at http://marshall.org/wp-content/uploads/2013/08/419.pdf. I am grateful to Jim David of the National Air and Space Museum for calling my attention to this report.
23. Michael D. Griffin, "The Legacy of the Space Shuttle," in Hale, *Wings in Orbit*, 514.
24. Columbia Accident Investigation Board, *Report*, 25.
25. For the NASA estimate of the cost of a shuttle launch, see NASA, "A White Paper on the Space Shuttle Cost History," July 25, 2012. I am grateful for Jonathan Krezel at NASA for providing me a copy of this paper. The total shuttle program costs include a number of elements not directly related to the actual operation of the shuttle, so it is probably not completely fair to make this simplistic calculation of program costs divided by number of flights. I am grateful to Dennis Jenkins for making this point in commenting on an earlier draft of this chapter. For the cost per pound to the space station, see NASA testimony to the Subcommittee on Space and Aeronautics, House Committee on Science, Space, and Technology, May 26, 2011, LSN/NHRC. It is also not really fair to compare the two cost-per-pound numbers, since they were based on very different calculations, but the stark difference between the two figures illustrates the high cost of shuttle operation.
26. Humbolt Mandell, "Space Shuttle Cost Analysis: A Success Story?" presentation to the annual conference of the International Cost Estimating and Analysis Association, Denver, CO, June 10, 2014. I am grateful to Dr. Mandell for sending me a copy of his presentation. Email from Robert Thompson, June 2, 2014.
27. Mark, *Space Station Decision*, 121–122; Ronald Reagan: "Address Before a Joint Session of the Congress on the State of the Union," January 25, 1984, http://www.presidency.ucsb.edu/ws/?pid=40205. Online by Gerhard Peters and John T. Woolley, *The American Presidency Project*. For an account of the space station decision, see Howard E. McCurdy, *The Space Station Decision: Incremental Politics and Technological Choice* (Baltimore, MD: Johns Hopkins University Press, 2007).
28. John M. Logsdon, "The Space Shuttle Program: A Policy Failure?" *Science*, May 30, 1986, 1099. For a discussion of the positive legacy of the space shuttle, see Wayne Hale, Executive Editor, *Wings in Orbit: Scientific and Engineering Legacies of the Space Shuttle, 1971–2010*, NASA SP-201-3409, available at www.nasa.gov/centers/johnson/wingsinorbit/
29. Columbia Accident Investigation Board, *Report*, 23.
30. Kraft is quoted in a story titled "Engineers Warned to Go Easy on Space Shuttle" in the *Arizona Republic*, May 22, 1970, LSN/NHRC. I am grateful to Andy Chaikin for bringing this article to my attention.
31. Email from Bill Anders, May 18, 2014.
32. For a fuller discussion of this point, see John Logsdon, "Retiring the Space Shuttle: What Next?" in Launius, *Space Shuttle Legacy*.

33. Committee on NASA's Strategic Direction, *NASA's Strategic Direction*, 1, 5.
34. Committee on Human Spaceflight, *Pathways to Exploration*, 1.
35. And even Kennedy in 1963 began to waver in his unconditional support for Project Apollo, proposing both that it become more military in rationale or perhaps a cooperative undertaking with the Soviet Union. See Logsdon, *John F. Kennedy and the Race to the Moon*, Chapter 13.
36. Committee on Human Spaceflight, *Pathways to Exploration*, 8.

Bibliography

Aldrin, Edwin E. "Buzz" Jr. with Wayne Warga. *Return to Earth*. New York: Random House, 1973.

Ambrose, Stephen. *The Triumph of a Politician*, Volume Two. New York: Simon and Schuster, 1989.

Borman, Frank with Robert J. Serling. *Countdown: An Autobiography*. New York: William Morrow, 1988.

Burns, James MacGregor. *Running Alone: Presidential Leadership—JFK to Bush II*. New York: Basic Books, 2006.

Chaikin, Andrew. *A Man on the Moon: the Voyages of the Apollo Astronauts*. New York: Viking, 1993.

Collins, Michael. *Carrying the Fire: An Astronaut's Journeys*. New York: Farrar, Straus and Giroux, 2009.

Committee on Human Spaceflight, National Research Council. *Pathways to Exploration—Rationales and Approaches for a U.S. Program of Human Space Exploration*. Washington, DC: National Academies Press, 2014.

Committee on NASA's Strategic Direction, National Research Council. *NASA's Strategic Direction and the Need for a National Consensus*. Washington, DC: National Academies Press, 2012.

Day, Dwayne and John Logsdon and Brian Latell, eds. *Eye in the Sky: The Story of the Corona Spy Satellites*. Washington, DC: Smithsonian Institution Press, 1999.

Dewar, James. *To the End of the Solar System: The Story of the Nuclear Rocket*, 2nd edition. Burlington, Ontario, Canada: Apogee Books, 2008.

Drew, Elisabeth. *Richard M. Nixon*. New York: Henry Holt and Company, 2007.

Ehrlichman, John. *Witness to Power: The Nixon Years*. New York: Simon & Schuster, 1982

Evans, Rowland Jr. and Robert D. Novak. *Richard Nixon in the White House: The Frustration of Power*. New York: Random House, 1971.

Ezell, Edward and Linda Ezell. *The Partnership: A History of the Apollo-Soyuz Test Project*, NASA SP-4209. Washington: Government Printing Office, 1978.

Fubini, David G. *Let Me Explain: Eugene G. Fubini's Life in Defense of America*. Santa Fe, NM: Sunstone Press, 2009.

Garment, Leonard. *Crazy Rhythm: My Journey from Brooklyn, Jazz, and Wall Street to Nixon's White House, Watergate, and Beyond...* New York: Random House, 1997.

Grey, Jerry. *Enterprise*. New York: William Morrow and Company, 1979.

Haldeman, H. R. *The Haldeman Diaries: Inside the Nixon White House*. New York: G.P. Putnam's Sons, 1994.

Hansen, James R. *First Man: The Life of Neil A. Armstrong.* New York: Simon & Schuster, 2005.

Heppenheimer, T. A. *The Space Shuttle Decision, 1965–1972.* Washington, DC: Smithsonian Institution Scholarly Press, 2002.

Jenkins, Dennis R. *Space Shuttle: The History of the National Space Transportation System.* Stillwater, MN: Voyageur Press, 2002.

Johnson, Lyndon B. *The Vantage Point: Perspectives of the Presidency, 1963–1969.* New York: Holt, Rinehart and Winston, 1971.

Kissinger, Henry. *White House Years.* Boston: Little, Brown and Company, 1979.

Krige, John, Angela Long Callahan, and Ashok Maharaj. *NASA in the World: Fifty Years of International Collaboration in Space.* New York: Palgrave Macmillan, 2013.

Lambright, W. Henry. *Powering Apollo: James E. Webb of NASA.* Baltimore, MD: The Johns Hopkins University Press, 1998.

Launius, Roger D. and Howard E. McCurdy, eds. *Spaceflight and the Myth of Presidential Leadership.* Urbana, IL: University of Illinois Press, 1997.

Launius, Roger D., John Krige, and James L. Craig, eds. *Space Shuttle Legacy: How We Did It and What We Learned.* Reston, VA, American Institute of Aeronautics and Astronautics, 2013.

Logsdon, John M. *John F. Kennedy and the Race to the Moon.* New York: Palgrave Macmillan, 2010.

Logsdon, John M., et al., eds. *Exploring the Unknown: Selected Documents in the History of the U.S. Civil Space Program*, Volume I: Organizing for Exploration, NASA SP-4407. Washington, DC: Government Printing Office, 1995.

Logsdon, John M., et al., eds. *Exploring the Unknown: Selected Documents in the History of the U.S. Civil Space Program,* Volume IV: Accessing Space, NASA SP-4407. Washington, DC: Government Printing Office, 1999.

McCurdy, Howard E. *The Space Station Decision: Incremental Politics and Technological Choice.* Baltimore, MD: Johns Hopkins University Press, 2007.

McCurdy, Howard E. *Space and the American Imagination*, 2nd edition. Baltimore: The Johns Hopkins Press, 2011.

Matthews, Christopher. *Nixon & Kennedy: The Rivalry That Shaped Postwar America.* New York: Simon & Schuster, 1997.

Mieczkowski, Yanek. *Eisenhower's Sputnik Moment: The Race for Space and World Prestige.* Ithaca, NY: Cornell University Press, 2013.

Nelson, Craig. *Rocket Men: The Epic Story of the First Men on the Moon.* New York: Viking, 2009.

Neufeld, Michael J. *Von Braun: Dreamer of Space, Engineer of War.* New York: Alfred A. Knopf, 2007.

Newell, Homer. *Beyond the Atmosphere: Early Years of Space Science*, NASA SP-4211. Washington, DC: Government Printing Office, 1980.

Nixon, Richard M. *The Memoirs of Richard Nixon.* New York: Grosset & Dunlap, 1978.

Nye, Joseph S. Jr. *The Paradox of American* Power. New York: Oxford University Press, 2002.

Osborne, John. *The First Two Years of the Nixon Watch.* New York: Liveright, 1971.

Osborne, John. *The Third Year of the Nixon Watch.* New York: Liveright, 1972.

Reeves, Richard. *President Nixon: Alone in the White House.* New York: Simon and Schuster, 2001.

Richelson, Jeffrey. *America's Secret Eyes in Space: The U.S. Keyhole Satellite Program.* Cambridge, MA: Ballinger, 1990.

Richelson, Jeffrey. *The Wizards of Langley.* New York: Basic Books, 2002.

Safire, William. *Before the Fall: An Inside View of the Pre-Watergate White House.* New Brunswick, NJ: Transaction Publishers, 2005.

Scott, David Meerman and Richard Jurek. *Marketing the Moon: The Selling of the Apollo Lunar Program.* Cambridge, MA: MIT Press, 2014.

Seamans, Robert. *Aiming at Targets,* NASA SP-4106. Washington, DC: Government Printing Office, 1996.

Sidey, Hugh. *John F. Kennedy, President.* New York: Atheneum, 1964.

Slotkin, Arthur. *Doing the Impossible: George E. Mueller & the Management of NASA's Human Space Flight Program.* London: Springer Praxis, 2012.

Temple, L. Parker III. *Shades of Grey.* Reston, VA: American Institute of Aeronautics and Astronautics, 2004.

Tribbe, Matthew D. *No Requiem for the Space Age: The Apollo Moon Landings and American Culture.* New York: Oxford University Press, 2014.

Tyson, Neil DeGrasse and Avis Lang. *Space Chronicles: Facing the Ultimate Frontier.* New York: W. W. Norton, 2012.

van Dyke, Vernon. *Pride and Power: The Rationale of the Space Program.* Urbana, IL: University of Illinois Press, 1964.

Wilford, John Noble. *Mars Beckons: The Mysteries, the Challenges, the Expectations of Our Next Great Adventure in Space.* New York: Viking, 1991.

Index

Advisory Council on Executive Management (Ash Council), 132, 133, 135–6
Aerospace Corporation, 119, 164, 190
Aerospace Industries Association, 178
aerospace unemployment, 2, 98–9, 236, 243, 253, 264–5
   1972 presidential election and, 137, 144, 145–6, 181, 232–5
   Apollo cancellations and, 180, 240–1
   full capability shuttle and, 261
   Low's "best case" essay and, 237
   NASA FY1972 budget and, 152, 153, 157–8
   NASA FY1973 budget and, 209, 233
   Rice and, 233–4, 260
   Rose and, 233, 234, 255, 266
   shuttle development's impact on, 266, 288–9, 291–2
   See also California Employment Project
Agnew, Spiro, 26
   Anders and, 52, 136, 200
   Apollo 13 and, 154–5
   Ehrlichman and, 137
   as NASC chair, 49, 50–2, 53, 107, 136–7
Agnew, Spiro as STG chair, 49–50, 52, 125
   Ehrlichman and, 77–8
   at final meeting, 77–8
   at first meeting, 56
   invited contributors meeting and, 60, 64
   Mars mission and, 59, 64, 65, 67–8, 70, 71, 73, 74, 77, 311n23
   Paine and, 59, 64, 65, 74, 75, 100–1
   at penultimate meeting, 75, 76
   program options and, 75, 76, 77–8, 79
   Seamans and, 171
   at second meeting, 58, 59
   at STG report press conference, 79–80
   See also Space Task Group (STG); Space Task Group (STG) report
Air Force, 164, 168, 290
   cross-range capability requirements from, 225
   Flax and, 202
   Mueller and, 170–1
   NASA/Air Force STS Committee, 171, 172
   payload bay size and, 192
   shuttle booster selection and, 272
Aldrin, Edwin "Buzz," Jr., 11, 13, 18, 26, 139
   Borman's profile of, 17
   "Giant Step" tour and, 28–9
   lunar landing and splashdown involvement of, 19–22, 24

Aldrin, Joan, 17, 29
American Institute of Aeronautics and Astronautics (AIAA), 60
"America's Next Decades in Space" (NASA input to STG). *See* Space Task Group (STG) report
Anders, Bill: on *Apollo 8*, 7, 51
   Apollo 13 and, 155, 156
   at Apollo review meeting, 122, 151
   Fletcher and, 187, 200, 226
   Low and, 151, 187, 199–200, 218, 226, 235
   NASA's budget and, 235, 273
   as NASC executive secretary, 51–2, 131, 136, 137, 199–200
   shuttle decision and, 261, 264, 288
   shuttle development and, 200, 213, 248, 253, 255, 261, 264, 269, 288, 298
   Weinberger and, 136, 151, 199, 218
Anderson, Clinton, 69, 262
Anderson, Robert, 234–5
Apollo 1 mission, 11
Apollo 8 mission, 7–8, 9, 33, 54, 110
   Anders and, 7, 51
   "Earthrise" photo taken during, 282, 283
   MacLeish poem on, 12
Apollo 9 mission, 10, 59
Apollo 10 mission, 10
Apollo 11 mission, 7–30, 93
   Borman and, 10–11, 12, 14, 16–18, 19, 21, 22–3
   dinner following landing, 26–7
   final preparations for, 16–18
   first anniversary of, 139
   "Giant Step" tour, 27–30
   launch and lunar landing of, 12–13, 19–20
   Luna-15 probe and, 17
   MacLeish poem on, 12, 22
   Nixon/Kennedy credit for, 13, 14–15, 16, 26, 30
   plaque commemorating, 13
   possibility of failure, preparations for, 18
   pre-launch dinner, 11, 13–14
   presidential preparation for, 8–9
   recovery of, 16
   risks of, 11
   as "soft power" exercise, 282–3
   splashdown of, 22–7
   success of, 1, 121, 125, 126, 137
   television conversation with Nixon during, 11, 15, 19, 20, 21–2
   von Braun and, 137

344 INDEX

Apollo 11 mission—*Continued*
  See also Aldrin, Edwin "Buzz," Jr.;
    Armstrong, Neil; Collins, Michael;
    Kennedy's moon landing goal; Nixon's
    involvement with *Apollo 11* mission
Apollo 12 mission, 90–3, 121, 282
Apollo 13 mission, 2, 10, 108, 117, 121, 122
  emotional impact of, on Nixon, 123, 144,
    154–7, 180, 241, 282, 322n28
Apollo 14 mission, 121, 122, 177
Apollo 15 mission, 121–2, 177, 211, 282
Apollo 16 mission, 121, 122, 158, 181, 214,
    284
Apollo 16 mission cancellation possibility, 146,
    177, 283
  FY1973 budget process and, 208, 210–11,
    242–4
  Nixon's suggestion for, 180, 187, 216–17,
    231, 235, 240, 242–4, 282
  rescheduling of, 244
  See also Apollo cancellations, Nixon and
Apollo 17 mission, 3, 121, 122, 214, 282
Apollo 17 mission cancellation possibility, 146,
    177, 283
  Apollo 13 accident and, 144, 154
  budget review as cause of, 150, 151, 152,
    157, 180
  decision to continue, 153, 159, 217, 242,
    244
  FY1973 budget process and, 208, 210–11
  Nixon's suggestion for, 144, 154, 157–9,
    180, 187, 216–17, 231, 235, 240, 242–4,
    282
  perceived risk as cause of, 144, 153–4, 157,
    158, 180, 217
  See also Apollo cancellations, Nixon and
Apollo 18 mission, 121, 122
Apollo 19 mission, 121, 122
Apollo 20 mission, 47, 98
Apollo cancellations, 119, 122, 177
Apollo cancellations, Nixon and, 123, 187,
    216–17, 231, 235, 282
  budget review and, 150, 151, 152, 157, 173,
    180, 204
  employment impact of, 158, 180, 240–1
  perceived risk and, 144, 153–4, 157, 158,
    180, 217
  vs. rescheduling, 158–9, 217, 242, 243, 244
Apollo hardware, use of by other missions, 70,
    285, 286
  "gap-filler" missions and, 214, 235
  for shuttle development, 2, 207
Apollo program, 47, 119, 120–1, 125
  employment and, 180, 240–1
  as exercise in "soft power," 8–9, 25, 282–3,
    304n4
  funding for, 123, 279
  launch rate of missions in, 87, 89, 93, 95, 96,
    106, 117
  Low's position on, 148
  priority of, 115, 279, 282
  retreat from, 123, 282
  review of, 121–2, 151
  spacecraft design in, 170
  See also Apollo cancellations, Nixon and;
    Apollo missions
Apollo-Soyuz Test Project, 17, 199, 289
Armstrong, Jan, 17, 29
Armstrong, Neil, 11, 13, 18, 26, 139
  Borman's profile of, 17
  "Giant Step" tour and, 28–9
  lunar landing and splash down involvement
    of, 1, 19–22, 24
Ash, Roy, 132
Ash Council, 132, 133, 135–6
Atomic Energy Commission (AEC), 56, 71
Augustine, Norm, 280
Aviation Week and Space Technology, 40

Bean, Alan, 91
Beggs, James, 294
Bell Laboratories, 135
Belson, James, 178
Bennett, Wallace, 174
Beresford, Spenser, 40
Berry, Charles, 14
Black, Shirley Temple, 60
Block I/Block II approach to shuttle
  development, 186, 208. *See also* Mark I/
  Mark II approach to shuttle development
BOB. *See* Bureau of the Budget (BOB)
Borman, Frank, 86
  Apollo 11 and, 10–11, 12, 14, 16–18, 19, 21,
    22–3
  Apollo 13 and, 154–5
  as candidate for NASA administrator, 53–4,
    141
  Christmas Eve lunar orbiting of, 7, 19
  European "goodwill" tour of, 9–10
  international space cooperation and, 9–10,
    22, 110, 111
  Nixon and, 9–10, 12, 14, 16–18, 22, 73,
    154–5
  Soviet Union visit by, 16–17
Branscomb, Lewis, 40, 57, 228
British Interplanetary Society, 37, 162
Buchanan, Patrick, 13, 16
budget process, FY1970, 38, 39, 57, 58, 73,
    118
budget process, FY1971, 74, 83–4, 90–101,
    126, 149
  balanced budget priorities in, 97
  confusion in, 83, 87, 103, 132, 278
  Congress and, 94, 97–8, 103, 117, 146
  Nixon space doctrine and, 105, 107–8, 114
  Nixon's explanation of decisions in, 100–1
  "Operation Paring Knife" and, 98
  space spending as low priority in, 83, 84–5,
    91, 95–6, 99, 103, 115, 278–80, 281
  See also NASA budget, FY1971
budget process, FY1972, 117, 120, 138, 144–
    53, 180, 209, 227–8, 248
  Congress and, 146, 160, 173
  Low and, 142, 144, 145, 146, 147–51, 153,
    157, 159–60, 173, 174, 186
  NASA submission in, 146–7, 173
  NASC and, 136
  Nixon's decisions in, 151–3, 157–60
  shuttle development goals in, 96, 122, 147,
    159–60, 186
  space spending as low priority in, 144
  See also NASA budget, FY1972
budget process, FY1973, 201–7, 240–1,
    244–5, 280
  Anders and, 200
  Apollo cancellations and, 208, 210–11,
    242–4
  civilian agencies' cuts in, 204

Flax committee and, 202–4
OMB hearings during, 214–15
shuttle development and, 204–5, 208–9, 235–8
See also NASA budget, FY1973
Bureau of the Budget (BOB), 46, 126, 133, 144
communication breakdown between White House and, 83, 90, 94–5, 96, 99
Mayo selected as director of, 47
NASA budget, FY1970, and, 38, 39, 57, 58
NASA budget, FY1971, and, 58, 74, 83, 84, 87–90, 91–101, 103
Nixon space doctrine and, 104, 106
space shuttle development and, 147, 188–9, 296
STG and, 56, 59, 75, 77, 84
See also Mayo, Robert; Office of Management and Budget (OMB)
Burns, Arthur, 14, 48, 80, 98, 109
Bush, George H. W., 85, 141
Butterfield, Alex, 155, 266

California, 145, 207
employment issues in, 180–1, 232–4, 241, 255, 264–5, 265–6, 291–2
shuttle development announcement in, 264–5
Vandenberg Air Force Base, 163, 168–9, 290
See also aerospace unemployment
California Employment Project, 232, 233, 241, 266, 288
Cannon, Howard, 262
Carter, Jimmy, 167, 275, 279, 291
Cash, Johnny, 156
Ceausescu, Nicolae, 13
Central Intelligence Agency (CIA), 13
Chaffee, Roger, 11
Challenger accident (1986), 291
Chapin, Dwight, 10–11, 12, 18, 77, 78
China, 13, 29, 242, 243, 244
Chou En-Lai, 13
Chrysler shuttle study, 191
Clarke, Arthur C., 137–8
Clauser, Francis, 40
Cold War, 8, 33, 45, 123, 281
Cole, Kenneth, 106, 108, 136
Collins, Michael, 11, 13, 14, 18, 22, 139
Apollo 13 and, 155, 156
Borman's profile of, 17
"Giant Step" tour and, 28–30
launch and splashdown and, 19, 24
Collins, Pat, 17, 29
Colson, Charles "Chuck," 264, 271, 288
Columbia Accident Investigation Board, 280, 293, 295
Columbia command and service module, 15, 19, 22, 24
Congress, U.S., 15, 82, 137, 283, 300
opposition in, to additional lunar missions, 119
opposition in, to Mars missions, 69, 74, 82
shuttle development and, 149, 194, 219, 224, 262, 290
SST funding and, 180, 186
STG information to, 59–60
tax increases and, 97–8
Connally, John, 234, 240
Conrad, Pete, 91, 93, 282

Corona (photo-intelligence satellite), 165, 167
Cortright, Edgar, 183, 200
Crabill, Don, 74
Cronkite, Walter, 175
cross-range capabilities, 162, 291
determination of requirements for, 168–9, 225
DOD/NASA report and, 166, 167–8
Fletcher/Packard meeting and, 224
of ILRV, 163
Mueller and, 170, 171
surveillance missions and, 165–6, 169
See also payload bay size/weight capacity; space shuttle development; space shuttle development, national security requirements for

David, Edward E., Jr., 145, 247, 265, 298
aerospace unemployment and, 181
Anders and, 200
Apollo 17 and, 158
Apollo cancellations and, 231
as DuBridge's successor, 131, 135
Flax committee and, 203–4
Fletcher and, 178, 186–7, 218–19
FY1972 budget process and, 148, 157
FY1973 budget process and, 205
glider preference of, 228
international space cooperation and, 198
"new NASA" discussions and, 200–1, 232
PSAC special panel and, 187, 202, 219
shuttle decision and, 207–8, 256, 258, 259, 260, 261
shuttle development and, 213, 226, 227, 228, 286–7
shuttle opposition by, 213, 218–19
See also Office of Science and Technology (OST)
Department of Defense (DOD), 56, 214
aerospace unemployment and, 232
Rice at, 133, 215
shuttle development and, 58, 162, 163, 184, 193, 194, 256, 262, 290, 291
shuttle support sought from, 224–6, 228–9, 237, 245–6
STG and, 48, 49, 56, 62, 75, 119
STG report and, 75
STS Committee and, 171, 172
See also DOD/NASA shuttle report; space shuttle development, national security requirements for
Department of State, U.S., 28, 126
desalination initiatives, 181–2
Dobrynin, Anatloy, 16
DOD. *See* Department of Defense (DOD)
DOD/NASA shuttle report, 70, 163–70, 225, 246, 263
cost analysis in, 166
mission areas in, 164–5
mission models in, 166, 190
payload bay size, weight and cross-range capabilities in, 166, 167–8
Domestic Council, 144, 149, 200–1
creation of, 131, 132–3
domestic policy, 45, 103, 108, 133, 247. *See also* foreign policy
Donlan, Charles, 177, 186, 193–4, 271
Dooley, Don, 164
Draper, Charles Stark, 99

Drew, Russell, 51, 122, 135, 148, 150
  STG and, 56, 66, 71, 75, 78, 79–80
"drop tank" design (expendable fuel tanks), 192–3, 205–6, 222
DuBridge, Lee, 15, 141, 283
  on Apollo review, 122
  as director of OST, 47
  Mars mission and, 71, 77
  NASC and, 51
  Paine and, 53, 57, 72, 122
  resignation of, 131, 135
  shuttle development and, 125–6
  STG and, 48–50, 51, 56, 58, 59–60, 70, 72
  STG report and, 75, 77, 78, 79–80
Dyna-Soar program, 168

Eagle lunar lander, 12, 15, 19, 21, 22
Earth-Moon space, Mueller's integrated plan and, 62–6, 72, 118, 140, 170
"Earthrise" photo, 282, 283
Ehrlichman, John, 274
  aerospace unemployment and, 152, 158, 232, 234, 240–1, 253, 288
  Agnew and, 77–8, 137
  antagonism between Mayo and, 94
  Apollo 11 and, 9, 12, 15, 22, 23
  Apollo 13 and, 154
  Apollo 17 and, 157, 158, 159
  Apollo cancellations and, 180, 240–1
  as Domestic Council director, 132–3
  Flanigan and, 45, 94, 132
  Foy and, 232
  FY1971 budget process and, 90, 91, 92, 93, 94, 96, 98, 99, 100
  FY1972 budget process and, 151, 152, 153, 157, 158, 159
  FY1973 budget process and, 240–1, 244–5
  Harper and, 144, 152
  international space cooperation and, 198
  Low and, 86, 227
  NASC and, 137
  "new NASA" discussions and, 181–2, 183
  Nixon space doctrine and, 113
  Paine and, 98, 140, 141
  roles and responsibilities of, 43–4, 77, 132
  SST funding and, 180
  STG report and, 77, 78
  Whitaker and, 145
  Witness to Power, 77
Ehrlichman, John, shuttle development and, 240–1, 244, 246, 248, 253, 287, 288
  David and, 187
  "new NASA" discussions and, 252
  shuttle decision and, 263, 266–9
Eisenhower, Dwight D., 15, 31–2
Electronics Research Center, 93
Elliott, David, 255
Ellsworth, Robert, 45
Europe, 9–10
  international space cooperation and, 82, 110–11, 148, 246, 268
  shuttle development and, 197–9
European Space Research Organization, 110
Executive Office of the President, 132, 133–4
exploration, 35, 66
  Agnew on, 64
  end of, 281–5, 300
  general goals of, 277, 278, 284

grand tour, 85, 97
  for its own sake, 65, 105, 150, 180
  lunar, 18, 41, 55, 58, 60–1, 106, 121, 123, 164, 166, 217
  as mark of great nations, 32, 124, 179
  Nixon space doctrine and, 105, 114–15, 281
  Nixon's call for "exploring the unknown," 124, 179–81, 278, 283–5
  Nixon's inaugural address and, 7
  planetary, 41, 77, 78, 82
  private sector and, 300
  shuttle development and, 231, 267–8, 277–8
  STG report and, 71, 75, 76, 77
  Townes report on, 41
  See also Mars mission(s)

Faget, Maxime (Max), 170, 184, 297
Finch, Robert, 145
Findlay, John, 121, 122
Fisk, James, 141
Flanigan, Peter, 126, 213, 277, 283
  aerospace unemployment and, 181, 233, 266, 288
  Anders and, 200
  antagonism between Mayo and, 94
  Apollo 11 mission and, 10–11, 12, 18, 27–8
  Apollo program review and, 121, 122
  bicentennial space spectacular and, 138
  Ehrlichman and, 45, 94, 132
  Fletcher and, 186–7, 205, 226, 248, 251
  Foy and, 232–3
  FY1971 budget process and, 74, 84, 90, 91–2, 93, 94, 95–7, 99, 100
  FY1972 budget process and, 148–9
  FY1973 budget process and, 200, 205
  "Giant Step" tour and, 27, 28
  international space cooperation and, 110, 111, 112, 113, 114, 198
  Low and, 85–6, 148–9, 237–8, 247
  NASA administrator selection and, 85–6, 141–2, 174–5, 177–8
  "new NASA" discussions and, 183
  Nixon space doctrine and, 104–5, 106, 107, 108, 112, 113–14
  OMB communication with, 226–7, 248, 251
  Paine and, 84, 98, 99, 139–40, 141
  role and responsibilities of, 45–6, 47, 73, 132
  STG and, 73–4, 75, 77, 84
  Whitehead's job change and, 131, 134, 198
Flanigan, Peter, shuttle development and, 186, 205, 213, 234–5, 247, 253
  employment impacts of, 233, 266, 288
  naming choice, 265–6
  NASA's "best case" essay for, 237
  national security potentials for, 246
  position of, 227
  presidential announcement of, 264–5
  size of shuttle and, 255, 256, 261
Flax, Alexander, 202, 219, 226–7, 228, 286, 298
Flax committee, 202–4, 213, 218, 219–21, 226–30, 261, 296
Fletcher, Fay, 178
Fletcher, James, 173, 174, 194, 298
  Anders and, 187, 200
  Apollo cancellations and, 231
  appointment of, 177–9
  David and, 178, 186–7, 218–19
  Flax committee and, 221

FY1973 budget process and, 205, 209, 245
Low, first meeting with, 177
"new NASA" discussions and, 183, 201
Rice and, 187–8, 193
support for NASA sought by, 224–7
von Braun and, 230
Weinberger and, 261–2
Whitehead and, 226, 247
Fletcher, James, shuttle development and, 253, 293
"Administrator's contingency" request by, 273–4
booster selection and, 288
glider approach to, 206
Heiss memorandum to, 222–4
Kissinger and, 246–7
naming choice, 265
national security requirements for, 169
NERVA and, 262
OMB shuttle design and, 250–1
phased approach of, 193, 194
presidential approval and, 264, 266–9, 274, 288
rationale for, 225–6, 245
shuttle decision and, 207–9, 255–7, 258, 261
Fletcher/Low, shuttle development and, 179, 208, 293
budget concerns of, 188, 194, 197, 205, 209, 273–4, 298
expendable fuel tank design, 193
Flax committee and, 221
fully reusable design abandonment, 193–4
presidential approval and, 261, 262, 263, 266–8, 287–8
Shultz and, 183
shuttle size and, 248–9, 250–1, 255–6, 261
support sought for full capability shuttle by, 213, 224–7, 256–8
Ford, Gerald, 174
foreign policy, 41, 45, 103, 247, 282–3. *See also* domestic policy; international space cooperation
Foster, Johnny, 56, 225, 228–9, 245, 246
Foy, Fred, 232–3
Freidman, Milton, 214
Fubini, Gene, 202–3, 221, 226, 228
full capability shuttle design, 185, 247, 251, 255, 286
approval of, 246, 252, 261, 262, 263, 274, 287
approval of, as policy mistake, 295–9, 299–300
approval of, reasons for, 288–9, 290, 292–4
NASA's insistence on, 172, 213, 241, 256, 257
OMB's alternatives to, 218, 220, 230, 260
See also NASA shuttle design; space shuttle development

Gambit (photo-intelligence satellite), 165
"gap-filler" missions, 214, 235
Garment, Leonard, 44–5, 141
Garwin, Richard, 203
Gemini Program Planning Board, 171
Gemini spacecraft, 37, 169, 170, 264, 293
modification of ("Big-G"), 205, 220–1
General Electric, 35, 139
"Giant Step" tour, 27–30
Gilruth, Robert, 40, 121, 154, 319n6

glider approach to shuttle development, 168, 206, 208, 214, 226–7, 285
cost analysis of, 228–9
Flax committee and, 221
Low's cost curve diagram and, 229, 236
NASA's lack of investigation into, 296
OMB director's review and, 218
OMB recommendation for, 235, 237
See also space shuttle development
Goldwater, Barry, 199
Gordon, Richard, 91
Grand Tour mission, 87, 95
Great Society programs, 34
Griffin, Mike, 293
Grissom, Gus, 11
Grumman Aerospace, 222, 272, 292
Grumman/Boeing study, 191–2

Haggerty, Patrick, 52–3
Haig, Alexander, 200, 246–7
Haise, Fred, 154, 156
Haldeman, Harry Robbins "Bob": advertising background of, 10, 43
Apollo 11 and, 10, 11, 14–15, 16, 18, 21, 23–4, 26, 27
Apollo 13 and, 154–5
Apollo cancellations and, 216–17
daily diary of, 23–4, 43, 86, 155
FY1971 budget process and, 100
Mars missions and, 100
NASA administrator selection and, 52–4, 174, 175
"new NASA" discussions and, 181–2
Nixon tapes and, 175–6
roles and responsibilities of, 43–4
shuttle development and, 217, 233
Hansen, Grant, 163, 192–3, 225
Harlow, Bryce, 15, 94, 96
Harper, Edwin, 144–5, 145–6, 152, 240, 252
Heiss, Klaus, 190, 226
memorandum by, 222–4
Henry the Navigator, Prince of Portugal, 35–6
Hess, Harry, 40
Hexagon (photo-intelligence satellite), 167–8
Hickham Air Force Base (Honolulu, HI), 156
Horowitz, Norman, 40
Hubble Space Telescope, 289
Huebner, Lee, 112
Humphrey, Hubert, 26, 39, 51

inflation, 89, 204
Ink, Dwight, 136
Institute for Defense Analysis, 202, 219
integral launch and reentry vehicle (ILRV), 162, 163
international space cooperation, 31–2, 246–7
Borman and, 9–10, 22, 110, 111
foreign astronaut participation, 109–12
as Nixon's pet idea, 109–12, 198–9, 289
Paine and, 22, 108, 110–11, 112, 113, 140
shuttle development and, 148, 172, 197–9, 215, 238, 268, 289
Skylab and, 148
STG and, 60, 76, 82
Townes report and, 41, 42, 48, 109
International Space Station (ISS), 289, 299
Invited Contributors meeting (public input to STG), 59–60, 64

J. Walter Thompson advertising agency, 10, 43
Jameson, Frank, 174, 176
Japan, 110, 113
Jet Propulsion Laboratory, 235
"John F. Kennedy" (*Apollo 11* spacecraft name suggestion), 14–15, 16
Johnson, Howard, 141
Johnson, Lyndon B., 30, 55, 176, 280
   NASA budget allocations by, 33–4, 38
   NASA budget reductions by, 32
   NASC and, 51, 136
   Webb and, 34–5, 52
Johnson, U. Alexis, 27, 56, 70, 80
Johnson administration, 11
"Joint DOD/NASA Study of Space Transportation Systems" report. *See* DOD/NASA shuttle report
Jones, Thomas, 141

Karth, Joseph, 149
Keldysh, Mstislav, 16, 17
Kendall, Donald, 141
Kennedy, David, 47
Kennedy, Edward, 68–9
Kennedy, John F., 25–6, 47–8, 300
   carte blanche for NASA budget by, 67
   NASC and, 50–1
   transition task force model of, 40
   Webb and, 52
Kennedy administration, 247
Kennedy Space Center, 12, 18, 90, 93, 123, 180
Kennedy's moon landing goal, 1, 121
   context of decision of, 31, 33, 47–8
   impact of, 275
   "John F. Kennedy" spacecraft naming rejection and, 14–15, 16
   Nixon's lack of credit for success of, 13, 14–15, 16, 26, 30
   as "soft power" exercise, 8–9, 25, 281–2, 304n4
   televised address to Congress announcing, 8, 104, 114, 124, 215–16, 269
   as top-down leadership initiative, 282
   See also Apollo 11 mission
Keogh, James, 108, 109
KH (Keyhole)-9 (photo-intelligence satellite), 167–8
KH (Keyhole)-10 (photo-intelligence camera) (Dorian), 169
Kissinger, Henry, 12, 16, 44, 226, 268
   Apollo 11 splashdown and, 22–3
   Apollo 13 and, 154, 155, 156
   "Giant Step" tour and, 28
   international space cooperation and, 110, 246–7
   Low and, 148
Klein, Herb, 15, 107
Klein, Milton, 80
Kraft, Chris, 155, 297
Kriegsman, Will, 134, 152

Laird, Melvin, 56, 224, 226
launch rate: of *Apollo* missions, 87, 89, 93, 95, 96, 106, 117
   Flax committee and, 203
   Mathematica study and, 190–1
   of shuttle missions, economic analysis of, 185, 189, 195, 292–3, 295
   of shuttle missions, in DOD/NASA report, 166, 167
Lenher, Samuel, 40
Lewis, Roger, 141–2
Life magazine, 156
Lilly, Bill, 87, 146, 151
Lindley, Robert, 189, 222
Lockheed, 190, 191, 222
Lovell, James, 7, 138, 154, 156
Low, George M., 300
   aerospace unemployment and, 153, 157
   Anders and, 151, 187, 199–200, 218, 226, 235
   Apollo cancellations and, 119, 123, 157–8, 159–60, 177
   appointment of, 85–6
   "best case" essay by, 237–8, 247
   booster selection and, 288
   cost-curve diagram of, 228–9, 236
   diary of, 85
   DuBridge and, 85–6
   Flanigan and, 85–6, 148–9, 237–8, 247
   Flax committee and, 202–4, 221
   Fletcher, first meeting with, 177
   Foster and, 245
   Fubini and, 203
   FY1972 budget process and, 142, 144, 145, 146, 147–51, 153, 157, 159–60, 173, 174, 186
   FY1973 budget process and, 205, 209, 227–8
   Kissinger and, 148
   as NASA acting administrator, 142, 143–4
   NASA FY1970 budget and, 99
   NERVA and, 262
   "new NASA" discussions and, 181, 183, 200–1, 231–2, 251–2
   NSC and, 246
   OMB meetings with, 145, 157, 214, 239
   OMB NASA hearings and, 147–50, 151
   Paine and, 84–5, 97, 131, 139, 141–2
   Rice and, 159–60, 236–7, 247
   on Saturn V production, 119
   Skylab and, 148, 157
   Soviet competition and, 153, 157
   support for NASA sought by, 147–51, 213, 224–7, 247
   von Braun and, 137, 230, 231
   See also Fletcher/Low, shuttle development and
Low, George M., shuttle development and, 120, 159–60, 177, 179, 207, 285–6
   economic analysis of, 188, 293
   glider approach and, 206, 208
   Mathematica study and, 190
   mini-shuttle approach and, 208
   OMB shuttle design and, 250–1
   orbiter/booster phased development and, 185–6
   phased approach and, 193
   presidential approval and, 266–9, 274, 287–8
   rationale for, 225–6
   shuttle decision and, 255–7, 258–9, 261
   shuttle/station development and, 298
Luna-15 robotic probe, 17
Lunar and Planetary Missions Board, 121, 122

MacGregor, Clark, 179, 261, 262
MacLeish, Archibald, 12, 22
Magruder, Jeb, 106
Magruder, William, 201, 231–2, 235–6, 242, 251–2
Malek, Fred, 174
Manned Orbiting Laboratory (MOL) program, 169
Manned Spacecraft Center (Houston, TX), 18, 26, 85, 156, 170
   Gilruth and, 40, 121, 154
   Grumman/Boeing study for, 191–2
   Kraft and, 297
   location choice of, 292
   shuttle development and, 191–2, 194, 206, 223, 272
Mansfield, Mike, 68–9
Mark, Hans, 167, 290, 294
Mark I/Mark II approach to shuttle development, 207, 209, 227, 230, 296
   aerospace unemployment and, 233
   Block I/Block II approach, 186, 208
   Flax committee on, 219
   Low's cost curve diagram and, 229, 236
   NASA FY1973 budget request and, 208
   OMB director's review and, 218
   See also space shuttle development
Marooned (film), 92–3
Mars mission(s), 64–9, 150, 284
   Agnew on, 59, 64, 65, 67–8, 70, 71, 73, 74, 77, 311n23
   ambitiousness of, 41, 54, 66
   budgeting for, 58, 75, 84, 88, 95, 96, 182, 247, 281
   Congressional opposition to, 69, 74, 82
   Newell's proposal to plan for, 62, 64
   Nixon space doctrine and, 105, 106, 114
   Nixon's desire for eventual, 2, 92, 100, 114, 180, 284
   Paine and, 59, 64–8, 69
   public opinion about, 68–9
   shuttle/station goals as steps to, 146, 148, 149
   STG discussions on, 30, 54, 58–9, 64–72, 74, 75, 88
   STG report and, 30, 71–2, 77, 78–9, 80, 81, 82, 84, 95, 96, 104, 125
   von Braun and, 66–8, 230
Mars mission(s), timing of, 62, 283
   after 1985, 71, 72, 75–6, 77, 79, 80–2, 84
   before 1985, 41, 71, 74, 75, 77, 79, 81, 88
   crash program, 66, 79, 80, 96
   eventual, 2, 92, 100, 114, 180, 284, 299
   Nixon space doctrine and, 104, 105, 106, 107, 114
   STG discussion and, 30, 64–8, 69, 71–2, 74, 75–6, 77, 125
   STG report and, 81, 82, 84, 104
   von Braun on, 66–8, 230
Marshall Space Flight Center (Huntsville, AL), 39, 66, 131, 137
   potential closing of, 216, 235
   shuttle development at, 191, 221, 236, 272
Mathematica, Inc. study, 189–91, 194, 195, 202, 221–4, 261
Mayo, Robert, 39, 57, 188
   balanced budget and, 97
   budget cuts and, 98–9

   NASA FY1971 and, 87, 88–90, 91, 92, 94, 96–9
   Nixon space doctrine and, 104, 105, 107
   Nixon's dislike of, 90, 94
   resignation of, 133
   selection of, as director of BOB, 47
   STG and, 56, 58, 59, 70–1, 73–4, 76, 83, 84, 126
   STG report and, 75, 77, 80
   See also Bureau of the Budget (BOB)
McCain, John, 23
McCracken, Paul, 157–8
McCurdy, Howard, 284
McDonald, Gordon, 84
McDonnell Douglas, 118, 184, 191, 192, 222, 272
McDougall, Walter, 278
McLucas, John, 225
McNamara, Robert, 171, 215
Meany, George, 155
Medal of Freedom, 26, 156
Mercury spacecraft design, 170
Mettler, Ruben, 40, 141
military, U.S.: NASA as separate from, 31–2
   satellites of, 149, 161, 238, 246, 288
   See also Air Force; Department of Defense (DOD); space shuttle development, national security requirements for
Miller, George, 69
mini-shuttle approach, 205–6, 208, 218, 296.
   See also space shuttle development
Mir (Soviet/Russian space station), 289–90
Mitchell, Edgar, 177
Mondale, Walter, 149, 271
Morgenstern, Oskar, 189, 222, 261
Moyers, Bill, 14, 15
Moynihan, Daniel Patrick, 14, 92
Mueller, George, 37, 85, 170–1, 296, 297
   "integrated plan" proposal by, 62–6, 72, 118, 140, 170
   Mars mission and, 64–6
   Myers as successor of, 120, 131
   shuttle development and, 162–3, 192
Muskie, Edward, 232, 271
Myers, Dale: Apollo cancellations and, 122, 157
   as Mueller's successor, 120, 131
   OMB shuttle design and, 250–1
   von Braun and, 230
Myers, Dale, shuttle development and, 188, 192, 193, 230
   Flax committee and, 203, 221, 228, 261
   Fletcher and, 177
   fully reusable design and, 173, 185, 186
   glider approach and, 206, 221, 227
   mini-shuttle and, 206
   national security requirements and, 184
   OMB shuttle and, 250–1
   phased approach and, 193
   space station and, 120, 131

NASA. *See* National Aeronautics and Space Administration (NASA)
NASA/Air Force STS Committee, 171, 172
NASA budget, 41, 55, 62–3, 66, 83–4
   Anders and, 235, 274
   anticipated, 187–8, 197, 207, 239
   carte blanche availability of, 67

NASA budget—*Continued*
 "constant budget" request of, 209
 employment impact of, 98–9
 FY1970, 38, 39, 57, 58, 99
 human space flight community and, 280–1
 Johnson and, 32, 33–4, 38
 Mars missions and, 58, 68, 70, 75, 84, 88, 95, 96, 182, 247, 281
 prioritization of, 277, 278–80
 reductions in, 32, 34, 74
 STG and, 73–4, 90
 Weinberger memorandum and, 210–11
 See also NASA budget, FY1971; NASA budget, FY1972; NASA budget, FY1973
NASA budget, FY1971, 111, 126, 316n1
 approval of, 93–4, 99, 103, 123, 146
 BOB and, 58, 74, 83, 84, 87–90, 91–101, 103
 FY1972 and, 160
 ratcheting down in, 94–9, 107–8
 Saturn V production and, 118–19
 shuttle/station program and, 117, 149
 STG report and, 76, 83, 84–5, 87, 103, 138
 See also budget process, FY1971
NASA budget, FY1972, 117, 120
 Domestic Council and, 144
 Flanigan's lack of involvement in, 148–9
 Low and, 142, 144, 145, 146, 147–51, 153, 157, 159–60, 173, 174, 186
 Nixon's decisions on, 157–60
 OMB and, 144, 146–51, 152, 159
 shuttle development language in, 159–60
 submission of, to OMB, 146–7, 173
 unemployment and, 152, 153, 157–8
 See also budget process, FY1972
NASA budget, FY1973, 201–7, 248
 cuts to, 204
 Flax committee and, 202–4
 Low and, 227–8
 Nixon's tentative approval of, 209–11
 OMB and, 213
 shuttle development and, 204–5, 208–9, 235–8
 unemployment and, 209, 233
 See also budget process, FY1973
NASA shuttle design, 213, 257–60, 285–86
 approval of, 246, 252, 261, 262, 263, 264, 274, 287–9, 293–4
 approval of as policy mistake, 295–9
 concerns about, 186–7, 193–4
 developing cost estimates of, 184, 188, 193, 195, 204, 228–9, 257, 286
 employment impact of, 266, 288–9, 291–2
 Flax committee and, 202–4, 219–21, 228–30
 glider approach to, 206, 208, 221, 227, 228–9
 Low's cost curve and, 228–9
 Mark I/Mark II approach to, 207, 208, 227, 229
 Mathematica study on, 189–91, 221–4
 mini-shuttle approach to, 205–6, 208
 naming of, 265–6, 267, 268–9
 national security uses of, 161–2, 163–72, 184, 194, 195, 207, 224–6, 245–7
 OMB's push for studies of alternatives to, 191, 218, 220, 230, 236–7, 255, 260
 operating costs of, 188–9, 190–1, 195, 209, 228–9, 257, 293–4
 payload bay size of, 170, 255–6, 257
 payload weight capacity of, 170, 255–6, 257
 Phase A studies of, 162, 166, 170, 184, 192
 Phase B studies of, 184–5, 190, 191, 192, 194, 206
 size of, 185, 191–3, 205–6, 247–8, 256
 TAOS configuration of, 222–4, 228
 See also full capability shuttle design; NASA-OMB shuttle design debates; OMB shuttle design; reusable shuttle development; space shuttle development
NASA-OMB shuttle design debates, 143, 147, 239–61, 286–7
 employment impact in, 233–5, 243, 253
 escalation of conflict in, 247–51
 NASA FY1972 wording and, 159–60
 national security issues and, 291
 OMB decision memorandum and, 242–5, 252
 OMB push for alternatives in, 191, 236–7, 242–5
 presidential decision in, 252–3, 261
 shuttle size and, 242–5, 247–8, 248–51
 trust issues in, 207–8
 See also NASA shuttle design; OMB shuttle design
NASA-OMB shuttle program debates, 213–38, 286–7
 David and, 218–19
 Flax committee and, 219–21
 OMB and White House players' positions in, 226–7
 OMB hearings and, 214–15
 Taft's paper and, 215–16
 Weinberger and, 216–18
NASC. *See* National Aeronautics and Space Council (NASC)
National Academy of Sciences, Space Science Board of, 60–1, 121, 122
National Advisory Committee on Aeronautics (NACA), 85
National Aeronautics and Space Administration (NASA), 26, 37–8
 2012 assessment of, 299
 creation of, 31–2
 division within program office, 34
 "Giant Step" tour and, 27–8
 international space cooperation and, 31–2, 110
 long-range planning conference, 137–9
 management of, 52–4, 85–6, 141–2, 143–4, 174–5, 176, 177–8
 Nixon space doctrine and, 106–7, 113
 at start of Kennedy and Nixon administrations, 47–8
 STG input by, 62, 79
 Townes report and, 32, 42, 57–8
 Webb and, 33–7, 52, 118, 123, 140, 308n10
 See also Fletcher, James; Low, George M.; "new NASA" discussions; Office of Manned Space Flight, NASA; Paine, Thomas O.
National Aeronautics and Space Council (NASC), 101
 Agnew as chair of, 49, 50–2, 53, 107, 136–7
 Anders as executive secretary of, 51–2, 131, 136, 137, 199–200
 composition of, 136
 proposals to eliminate, 131, 135–7, 200

National Reconnaissance Office (NRO), 164, 168, 290–1
National Security Council (NSC), 112, 132, 246–7
Nelson, Craig, 16
NERVA (nuclear rocket engine) program, 151, 210, 211, 217
  cancellation of, 153, 204, 262
  restoration of, 157, 240
"new NASA" discussions, 70, 173, 181–3, 200–1, 231–2
  as "New Technology Opportunities" effort, 201, 232–5, 251–2
"New Technology Opportunities" effort, 201, 232–5, 251–2
New York Times, The, 12, 15–16, 22, 97, 271
  Mars mission criticism by, 69
  Nixon reaction to criticism by, 15–16
  STG report and, 80, 82
Newell, Homer, 37–8, 39, 82
  Paine and, 61–2, 64, 140
  STG and, 56, 75
Newsday (newspaper), 14
Newsweek magazine, 92, 158
Niskanen, William, 159, 214–15
Nixon, Pat, 17, 29, 90
Nixon, Richard Milhous, 46, 69, 335n1
  antagonism between Mayo and, 90, 94
  Apollo 8 and, 7–8
  at *Apollo 12* launch, 90–1
  Apollo 13's impact on, 123, 144, 154–7, 180, 241, 282, 322n28
  Apollo 17 risks and, 144, 153–4, 157, 158, 180, 217
  Ash Council and, 132
  California contracts and, 207
  credit to Kennedy, for moon landing, 13, 14–15, 16, 26, 30
  desalination initiatives and, 181–2
  "exploring the unknown" and, 124, 179–1, 278, 283–5
  FY1971 budget explanations by, 100–1
  FY1971 budget process and, 90, 91–101
  FY1972 budget process and, 151–3, 157–60
  FY1973 budget process and, 209–11
  image of, 10–11
  inaugural address of, 7–8
  international space cooperation and, 109–12, 198–9, 289
  Low's appointment by, 86
  Mars mission, desire for eventual, 2, 92, 100, 114, 180, 284
  Memoirs of, 7, 32, 50
  NASA administrator selection by, 53–4, 174–5, 177–8
  national security potentials of shuttle and, 246
  Paine's meeting requests with, 73, 93, 100, 139–40, 146, 160
  reelection of, 145
  relationship with Borman, 9–10, 12, 14, 16–18, 22, 53–4, 73, 154–5
  Rice memorandum to, 238, 239, 242–5, 253
  shuttle approval announcement by, 265–9
  shuttle decision involvement of, 252–3, 261, 262–3, 266
  STG meeting with, 79–80
  Weinberger's memorandum to, 210–11, 213, 216–17, 240
Nixon, Tricia, 90
Nixon administration, 103, 277–8
  communication breakdown between BOB and, 83, 90, 94–5, 96, 99
  early confusion of operations in, 42–3
Nixon space doctrine, 104–15, 120, 134, 230, 281
  acceleration of schedule of, 105–7
  content of, 105, 114–15, 126
  delay of, 107–8
  drafting of, 84, 104–5
  impact of, 104, 279–80
  release of, 113–15
  revision of, 106, 109–12, 113
  timing of, 84–5, 105–7, 107–8
Nixon tapes, 175–6
Nixon's involvement with *Apollo 11* mission, 1, 9, 11–12, 31
  "Giant Step" tour, 27–30
  signature on plaque, 13, 15
  splashdown attendance, 22–7
  televised phone conversation with Armstrong, 11, 15, 19, 20, 21–2
  See also Apollo 11 mission
North American Rockwell, 118, 184–5, 234–5, 250, 291–2
North Atlantic Treaty Organization (NATO), 45
NRO (National Reconnaissance Office), 164, 168, 290–1

O'Dell, Charles, 40
Office of Management and Budget (OMB), 207–8, 209, 257–60, 294
  Anders and, 200
  creation of, 131, 133–4
  decision memorandum of, 242–5, 252, 253
  director's review of, 217–18, 239
  Evaluation Division of, 159, 214
  Flanigan and, 226–7, 248, 251
  Flax committee and, 202
  glider recommendation by, 235, 237
  Heiss memorandum and, 223
  Lindley's study and, 189
  Low and, 145, 157, 214, 239
  Mathematica study and, 190–1
  NASA FY1972 budget and, 144, 146–51, 152, 159
  NASA FY1973 budget and, 213
  NASA hearings with, 214–15
  NASC and, 136–7, 199
  paper requested by Weinberger from, 152
  presidential input solicitation by, 151–3
  Shultz as first director of, 133, 134
  shuttle booster selection and, 274
  space strategy paper collaboration with OST, 150
  Weinberger and, 133, 134, 152, 274
  See also NASA-OMB shuttle design debates; NASA-OMB shuttle program debates; NASA shuttle design; OMB shuttle design
Office of Manned Space Flight, NASA, 34, 37–8, 208
  Mathematica study and, 189–91
  Mueller as associate administrator of, 37, 62, 85, 162
  Myers as associate administrator of, 120, 131, 173, 184, 221
  See also National Aeronautics and Space Administration (NASA)

INDEX

Office of Science and Technology (OST), 46, 49, 143, 286
  Anders and, 200
  David as director of, 131, 135
  DuBridge as director of, 47
  Ehrlichman and, 133
  Nixon space doctrine and, 106
  shuttle decision and, 257–60
  space strategy paper collaboration with OMB, 150
  See also David, Edward E., Jr.
Office of Telecommunications Policy, 134, 198
OMB shuttle design, 248–51, 286, 298
  contractor assistance with, 249–50, 259
  cost of, 243, 248–9, 251
  NASA's evaluation of, 255, 256, 260–1
  NASA's reaction to, 250–1
  operation costs of, 251
  size of, 243, 247–8, 248–9
  See also NASA-OMB shuttle design debates; NASA shuttle design; Office of Management and Budget (OMB)
"Operation Paring Knife," 98
orbiter/booster development phasing, 177, 185–6, 193–5
  cost of, 188
  expendable fuel tanks and, 192
  Flax committee and, 203
orbiter design (MSC-040), 206–7, 221, 226–7
Outer Space Treaty (1967), 166

Packard, David, 49, 169, 224–5, 245, 246, 290
Pagnotta, Frank, 80
Paine, Thomas O., 10–11, 14, 38–40, 91, 282, 296
  Agnew and, 59, 64, 65, 74, 75, 100–1
  Apollo program review and, 121–2
  DOD/NASA report and, 163
  DuBridge and, 53, 57, 72, 122
  Flanigan and, 84, 92, 98, 99, 139–40, 141
  "Giant Step" tour and, 27, 28, 29
  international space cooperation and, 22, 108, 110–11, 112, 113, 140
  Low and, 84–5, 97, 131, 139, 141–2
  NASC and, 51–2
  Newell and, 61–2, 64, 140
  Nixon space doctrine and, 107, 108, 113–14, 117
  resignation of, 131, 139–41, 143, 173, 177
  roles and responsibilities of, 52–4
  shuttle development and, 171, 172, 188
  space goals proposals by, 35–6, 39–40, 55
  Townes report and, 42
  von Braun and, 66–8, 137–8, 141, 230
Paine, Thomas O., NASA budget and, 118–19
  employment impact and, 98–9
  FY1971, 92, 96–7
  long-range planning conference called by, 66–8, 137–9
  Nixon meetings on, 73, 93, 100, 139–40, 146, 160
  requests by, 38, 88–90
Paine, Thomas O., on STG, 48–9, 55–6, 58–9, 69–70
  FY1971 budget process and, 83–4
  Mars mission and, 59, 64–8, 69
  Mueller's integrated plan and, 63–4
  Newell and, 61–2, 64, 140

STG report and, 71–2, 75, 77–8, 79–80, 101, 110–11
Palley, Nevin, 56, 75
payload bay size/weight capacity, 163, 262
  Air Force requirements for, 192
  DOD/NASA report and, 166, 167
  Flax committee and, 203
  Fletcher/Packard meeting and, 224
  of glider, 206, 221
  Hexagon and, 167–8
  of ILRV, 162
  Low and, 229, 255
  of mini-shuttle, 205–6
  Mueller and, 170, 171
  NASA full-capability shuttle and, 170, 255–6, 257, 261
  OMB shuttle design and, 249
  shuttle decision and, 256–7
  See also cross-range capabilities; space shuttle development
Pegasus (proposed shuttle name), 265
Peterson, Peter, 227, 253
phased approach to shuttle development, 193–5, 206–7, 236
Pickering, William, 147–8
Portuguese "Research Institute for Navigation," 35–6
"Post-Apollo Advisory Group," 36–7
Pratt Whitney, 264
presidential election (1972), 2, 158
  employment issues and, 137, 144, 145–6, 181, 232–5
President's Science Advisory Committee (PSAC), 49, 70, 75, 175, 176
  shuttle review panel of, 40, 57–8, 187, 202, 219
Price, Ray, 21, 106
"Problems and Opportunities in Manned Space Flight" (Paine February 1969 memorandum), 57
Project Apollo. *See under* Apollo missions; Apollo program
Proxmire, William, 149, 271
PSAC. *See* President's Science Advisory Committee (PSAC)
Puckett, Alan, 40

quarantine of *Apollo 11* astronauts, 11–12, 24–5, 26

Ramo, Simon, 52
Rand Corporation, 215
Reagan, Ronald, 133, 233, 294
Reagan administration, 291
Reeves, Richard, 155, 322n28
reusable shuttle development, 2, 37
  abandonment of fully reusable design, 173–4, 186, 193–4, 204, 205, 222, 236, 296
  AIAA recommendation of, 60
  alternatives to, 174, 193–4, 195
  cost of, 188
  DOD/NASA report and, 163–70
  early stages of, 147
  expendable fuel tanks and, 192, 195
  Fletcher on, 178–9
  Heiss memorandum and, 222–4
  Low's cost curve diagram and, 229
  Mathematica study and, 191, 195, 221–4

Mueller and, 62, 63, 162, 170
Phase B studies on, 184–5, 194
phased approach to, 185–6, 193–5, 206
space station and, 216
STG discussion of, 70
See also NASA shuttle design
Rice, Donald, 133–4, 149, 207
 aerospace unemployment and, 233–4, 260
 Anders and, 200
 Apollo cancellations and, 180
 Flax committee and, 202–3
 Fletcher and, 187–8, 193
 Low and, 159–60, 236–7, 247
 Mathematica study and, 191
 memorandum to Nixon by, 238, 239, 242–5, 253
 NASC and, 136
 "new NASA" discussions and, 252
 OMB Evaluation Division and, 159
 OMB hearings and, 214–15
 OMB shuttle design and, 249–51
 opposition to space shuttle by, 194–5, 213, 226, 239, 286–7
 Rockwell/Anderson meeting and, 234–5
 roles and responsibilities of, 133, 145
 shuttle booster selection and, 272
 shuttle decision and, 255, 256, 257, 258, 259–60, 261
 "whole systems" approach of, 215
Roberts, Walter, 40
rocket-assisted takeoff (RATO) shuttle design, 222
Rocketdyne Corporation, 207, 234, 243, 264
Rockwell, Willard "Al," Jr., 234–5, 250, 263, 292
Rogers, William, 9, 22–3, 29, 48, 109, 110
Romney, George, 98, 132
Roosa, Stuart, 177
Rose, Jonathan, 18, 213, 227, 247
 aerospace unemployment and, 233, 234, 255, 266
 Low's "best case" essay for, 237–8
 roles and responsibilities of, 134
 shuttle approval and, 264
 shuttle decision and, 255, 256, 260, 288
Rosen, Milton, 56, 66, 82, 104
Russell, Newton, 233
Ruwe, Nicholas, 28
Ryan, John, 171

Safeguard antiballistic missile system, 175
Safire, William, 12, 13, 18, 182–3, 265, 306n30
Sato, Eisaku (Japanese Prime Minister), 264
Saturn V rockets, 2, 19, 34, 41, 55, 57, 58, 90, 122
 Apollo program review and, 121
 integrated plan and, 63
 Mars mission and, 65
 suspension/abandonment of production of, 89, 91, 93, 106, 118–19, 143, 283–4
Scheer, Julian, 11–12, 27–8
Schlesinger, James, 59, 74, 95–6
Schmitt, Harrison "Jack," 282
Schorr, Daniel, 175
Schriever, Bernard, 52, 141, 181
Science magazine, 82
Seaborg, Glenn, 56, 71, 76, 77

Seamans, Robert C., 35, 40, 49, 125, 308n10
 shuttle development and, 163, 168, 171–2, 194, 225, 297
 STG and, 56, 70, 75–6, 79–80, 171, 297
Shapley, Willis, 33, 56, 186
 "constant budget" idea of, 209
 Fletcher and, 177
 Low's "best case" essay and, 237
 NASA FY1972 budget and, 146
 OMB shuttle design and, 250
 Rice and, 215
 STG report evaluation and, 84
Shepard, Alan, 47, 177
short-duration orbital missions, 165–6, 168
Shultz, George, 145, 149–50, 190, 246, 253
 "Administrator's contingency" and, 274
 aerospace unemployment and, 234
 Apollo cancellations and, 180
 as first OMB director, 133, 134
 FY1972 budget process and, 151, 157, 158
 FY1973 budget process and, 240–1, 244
 NASC elimination proposal to, 136
 "new NASA" discussions and, 182, 183
 roles and responsibilities of, 133
 shuttle approval and, 264, 265
 shuttle booster selection and, 273
 shuttle decision and, 256, 257, 258, 259, 261, 262–3, 287
 Weinberger memorandum and, 211
shuttle booster selection process, 272–5, 288
shuttle reference missions, 168–9
shuttle/station development, 117, 120, 147, 149, 298
 abandonment of simultaneous development of, 195
 DOD/NASA report and, 165
 Mueller and, 162
 OMB and, 294
Sidey, Hugh, 9, 156
Skylab workshop, 47, 131, 187, 200
 cancellation discussions of, 144, 145, 148, 151, 152, 153
 launch of, 118
 Low's position on, 148, 157
 priority of, 214
Smart, Jacob, 245–6
"soft power," 8–9, 25, 281–2, 304n4
Soviet Academy of Sciences, 16
Soviet Union, 13, 242
 Apollo-Soyuz Test Project docking mission and, 17, 199, 214, 243, 268, 289
 Borman's visit to, 16–17
 competition/cooperation with U.S., 16–17, 45, 55, 57, 60, 123, 153, 157, 216, 238
 Foster and, 229
 Low and, 153, 157
 Luna-15 and, 17
 Mir and, 289–90
 Nixon/*Apollo 11* conversation not aired in, 21
 Outer Space Treaty and, 166
 shuttle development and, 163, 229
 "soft power" U.S. dominance over, 281
 Sputniks 1 and *2* and, 31
 STG report sent to, 111
Space Act (1958), 50
Space Clipper (proposed shuttle name), 265, 266, 267, 268–9

Space Council. *See* National Aeronautics and Space Council (NASC)
space glider. *See* glider approach to shuttle development
Space Science Board (SSB), 60–1, 121, 122
space shuttle approval, 186, 208, 245
  early indication of, 151, 210, 211, 213, 217
  employment impact of, 217, 233, 241, 264
  of full capability design, 246, 252, 261, 262–3, 264, 274, 286–9, 293–4
  Nixon's announcement of, 264–9
  as policy mistake, 294–5, 295–300
  reasons for, 288–9, 290, 292–4
  shuttle/station as single program, 117, 119–20, 147, 149, 162, 165, 195, 295
space shuttle development, 2–3, 92, 125–6, 147–8, 178–9, 186–7, 267–8
  airframe development and, 151, 153, 157, 160
  alternatives to, 214, 219–21, 236–7, 239–40, 242
  BOB and, 147, 188–9, 296
  booster selection for, 272–3, 274–5, 288
  Congress and, 149, 194, 219, 224, 262, 290
  contractor studies of, 58, 93, 160, 191–2, 194, 207–8
  deferral of, 204, 209, 216, 220, 285
  delta-shaped wing design in, 184, 206, 225
  employment impact of, 233
  expendable fuel tanks in, 192, 193, 205–6, 222
  Flax committee and, 202–4, 213, 218–21, 226–30, 296
  FY1971 and, 149
  FY1972 and, 96, 122, 147, 159–60, 186
  FY1973 and, 204–5, 208–9, 235–8
  impact of, 289–94
  influence of, on U.S. space program, 285–7
  "integrated plan" for, 62–3
  international participation in, 148, 172, 197–9, 215, 238, 268, 289
  Mars goal and, 146
  mini-shuttle and, 205–6, 208, 218, 296
  Mueller and, 162–3, 192
  naming of program, 264, 265, 266, 267
  NASA trying to get Nixon to commit to, 112, 113, 114
  NASA's best case for approval of, 235–8
  necessity of for future of space program, 117–18
  Niskanen as opponent to, 214–15
  Nixon space doctrine and, 112, 113
  as Nixon's legacy, 285–301
  orbiter size and, 248–9, 250–1, 255–6, 261
  Phase A studies of, 162–3, 166, 170, 184, 192
  Phase B studies of, 184–5, 190, 191, 192, 194, 206–7
  priority of, 214
  program as policy failure and, 295
  rationale for, 225–6, 229, 235, 245, 247, 277–8, 285
  STG report and, 75, 105
  straight-wing orbiter design in, 184
  systems analysis of, 215
  transition in NASA's thinking on, 173–4
  von Braun's absence in advocating, 230–1
  See also cross-range capabilities; glider approach to shuttle development; Mark I/Mark II approach to shuttle development; NASA-OMB shuttle design debates; NASA-OMB shuttle program debates; NASA shuttle design; OMB shuttle design; payload bay size/weight capacity
space shuttle development, costs of, 161, 184–5, 188–91, 239, 258, 295–7
  booster development and, 272–3
  budget commitment levels and, 273–4
  contractor studies and, 191–2, 227
  controversy around, 147
  economic analysis and, 195, 219–20
  Flax committee and, 203, 204
  Fletcher/Low and, 188, 194, 197, 205, 209, 273–4, 298
  fully reusable design abandonment and, 173–4, 186, 193–4
  Heiss memorandum and, 222–4
  Low's cost curve diagram and, 228–9, 236
  Mathematica study and, 189–91, 194, 202, 221–4
  OMB memorandum and, 252–3
  OMB shuttle and, 249, 252–3
  vs. operating costs, 228, 236, 243, 251, 293–4, 337n25
  phased approach to, 197, 206–7
  Rice's "whole systems" approach and, 215
  shuttle decision and, 256–7
  Weinberger and, 218
space shuttle development, national security requirements for, 161–72, 195, 219, 237, 263, 266, 267–8
  Air Force and, 164, 168, 170–1, 192, 202, 225, 272, 297
  cross-range capabilities in, 165–6, 168–9, 170, 171, 184, 224
  DOD and, 58, 162, 163, 184, 193, 194, 256, 262, 290, 291
  DOD/NASA 1969 study and, 70, 163–70, 190, 225, 246, 291
  economic analysis of, 189
  European participation and, 198
  evaluation of, 290–4
  Fletcher/Packard meeting and, 224–6
  ILRV and, 162–3
  Mark II orbiter and, 207
  Mueller and, 170–1
  Nixon's interest in, 288, 289
  NRO and, 164
  payload bay size and, 166, 167–8, 170, 171, 192–3, 224
  payload weight and, 166, 170, 171, 184, 224
  Phase A studies of, 184
  Phase B studies of, 184–5, 190
  reference missions for, 168–9
  Smart's memorandum on, 245–6
space station development, 2, 3, 131, 216, 300
  Apollo hardware reused in, 47, 118, 205, 297
  bicentennial plans for, 138
  budget additions for, 57
  deferral of, 125, 173
  funding for studies of, 58
  Low on, 205
  Mars goal and, 146, 148
  NASA trying to get Nixon to commit to, 112, 114
  NASA's uncertainty about, 39–40
  Newell and, 39, 61–2

Paine and, 55, 92
"Post-Apollo Advisory Group" and, 37
STG discussion about, 70
STG report and, 75, 79, 105
studies of, 58, 93, 143
Townes report on, 41
See also space shuttle development
Space Task Group (STG), 83, 126, 171, 297
  alternatives to Mars mission considered by, 69–72
  BOB and, 56, 59, 75, 77, 84
  Congressional information session, 59–60
  creation of, 49–50, 80, 109
  DOD and, 48, 49, 56, 62, 75, 119
  DOD/NASA joint study directed by, 163, 164, 170
  early decisions of, 58–9
  final meeting of, 75–9
  first meeting of, 56–8
  Mars discussions in, 30, 54, 58–9, 64–72, 74, 75–6, 77, 88, 125
  NASA's input to, 62, 79
  Nixon's meeting with, 79–80
  outside inputs to, 59–61
  penultimate meeting of, 75–7
  public input to, 59–60, 64
  second meeting of, 58–9
  shuttle development and, 213, 284
  Staff Directors Committee of, 56, 58, 72, 75
  White House involvement in, 72–4
  See also Agnew, Spiro as STG chair; Paine, Thomas O., on STG
Space Task Group (STG) report, 195
  alterations of, to suit White House budgeting needs, 77
  comparative milestones in various options of, 81
  elimination of Mars mission options from, 77–8
  evaluation of, 84–5
  extended preparation time for, 75–9
  late 70s forecasts for NASA budget in, 146–7
  as marketing document, 83
  Mars missions and, 30, 71–2, 77, 78–9, 80, 81, 82, 84, 95, 96, 104, 125
  NASA FY1971 budget and, 76, 83, 84–5, 87, 103, 138
  Nixon space doctrine and, 104
  options of, 71–2, 75–6, 77–9, 81, 84–5, 88–9, 104, 105
  Paine and, 71–2, 75, 77–8, 79–80, 101, 110–11
  press reaction to, 80–2
  public release of, 79–80
  rejection of, 278, 285
  relabeling of options in, 78–9
Space Transportation System (STS), 164, 171, 172
space "tugs," 76, 81
  development of as step to Mars goal, 65
  European participation in design of, 198
  Mueller on, 62, 63
  shuttle development and, 165, 167, 189, 190, 198, 223, 290
Spacelab program, 199
Speier, Richard, 87
"Spirit of Apollo" tour, 13, 24–5, 26
Sputniks 1 and *2*, 31, 32
Starlighter (proposed shuttle name), 265

State of the Union Address (1972), 201
supersonic transport (SST), 100, 179–80, 186, 187, 201, 288
Suter, Luis, 176
Swigert, Jack, 154, 156

Taft, Dan, 150, 215–16
TAOS (thrust-assisted orbiter shuttle), 222–4, 228
tax reform, 89, 97–8
television, *Apollo 11* and, 11, 19, 21–2, 1520
Thompson, Floyd, 37
Thompson, Robert, 194, 271, 294
thrust-assisted hydrogen-oxygen (TAHO) takeoff shuttle design, 222
Titan III booster, 192–3, 221, 226, 229, 243, 333n14
  OMB memorandum and, 252
Titan IV booster, 291
Townes, Charles, 40, 70, 121, 141
Townes report (transition task force on space, 1969), 30, 40–2, 47, 70
  Burns's distillation of, 48, 109
  NASA's status and, 32, 42, 57–8
  NASC discussions in, 51
TRW, 52
Tyson, Neil DeGrasse, 281

"U.S. Civilian Space Program, The: a Look at the Options" (Taft paper), 215–16
U.S. Information Agency, 28
U.S. Treasury Department, 97
unemployment levels, 204. *See also* aerospace unemployment
United Nations, 109
USS Arlington, 22, 23
USS Hornet aircraft carrier, 16, 23, 24
USS John F. Kennedy aircraft carrier, 16

van Allen, James, 40, 194
Vandenberg Air Force Base (CA), 163, 168–9, 290
von Braun, Wernher, 39–40, 70, 71
  FY1973 budget process and, 205
  glider and, 206
  move to Washington of, 131, 137
  Paine and, 66–8, 137–8, 141, 230
  resignation of, 231
  shuttle development and, 230–1
von Neumann, John, 189
"Voyage to the Moon" (MacLeish poem), 22

Washington Post, The (newspaper), 15, 35, 80, 96
Washington Star, The (newspaper), 99
Watergate scandal, 43, 45
Webb, James, 33–7, 52, 118, 123, 140, 308n10
Weber, Arnold, 136
Weinberger, Caspar "Cap," 134, 145, 246
  aerospace unemployment and, 158, 181, 233
  Anders and, 136, 151, 199, 218
  Apollo cancellations and, 231
  director's review meeting and, 239
  "director's review" with Rice and, 150–1
  Fletcher and, 261–2
  FY1973 budget process and, 204, 210, 244, 248
  Low's correspondence with, 149, 153, 157
  meeting with Paine and Low, 145

Weinberger, Caspar "Cap,"—*Continued*
 memorandum to Nixon by, 210–11, 213, 216–17, 240
 OMB and, 133, 134, 152, 274
 OMB memorandum and, 244, 253
 OMB paper requested by, 152
 OMB shuttle design and, 248–9, 250
 Rockwell/Anderson meeting and, 234–5
 roles and responsibilities of, 133
 shuttle booster selection and, 273, 288
 shuttle decision and, 255, 256, 258, 261, 263, 287
 shuttle support by, 217–18, 226
Whitaker, John, 145
White, Ed, 11
White House organization: Ash Council and, 132, 133, 135–6
 Domestic Council creation and, 131, 132–3
 OMB creation and, 133
 proposed NASC elimination, 135–7
Whitehead, Clay Thomas "Tom," 44, 45–6, 126–7, 213, 283
 Anders and, 200
 BOB and OST and, 47
 budget cutting options demanded by, 75
 Fletcher and, 226, 247
 international cooperation and, 112, 113, 198
 NASA FY1971 budget, 94–5, 96
 Nixon space doctrine and, 104–5, 106, 107, 112, 113, 134
 shuttle decision and, 255
 shuttle development position of, 227
 STG review and, 73–4
Wilson, Harold, 54
Witness to Power (Ehrlichman), 77
Wolff, Jerome, 56, 60, 75, 80
Woods, Rose Mary, 18
Wyatt, DeMarquis, 75, 78

Yarymovych, Michael, 163

Ziegler, Ron, 154
Zwick, Charles, 38